RECENT ADVANCES IN LIQUID CRYSTALLINE POLYMERS

Based on the Proceedings of the European Science Foundation Sixth Polymer Workshop on Liquid Crystal Polymer Systems held at Lyngby, Denmark, 12–14 September 1983.

RECENT ADVANCES IN LIQUID CRYSTALLINE POLYMERS

Edited by

L. LAWRENCE CHAPOY

Instituttet for Kemiindustri,
The Technical University of Denmark,
Lyngby, Denmark

ELSEVIER APPLIED SCIENCE PUBLISHERS
LONDON and NEW YORK

ELSEVIER APPLIED SCIENCE PUBLISHERS LTD
Crown House, Linton Road, Barking, Essex IG11 8JU, England

Sole Distributor in the USA and Canada
ELSEVIER SCIENCE PUBLISHING CO., INC.
52 Vanderbilt Avenue, New York, NY 10017, USA

British Library Cataloguing in Publication Data

Recent advances in liquid crystalline polymers.
1. Polymers and polymerization 2. Liquid crystals
I. Chapoy, L. Lawrence
547.7 QD381

ISBN 0-85334-313-6

WITH 30 TABLES AND 176 ILLUSTRATIONS

Reprinted 1986

Printed in Great Britain by Galliard (Printers) Ltd, Great Yarmouth

PREFACE

This volume contains an eclectic collection of 22 papers on liquid crystalline polymers presented at the Sixth Polymer Workshop, in the series sponsored by the European Science Foundation, entitled: 'Liquid Crystal Polymer Systems', in Gentofte, Denmark, 12–14 September 1983. Since a contribution to this volume was strictly voluntary, and in some cases represents a considerably expanded version of that which was presented, it is strictly speaking not correct to term this a 'proceedings'. A description of the aims and purposes of the European Science Foundation with respect to the polymer area has been presented in: *Shell Polymers*, Vol. 5, No. 2, pp. 34–35, 1981.

The papers given here represent a cross-section of current research interests in liquid crystalline polymers in the areas of theory, synthesis, characterization, structure–property relationships and applications. At least some of the current interest is motivated by attempts to practically exploit the novel properties of these materials in the developing technologies of high strength fibres and advanced materials for constructional purposes, but also for functional materials in the areas of information retrieval, electronics and opto-electronics applications.

The editor wishes to thank all those involved for their courtesy and co-operation.

L. LAWRENCE CHAPOY

CONTENTS

PART III: CHARACTERIZATION

PART IV: APPLICATIONS

LIST OF CONTRIBUTORS

G. Baur

Fraunhofer Institut für Angewandte Festkörperphysik, D-7800 Freiburg i Br., Federal Republic of Germany

C. Bhaskar

Division of Polymer Chemistry, National Chemical Laboratory, Poona 411 008, India

D. Biddle

Institute of Physical Chemistry, University of Gothenburg and Chalmers University of Technology, 412 96 Gothenburg, Sweden

A. Blumstein

Polymer Program, University of Lowell, Lowell, Massachusetts 01854, USA

R. B. Blumstein

Polymer Program, University of Lowell, Lowell, Massachusetts 01854, USA

L. L. Chapoy

Instituttet for Kemiindustri, The Technical University of Denmark, DK-2800 Lyngby, Denmark

E. Chiellini

Istituto di Chimica Generale (Facoltà di Ingegneria), Università di Pisa, Via Diotisalvi 2, 56100 Pisa, Italy

F. N. Cogswell
Imperial Chemical Industries plc, Research and Technology Department, PO Box 90, Wilton, Middlesbrough, Cleveland TS6 8JE, UK

H. J. Coles
Liquid Crystal Group, Schuster Laboratory, Department of Physics, University of Manchester, Manchester M13 9PL, UK

P. Dais
Department of Chemistry, McGill University, Montreal, Quebec, Canada H3A 2A7

E. J. Diekmann
Instituttet for Kemiindustri, The Technical University of Denmark, DK-2800 Lyngby, Denmark

A. M. Donald
Cavendish Laboratory, Madingley Road, Cambridge CB3 0HE, UK

J. W. Emsley
Department of Chemistry, University of Southampton, Highfield, Southampton SO9 5NH, UK

P. J. Flory
Department of Chemistry, Stanford University, Stanford, California 94305, USA

V. Frosini
Dipartimento de Ingegneria Chimica, Centro CNR Processi Ionici, Università di Pisa, 56100 Pisa, Italy

G. Galli
Istituto di Chimica Organica Industriale, Università di Pisa, Via Risorgimento 35, 56100 Pisa, Italy

B. Hisgen
Institute of Organic Chemistry, University of Mainz, D-6500 Mainz, Federal Republic of Germany

R. KIEFER
Fraunhofer Institut für Angewandte Festkörperphysik, D-7800 Freiburg i Br., Federal Republic of Germany

J. KOPS
Instituttet for Kemiindustri, The Technical University of Denmark, DK-2800 Lyngby, Denmark

G. KOTHE
Institute of Physical Chemistry, University of Stuttgart, D-7000 Stuttgart 80, Federal Republic of Germany

R. W. LENZ
Department of Chemical Engineering, University of Massachusetts, Amherst, Massachusetts 01003, USA

J. A. LOGAN
IBM Research Laboratory, San Jose, California 95193, USA

J. R. LUCKHURST
Department of Chemistry, University of Southampton, Highfield, Southampton SO9 5NH, UK

B. MARCHER
Instituttet for Kemiindustri, The Technical University of Denmark, DK-2800 Lyngby, Denmark

A. F. MARTINS
S A Ciência des Materiais, FCT/UNL, and Centro de Física da Matéria Condensada, INIC, Av. Prof. Gama Pinto 2, 1699 Lisboa Codex, Portugal

K. MUELLER
Institute of Physical Chemistry, University of Stuttgart, D-7000 Stuttgart 80, Federal Republic of Germany

J. MÜGGE
Institut für Physikalische Chemie der TU Clausthal, Adolf-Römer-Strasse 2a, D-3392 Clausthal-Zellerfeld, Federal Republic of Germany

D. K. MUNCK
Instituttet for Kemiindustri, The Technical University of Denmark, DK-2800 Lyngby, Denmark

C. NOËL
Laboratoire de Physicochimie Structurale et Macromoléculaire, ESPCI, 10 rue Vauquelin, 75231 Paris Cedex 05, France

M. G. NORTHOLT
Akzo Research Laboratories, Corporate Research Department, PO Box 60, 6800 AB Arnhem, The Netherlands

C. M. PALEOS
Nuclear Research Center 'Demokritos', Aghia Paraskevi, Attikis, Greece

S. DE PETRIS
Dipartimento de Ingegneria Chimica, Centro CNR Processi Ionici, Università di Pisa, 56100 Pisa, Italy

K. H. RASMUSSEN
Instituttet for Kemiindustri, The Technical University of Denmark, DK-2800 Lyngby, Denmark

H. RINGSDORF
Institute of Organic Chemistry, University of Mainz, D-6500 Mainz, Federal Republic of Germany

H.-W. SCHMIDT
Institute of Organic Chemistry, University of Mainz, D-6500 Mainz, Federal Republic of Germany

R. K. SETHI
Instituttet for Kemiindustri, The Technical University of Denmark, DK-2800 Lyngby, Denmark

R. SIMON
Liquid Crystal Group, Schuster Laboratory, Department of Physics, University of Manchester, Manchester M13 9PL, UK

H. SPANGGAARD
Instituttet for Kemiindustri, The Technical University of Denmark, DK-2800 Lyngby, Denmark

J. SPRINGER
Institut für Technische Chemie der Technischen Universität Berlin, Fachgebiet Makromolekulare Chemie, Strasse des 17 Juni 135, 1000 Berlin 12, Federal Republic of Germany

G. STROBL
Institute of Organic Chemistry, University of Mainz, D-6500 Mainz, Federal Republic of Germany

F. W. WEIGELT
Institut für Technische Chemie der Technischen Universität Berlin, Fachgebiet Makromolekulare Chemie, Strasse des 17 Juni 135, 1000 Berlin 12, Federal Republic of Germany

A. H. WINDLE
Department of Metallurgy and Materials Science, University of Cambridge, Pembroke St, Cambridge CB2 3QZ, UK

F. VOLINO
CNRS and Département de Recherche Fondamentale, Centre d'Etudes Nucléaires de Grenoble, 85X, 38041 Grenoble Cedex, France

A. E. ZACHARIADES
IBM Research Laboratory, 5600 Cottle Road, San Jose, California 95193, USA

R. ZENTEL
Institute of Organic Chemistry, University of Mainz, D-6500 Mainz, Federal Republic of Germany

P. ZUGENMAIER
Institut für Physikalische Chemie der TU Clausthal, Adolf-Römer-Strasse 2a, D-3392 Clausthal-Zellerfeld, Federal Republic of Germany

PART I

SYNTHESIS, STRUCTURE, PROPERTIES RELATIONSHIPS

1

SYNTHETIC ROUTES TO LIQUID CRYSTALLINE POLYMERS

ROBERT W. LENZ

Chemical Engineering Department, University of Massachusetts, Amherst, USA

INTRODUCTION

This chapter is concerned with the synthesis of main chain, liquid crystal polymers, but only with the reactions used to prepare polyesters and polyamides. At the time of this review, polyesters have been of interest only as thermotropic polymers and polyamides for lyotropic behaviour, but this situation will, likely, soon change. There are undoubtedly aromatic polyesters which show liquid crystallinity in solution, and a thermotropic polyurethane has recently been reported.[1]

The practical application of lyotropic polyamides preceded that of thermotropic polyesters by at least ten years. Indeed, the development of the Kevlar® fibre by the du Pont Company, by the solution spinning of an aromatic polyamide in the lyotropic state, probably provided most of the early incentive for research on thermotropic polyesters.[2] That is, the goal of melt spinning a fibre, with comparable high strength and high modulus properties to 'Kevlar', led to the initial investigations of aromatic polyesters, which were closely related in structure to the 'Kevlar' polyamide. As yet this goal has not been realized commercially, but an aromatic copolyester containing biphenol terephthalate and *p*-oxybenzoate units which is capable of forming a thermotropic melt,[3] is marketed for ceramic-like plastics, and a family of oxybenzoate–ethylene terephthalate copolyesters has been intensively studied for use in injection moulded thermoplastics.[4]

AROMATIC POLYAMIDES

Morgan has reviewed the historical development of the preparation of fibre-forming aromatic polyamides.[5] He reported that the first polyamide of the non-peptide type, which was found to form liquid crystalline solutions, was poly(1,4-benzamide). Subsequent investigations revealed that other molecular structures, which retained the linearity and rigidity of the *p*-phenylene group, could also lead to lyotropic polyamides, including the 2,6-naphthalene, *trans*-1,4-cyclohexylene, and *p,p'*-biphenylene. In addition, *p*-phenylene units can be connected by *trans*-azo or *trans*-vinylene groups and still retain their linear, rigid conformations. The molecular structures of such units, as compiled by Morgan, are shown in Table 1.[5]

Four types of reactions have been applied to the preparation of polyamides in general, but only one of these, the reaction of an acid chloride with an amine, has been broadly used to prepare lyotropic polyamides. The direct amidation of acids and amines through a crystalline salt intermediate, as used for the preparation of aliphatic polyamides[6] (termed a 'nylon salt'), is not applicable for aromatic amine–acid monomer pairs, but three other types of reactions have recently been developed for polyamides; these are:

(1) The oxidative amidation reaction of aromatic amines and acids with phosphorous compounds, which is generally carried out in an amine or amide solvent with organic chlorine compounds as co-reactants, as in the following reaction scheme:[7]

$$\mathrm{R\overset{\overset{O}{\|}}{C}OH + R'NH_2 + (C_6H_5)_3P + C_2Cl_6 \xrightarrow{\text{pyridine}}}$$

$$\mathrm{R\overset{\overset{O}{\|}}{C}NHR' + (C_6H_5)_3PO + Cl_2C{=}CCl_2 + C_5H_5N^{\oplus}{-}HCl^{\ominus}}$$

Lithium salts may be added to enhance the solubility of the polymer, and polymers with inherent viscosities well above 1·0 can be prepared by this route.[8] Unlike direct amidation, this reaction is essentially irreversible, and it is capable of very high yields, so high molecular weight polymers can be readily obtained.[9]

(2) The aminolysis of an active ester, such as those of substituted phenols and hydroxytriazoles, can be used to prepare aromatic and olefinic

TABLE 1
Repeating Units of Polyamides which form Lyotropic Solutions[5]

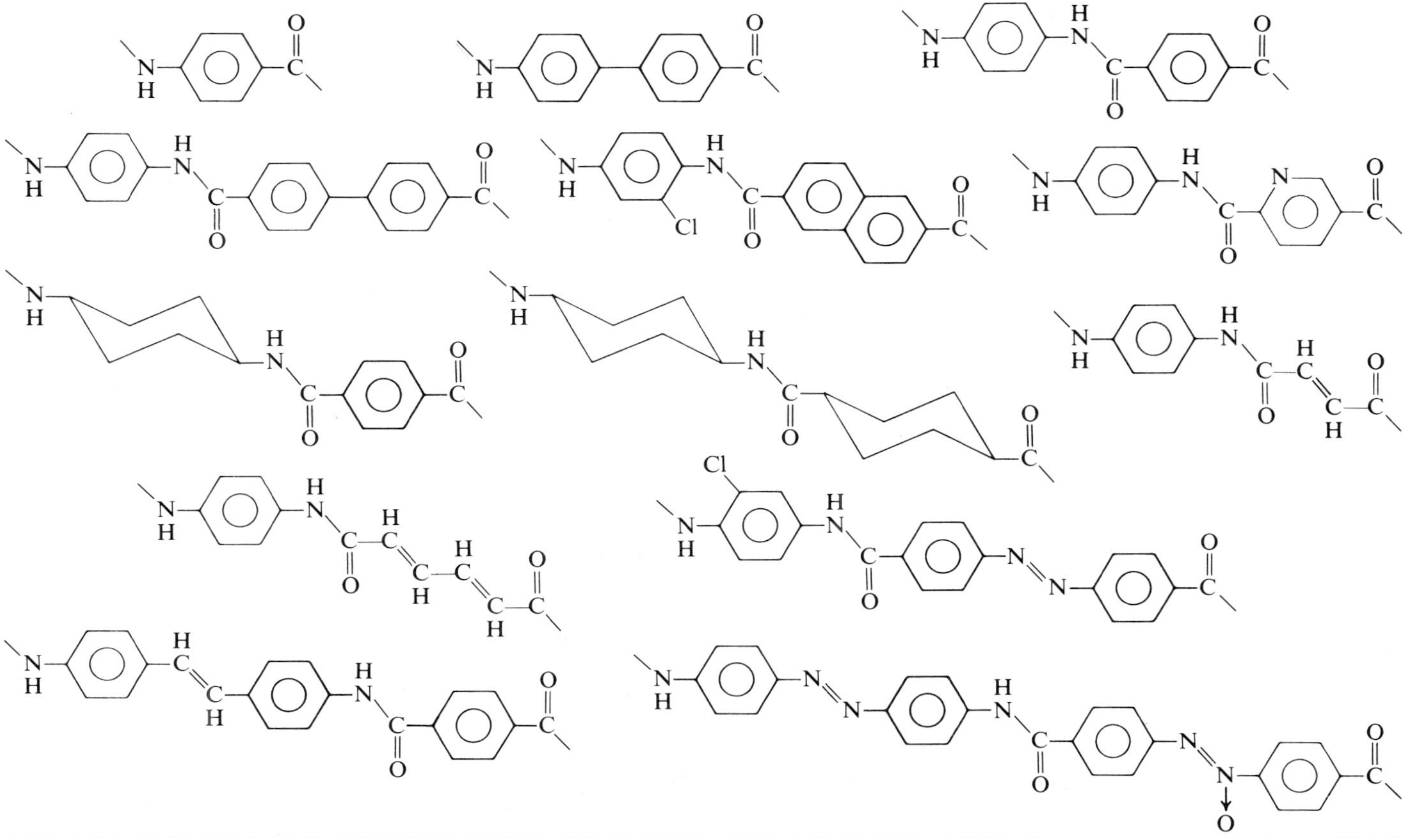

polyamides under relatively mild conditions in reasonably high molecular weights.[10]

(3) Amidation reactions in the presence of imidazole or its derivatives, such as in the following reaction, have been used by organic chemists for some time to prepare amides in high yields, but only recently has advantage been taken of this reaction for the preparation of polyamides:[11]

$$\mathrm{R\overset{O}{\overset{\|}{C}}OH + N\overbrace{\quad\;}N{-}X + R'NH_2 \longrightarrow R\overset{O}{\overset{\|}{C}}NHR' + N\overbrace{\quad\;}NH + XOH}$$

in which X is $\mathrm{{-}H,\ N\overbrace{\quad\;}N{-}\overset{O}{\overset{\|}{C}}{-}}$ and other groups.

While it is likely that these reactions will find increasing utility, the low temperature polycondensation of diamines with bis-acid chlorides is still the simplest and most direct route to the preparation of high molecular weight polyamides. For this approach, two reaction systems can be used: (1) interfacial polycondensation in a two-phase reaction system,[12] and (2) solution polymerization, either in an amide solvent (with or without lithium salts to increase polymer solubility), or in chlorocarbon salts with amines present to form ammonium hydrochloride salts.[13] Salts of the latter type

TABLE 2
Solvents for Aromatic Polyamides

Solvent	*Solvent activity*
Amide solvents (with or without Li or Ca salts):	
N,N-dimethylacetamide	Strong (toxic)
Hexamethylphosphoramide	Strong (toxic)
N-methylpyrrolidone	Strong with salts
N,N,N′,N′-tetramethylurea	Strong
N methylcaprolactam	Weak
N-acetylpyrrolidone	Weak
N,N-dimethylpropionamide	Weak
N-methylpiperidone	Weak
1,3-Dimethylimidazolidinone	Strong
N,N,N′,N′-tetramethylmalonamide	
Acid solvents:	
Sulphuric Acid (100·6%, oleum)	Very strong
Hydrofluoric Acid	Strong
Chlorosulphonic Acid	Strong

may also be added to the reaction to increase polymer solubility,[5] and reaction temperature and solvent are particularly important in preventing premature precipitation, which could make the end groups of the aromatic polyamides inaccessible before high molecular weights can be achieved. Solvents for aromatic (or other rigid-rod) polyamides which may be used to process these polymers for films or fibres are compiled in Table 2.[14]

AROMATIC POLYESTERS

The problem of polymer solubility is particularly important, and can be a limiting factor, in the preparation of high molecular weight, aromatic polyesters because such rigid-rod polymers have, in general, both very high melting points and very low solubilities in all presently available reaction solvents. As a result of these characteristics, and in order to achieve very high molecular weights, aromatic polyesters must be prepared by a two-step sequence involving first either a homogeneous solution polymerization reaction or a melt polymerization reaction, to form a polymer of intermediate molecular weight, followed by a final reaction of the polymer in the solid state. This latter reaction, often referred to as 'solid stating', occurs presumably in the non-crystalline phase of the semi-crystalline polymer at a temperature slightly below its melting point and, if carried out under high vacuum, is capable of generating very high molecular weight polymers.[15]

A great many aromatic polyesters and copolyesters which show thermotropic behaviour have now been reported, particularly in the patent literature.[16] The copolyesters may contain either combinations of different types of mesogenic units or combinations of mesogenic and non-mesogenic (non-linear) units. The structures of many of the different types of monomers used to form liquid crystal aromatic polyesters and copolyesters are shown in Table 3.[16]

There are four basic reactions which can be used for the synthesis of aromatic polyesters of intermediate molecular weights, as follows:[17]

(1) The Schotten–Baumann reaction of an aromatic acid chloride with a phenol, which can be carried out either in solution at elevated temperature (generally in a chlorocarbon solvent with a tertiary amine present to react with the HCl liberated to form an ammonium salt), or by interfacial polycondensation at room temperature (again a chlorocarbon solvent is generally used for the phase containing the acid chloride and a base is present as an 'HCl-acceptor' in the aqueous phase containing the difunctional phenol), or as a melt reaction, in which the HCl liberated is

TABLE 3
Monomers used in the Synthesis of Liquid Crystal Aromatic Polyesters

Aromatic diol	*Aromatic dicarboxylic acid*	*Hydroxyacid*
HO— —OH	HOC(=O)— —COH(=O)	HO— —COH(=O)
HO— —OH, X, Y (X, Y = halogen or alkyl)	HOC(=O)— —COH(=O), X (X = halogen, alkyl)	HO— —COH(=O), X (X = halogen, alkyl)
HO, OH	HOC(=O)— — —COH(=O)	HO, COH(=O)
HO— — —OH	HOC(=O), COH(=O)	HO— —CH═CHCOH(=O)
HO— —OH, X (X = H, halogen, alkyl)	HOC(=O)— —O— —COH(=O)	

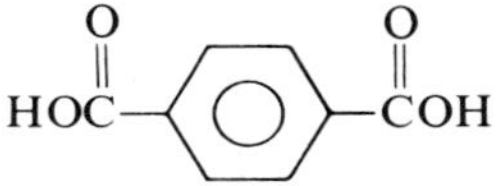

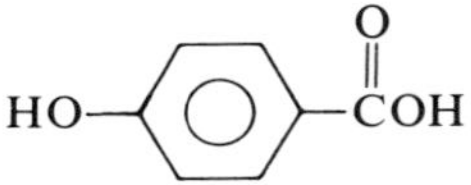

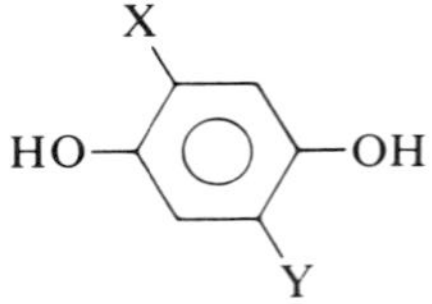

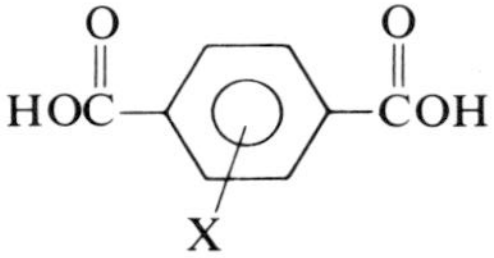

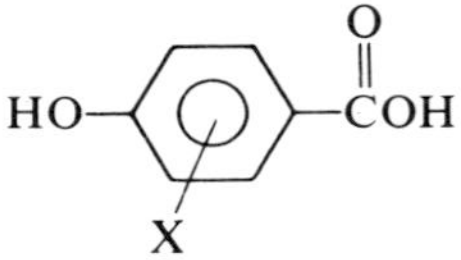

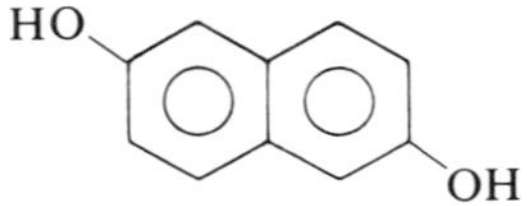

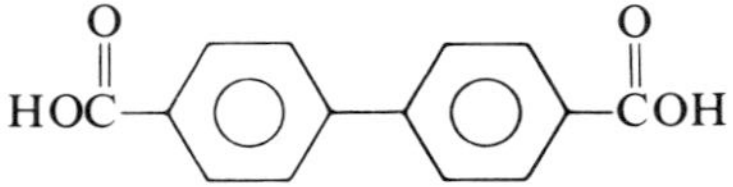

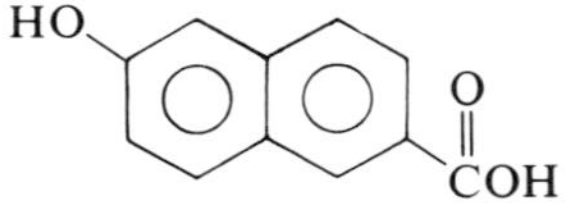

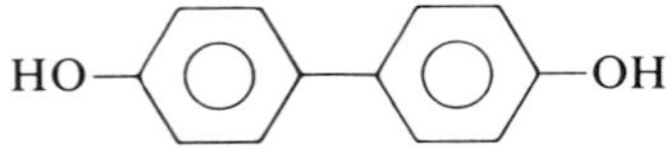

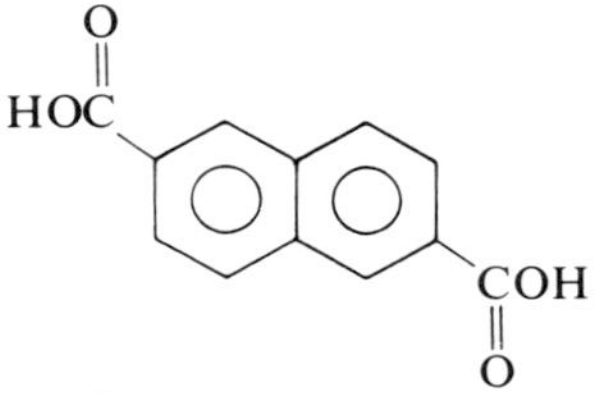

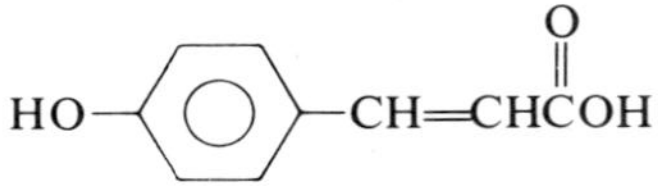

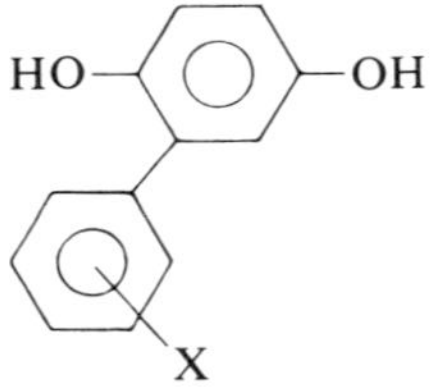

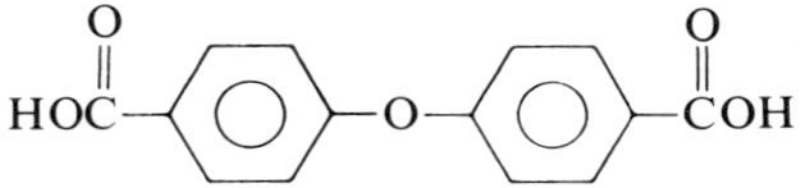

removed by an inert gas stream or by maintaining the reaction mixture under vacuum.[18]

(2) Ester interchange reactions in the melt at high temperatures of either of two different types: (a) reaction of a diphenyl ester of the aromatic dicarboxylic acid monomer with the difunctional phenol monomer to liberate phenol, which is removed under vacuum,[19] or (b) reaction of a diacetate ester of the difunctional phenol monomer with the dicarboxylic acid monomer by an 'acidolysis' reaction to liberate acetic acid, which is readily removed under vacuum:[20]

$$CH_3\overset{O}{\overset{\|}{C}}OArO\overset{O}{\overset{\|}{C}}CH_3 + HO\overset{O}{\overset{\|}{C}}Ar'\overset{O}{\overset{\|}{C}}OH \longrightarrow \left[OArO\overset{O}{\overset{\|}{C}}Ar'\overset{O}{\overset{\|}{C}} \right] + CH_3COOH$$

Both of these reactions may, of course, be equally applied to the polymerization of hydroxyacid monomers.

(3) The oxidative esterification reaction of an aromatic carboxylic acid with a phenol in the presence of a phosphorous compound and a chlorocarbon, as illustrated below:[21]

$$ArOH + Ar'COOH + (C_6H_5)_3P + C_2Cl_6 \longrightarrow Ar'\overset{O}{\overset{\|}{C}}OAr + (C_6H_5)_3PO + C_2Cl_4 + HCl$$

This type of reaction was first applied to the preparation of polyamides, as discussed above, for which it is somewhat more effective in forming high molecular weight polymers, but it has now been used for the synthesis of a wide variety of aromatic polyesters, either by the self-condensation of hydroxyacids or by the co-condensation of dicarboxylic acids and difunctional phenols.[22] A fairly wide variety of phosphorous compounds can be used as reducing or dehydrating agents in these reactions, in addition to phosphines, including phosphites, chlorophosphates, phosphates, polyphosphates and phosphazenes.[23] In most cases, lithium chloride is added and the reaction is run in either pyridine or an amide solvent system. The reaction has also been found to be catalyzed by tertiary amine salts.[24]

(4) Polymerization reactions involving the use of mixed anhydrides, formed *in situ*, have been reported to give fairly high molecular weight polymers, as for example in the recent report of the use of arylsulphonyl chlorides to effect the direct polycondensation of aromatic dicarboxylic

acids with biphenols, presumably through the intermediate formation of the mixed sulphonate–carboxylate anhydride, as follows:[25]

$$\mathrm{R\overset{O}{\overset{\|}{C}}OH + ArSO_2Cl \longrightarrow R\overset{O}{\overset{\|}{C}}OSO_2Ar \xrightarrow{R'OH} R\overset{O}{\overset{\|}{C}}OR' + ArSO_3H}$$

Copolyesters

All of the reactions listed above can be equally applied to the preparation of aromatic copolyesters from almost any combination of hydroxyacids, dicarboxylic acids and biphenols of the types listed in Table 3. In general, the copolymers so obtained are random or statistical in composition. Furthermore, the acidolysis reaction has been successfully applied to preparation of random copolyesters by interchanging monomers and high molecular weight polymers, most notably in the preparation of poly(ethylene terephthalate–co-oxybenzoate) by the insertion reaction of *p*-acetoxybenzoic acid with preformed poly(ethylene terephthalate), as follows:[26]

$$\mathrm{\left[OCH_2CH_2O\overset{O}{\overset{\|}{C}}\langle\bigcirc\rangle\overset{O}{\overset{\|}{C}}\right] + CH_3\overset{O}{\overset{\|}{C}}O\langle\bigcirc\rangle\overset{O}{\overset{\|}{C}}OH \longrightarrow}$$

$$\mathrm{CH_3\overset{O}{\overset{\|}{C}}OH + \left[OCH_2CH_2O\overset{O}{\overset{\|}{C}}\langle\bigcirc\rangle\overset{O}{\overset{\|}{C}}\right]\!\sim\!\sim\!\left[O\langle\bigcirc\rangle\overset{O}{\overset{\|}{C}}\right]}$$

A problem can arise, however, if this type of high melting, semicrystalline aromatic copolyester is maintained at temperatures at or near its melting point for a prolonged period of time. Under these conditions, the initially random copolymers can slowly reorganize to multiple block structures by a process termed the 'crystallization induced reaction' of the copolymer.[27] This reaction has been demonstrated to occur for the specific oxybenzoate copolymer mentioned above and for several other types of aromatic copolyesters studied in the author's laboratory.[28] The ester interchange reorganization reaction can apparently occur either in the solid state at temperatures close to the melting point or in the liquid crystalline phase of thermotropic copolymers.

Solvents

The rigid rod structures and high melting points of aromatic polyesters make it difficult to dissolve these polymers at lower temperatures in concentrations of more than a few per cent. Some of the solvents and solvent combinations which have been used for this purpose are listed in Table 4 with a qualitative indication of their solvent ability. Most of these

TABLE 4
Solvents for Aromatic Polyesters

Solvent	*Solvent activity*
p-Chlorophenol	Strong
p-Chlorophenol/tetrachloroethane	Strong
Trifluoromethanesulphonic acid	Very strong
o-Dichlorobenzene/*p*-chlorophenol (50/50)	Strong
Phenol/tetrachloroethane/*p*-chlorophenol (25/35/45)	Strong
Trifluoroacetic acid (TFA)	Weak
TFA/methylene chloride (60/40 or 30/70)	Strong
Phenol/tetrachloroethane (60/40)	Strong
Phenol/chloroform	Weak
Hexafluoroisopropanol	Weak
m-Cresol	Weak
o-Chlorophenol	Weak
Pentafluorophenol	Strong
p-Fluorophenol	Strong
1,3-Dichloro-1,1,3,3-tetrafluoroacetone hydrate (DCTFAH)	Strong
DCTFAH/perchloroethylene (50/50)	Strong
TFA/DCTFAH/methylene chloride/perchloroethylene (15/25/35/25)	Strong
Tetrachloroethane	Weak
N,*N*-dimethylformamide	Weak
Dioxane	Very weak

solvents are acids or phenols themselves, so care must be taken to prevent hydrolytic or exchange reactions which can degrade the polymers, and this problem is a particular concern in the measurement of the molecular weights of these polymers. Nevertheless, by trial and error, a particular solvent can usually be found for a particular polymer with which molecular weights can be either estimated by the use of either solution viscosity or gel permeation chromatography, and weight-average molecular weights can be directly determined by solution light scattering in such solvents.[29]

ACKNOWLEDGEMENT

The author wishes to express his appreciation to the Office of Naval Research, which has financially supported his research on liquid crystal polymers at the University of Massachusetts.

REFERENCES

1. Iimura, K., Koide, N. and Katahira, A., *Progr. Polym. Phys., Japan*, 1982, **25**, 295.
2. Kwolek, S. L., US Patent 3 671 542 (1972); Kwolek, S. L., Morgan, P. W., Schaefgen, J. R. and Gulrich, L. W., *Macromolecules*, 1977, **10**, 1390.
3. *National Tech. Conf. SPI Preprints*, October 1976, p. 229; Cottis, S. G., Economy, J. and Nowak, B. E., US Patent 3 637 595 (1972).
4. McFarlane, F. E., Nicely, V. A. and Davis, T. G., in *Contemporary Topics in Polymer Science*, Vol. 2, Pearce, E. M. and Schaefgen, J. R. (Eds), Plenum, New York, 1977, p. 109.
5. Morgan, P. W., *Macromolecules*, 1977, **10**, 1381.
6. Lenz, R. W., *Organic Chemistry of Synthetic High Polymers*, Interscience Publishers, New York, 1967, p. 103.
7. Yamazaki, N., Matsumoto, M. and Higashi, F., *J. Polym. Sci., Polym. Chem. Ed.*, 1975, **13**, 1373; Wu, G.-C., Tanaka, H., Sanui, K. and Ogata, N., *Polymer J.*, 1982, **14**, 571.
8. Preston, J., Krigbaum, W. R. and Kotek, R., *J. Polym. Sci., Polym. Chem. Ed.*, 1982, **20**, 3241.
9. Yamazaki, N., Niwano, N., Kawabata, J. and Higashi, F., *Tetrahedron*, 1975, **31**, 665; Ogata, N. and Tanaka, H., *Polymer J.*, 1974, **6**, 461.
10. Jacovic, M. S., Djonlagic, J. and Lenz, R. W., *Polym. Bull.*, 1982, **8**, 295.
11. Ogata, N., Sanui, K. and Hasada, M., *J. Polym. Sci., Polym. Chem. Ed.*, 1979, **17**, 2401; Staab, H. A., *Chem. Ber.*, 1957, **90**, 1326.
12. Morgan, P. W., *Condensation Polymers by Interfacial and Solution Methods*, Interscience, New York, 1965.
13. Bair, T. I., Morgan, P. W. and Killian, F. L., *Macromolecules*, 1977, **10**, 1396.
14. Preston, J., *Angew. Makromol. Chemie*, 1982, **109**/**110**, 1.
15. Kleinschuster, J. J., Pletcher, T. C., Schaefgen, J. R. and Luise, R. R., German Offen. 2 520 820 (1975); Dicke, H.-R. and Lenz, R. W., *J. Polym. Sci., Polym. Chem. Ed.*, 1983, **21**, 2581.
16. Jin, J.-I., Antoun, S., Ober, C. and Lenz, R. W., *Brit. Polym. J.*, 1980, **12**, 132.
17. Lenz, R. W., Ref. 6, p. 83.
18. Majnusz, J., Catala, J. M. and Lenz, R. W., *Eur. Polym. J.*, 1983, **19**, 1043.
19. Kricheldorf, H. R. and Schwarz, G., *Makromol. Chem.*, 1983, **184**, 475; Blaschke, F. and Ludwig, W., US Patent 3 395 119 (1968); Cottis, S. G., Nowak, B. E. and Economy, J., German Offen. 2 025 972 (1970).

20. Hamb, F. L., *J. Polym. Sci. A-1*, 1972, **10**, 3217; US Patent 3772405 (1973).
21. Higashi, F., Kubota, K. and Sekizuka, M., *Makromol. Chem., Rapid Commun.*, 1980, **1**, 457.
22. Higashi, F., Kokubo, N,. and Goto, M., *J. Polym. Sci., Polym. Chem. Ed.*, 1980, **18**, 2879.
23. Yamazaki, N. and Higashi, F., *Advances in Polymer Science, Vol. 38*, Cantow, H.-J. (Ed.), Springer-Verlag, New York, 1981, p. 1.
24. Higashi, F. and Mochizuki, A., *J. Polym. Sci., Polym. Chem. Ed.*, 1983, **21**, 3337.
25. Higashi, F., Akiyama, N. and Koyama, T., *J. Polym. Sci., Polym. Chem. Ed.*, 1983, **21**, 3233.
26. Jackson, W. J., Jr and Kuhfuss, K. H., *J. Polym. Sci., Polym. Chem. Ed.*, 1976, **14**, 2043.
27. Lenz, R. W. and Schuler, A. N., *J. Polym. Sci., Polym. Symp.*, 1978, **63**, 343.
28. Lenz, R. W., Jin, J.-I. and Feichtinger, K. A., *Polymer*, 1983, **24**, 327.
29. Jackson, W. J., Jr, *Brit. Polym. J.*, 1980, **12**, 154.

2

CHIRAL THERMOTROPIC LIQUID CRYSTAL POLYMERS

EMO CHIELLINI

Istituto di Chimica Generale (Facoltà di Ingegneria), Università di Pisa, Italy

and

GIANCARLO GALLI

Istituto di Chimica Organica Industriale, Università di Pisa, Italy

INTRODUCTION

Among the several liquid crystal polymers that have been studied in recent years those containing intrinsically chiral elements with a prevalent chirality hold a particular position. Some of these, in fact, by virtue of their structural characteristics, assume a spatial array with nematic planes stacked in a superhelical structure characterized by a prevalent screw sense and are known as *cholesteric phase*. This kind of order can be controlled by either concentration in solution (*lyotropic* systems) or temperature in bulk (*thermotropic* systems).

The cholesteric structure is generated by the rotation of the director of each successive nematic plane placed away from the previous one by a certain angle of twist, which determines the length of the helical pitch. A preferential twisting of a homogeneously oriented nematic phase between two plane-parallel plates is able to produce a preferentially chiral helical conformation,[1] whose helicity is however rather weak. More significantly, stronger effects can be induced by simply mixing an optically active substance with a 'chemically compatible' nematic phase,[2] or, better, by introducing preferential chirality into the molecule of a nematogen.[3] The molecular structure of the nematogen, its physical state and external parameters such as temperature, pressure, and magnetic and electric fields affect the angle of twist, and hence the general helical properties.[4-7] The helical twisting power ($\Phi = 2\pi d/p$, where d is the distance between two successive nematic layers, and p is the length of the pitch) may vary in thermotropic binary systems with the content of one component in either a linear[8,9] or non-linear fashion.[10–12]

The influence of temperature on the helical twisting power and pitch may result in a contraction of the helical pitch (which corresponds to an increase of the twisting power) in thermotropic liquid crystals,[5] whereas an expansion is often observed in lyotropic systems.[4,7] Cholesterics reflect the light according to a Bragg-type scattering of incident light and, when the length of the pitch is comparable to the wavelength of visible light, they exhibit typical brilliant colours.[13]

The peculiar feature related to temperature-sensitive responses has led to the utilization of cholesteric materials in various kinds of optical applications such as thermal indicators and radiation sensors.[14–17] Other applications connected with the possibility of storing thermal and electromagnetic inputs for cholesteric liquid crystals with suitably elevated relaxation times have been proposed.[18,19] Particularly attractive along these lines are polymeric liquid crystals that may retain the conformational properties after removal of the perturbing factors. A freezing of the cholesteric organization, that was claimed in work of French researchers,[20] has been realized more recently and very elegantly in lyotropic systems by polymerization of solutions of vinyl monomers containing polypeptides.[21,22] Experimental conditions were selected such that the bulk conformational properties of the dissolved high molecular weight cholesterogen were not altered. The insensitivity of selective reflection to temperature makes the reported systems possibly suitable for the fabrication of passive optical devices such as polarizers, electro-magnetic filters and reflective displays. A tuning of the viscoelastic properties of one-component thermotropic liquid crystal polymers can, likely, lead to potentially reversible systems with different reflective properties simply selected on the basis of the manufacturing temperature. In this area of liquid crystal polymers it is the commonly accepted opinion that ‘the best is yet to come’ for both practical application and speculative advances.[23] From these points of view, cholesteric polymers appear particularly promising.

The importance of the liquid crystal order in living systems and in model substances is well documented.[24–28] The lyotropic mesomorphism usually observed in these systems is generally related to the presence of amphiphilic moieties embedded in an intrinsically chiral environment that may cause chiral smectic or, more frequently, cholesteric states. However, any comment on this special area is outside the scope of the present paper, which is focused on thermotropic liquid crystal polymers characterized by the presence of intrinsically chiral elements.

In Part I we present a survey of the work done in the preparation and characterization of synthetic and semi-synthetic chiral thermotropic liquid crystal polymers. For convenience, we have grouped the polymeric materials (until now reported) according to the nature of the repeat unit and relevant position of the mesogen, *side chain* and *main chain* polymers. In Part II we report on the results obtained in our laboratories on optically active thermotropic polyesters containing mesogenic aromatic dyads or triads based on *p*-oxybenzoic acid.

PART I. LITERATURE SURVEY OF VARIOUS HOMOPOLYMERS AND COPOLYMERS

1. Side Chain Chiral Thermotropic Liquid Crystal Polymers

Polymers containing chiral groups with a prevalent chirality in the side chain have been studied rather extensively by several research groups.[20,29–45] Homopolymer and copolymer materials are grouped in Tables 1 and 2, respectively.

A. Homopolymers

The reported homopolymers[29–41] have been obtained by polymerization of intrinsically chiral unsaturated vinyl or vinylidenic monomers, such as acryloyl or methacryloyl derivatives, containing a cholesteryl group or, in one case, an (*S*)-2-methylbutyl residue. The cholesterol-containing monomers are usually liquid crystalline in themselves and incorporation into a polymer backbone does not disrupt their mesogenic character. By starting from either a cholesteric or smectic monomer, a smectogenic polymer is obtained. This indicates that the polymerization process of an anisotropic monomer tends to stabilize the liquid crystalline order.[32,40] As a consequence, if twisted nematic polymers have to be prepared, isotropic optically active monomers with mesogenic propensity should be necessarily used.[40]

In all cases the stereoirregular polymers obtained possess rather high glass transition temperatures and normally, above T_g, a marked anisotropic character associated with the establishment of a smectic order, as anticipated. Wider temperature ranges of mesophase existence are recorded for polymers with higher flexibility connected with either less main chain rigidity or higher conformational freedom of the side chain. The values of the isotropization temperature in corresponding samples

TABLE 1
Chiral Thermotropic Homopolymers with Mesogenic Units in the Side Chain

Chiral component	Anchoring repeating unit		Mesophase				Ref.
	Type	*n*	*Temperature* T_i (°C)[a]	ΔT[b]	*Structure*	*Remarks*	
Cholesteryl	—CH(—CH$_2$—)—COO—	—	n.r.	—	Smectic?	Intermediate order	29, 30
Cholesteryl	—CH(—CH$_2$—)—COO—C_6H_4—COO—	—	n.r.	—	Smectic		29
Cholesteryl	—CH(—CH$_2$—)—CONH$(CH_2)_n$COO—	2–11	220	75	Smectic ($n \geq 5$)	Not liquid crystalline ($n < 5$)	31, 32
Cholesteryl	—CH(—CH$_2$—)—COO$(CH_2)_n$COO—	5	n.r.	—	Cholesteric[c]	Reflection of UV light	33
Cholesteryl	CH_3—C(—CH$_2$—)—COO—	—	>200[d]	—	—		34, 35
		—	n.r.	—	Smectic		29, 30

Cholesteryl	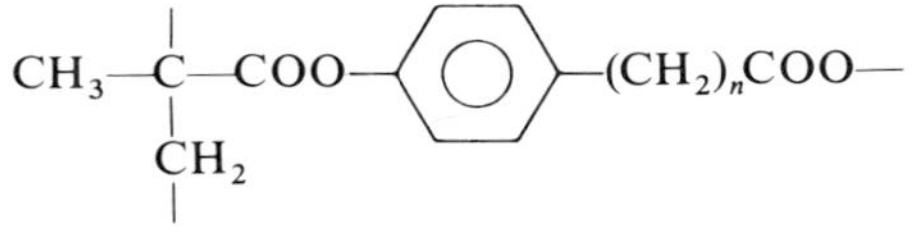 $CH_3-C(-CH_2-)-COO-C_6H_4-(CH_2)_nCOO-$	2–12	182	n.r.	Smectic		36
Cholesteryl	$CH_3-C(-CH_2-)-CONH(CH_2)_nCOO-$	2–11	220	90	Smectic ($n \geq 5$)	Not liquid crystalline ($n < 5$)	34, 35
Cholesteryl	$CH_3-C(-CH_2-)-COO(CH_2)_nCOO-$	5–14	210	125	Smectic		32, 37
Cholesteryl	$CH_3-Si(-O-)-(CH_2)_nCOO-$	3	115	70	Smectic		38, 39
(*S*)-2-methylbutyl	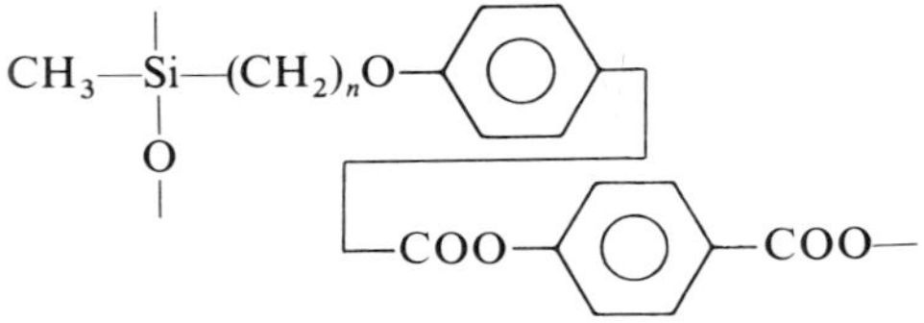 $CH_3-Si(-O-)-(CH_2)_nO-C_6H_4-COO-C_6H_4-COO-$	3–5	311	41	Smectic and/or cholesteric	Reflection of IR and VIS light	40, 41

[a] Maximum value of the isotropization temperature in the series.
[b] Maximum value of the temperature range of mesophase existence.
[c] Extrapolated from mesophase behaviour of copolymers.
[d] Glass transition temperature.
n.r., Not reported.

TABLE 2
Chiral Thermotropic Copolymers with Mesogenic Units in the Side Chain

Residue from chiral component	Residue from achiral component[a]: Type	n	Content (mol %)	Mesophase: Temperature $\bar{T}_i$ (°C)[b]	ΔT[c]	Structure	Remarks	Ref.
$-CH(-CH_2-)-COO(CH_2)_5COO-$cholesteryl	$-CH(-CH_2-)-COO(CH_2)_nCOO-C_6H_4-COO-C_6H_4-OCH_3$	5	55–86	110	78	Cholesteric	Reflection of VIS light	33, 42
	$-CH(-CH_2-)-COO(CH_2)_nCOO-C_6H_4-C_6H_4-CN$	5	35–66	105	52	Cholesteric	Reflection of VIS and IR light	33
$-CH(-CH_2-)-COO-$cholesteryl	$-CH(-CH_2-)-COO-C_6H_4-CH{=}N-C_6H_4-N{=}CH-C_6H_4-OCO-CH(-CH_2-)-$ + $-CH(-CH_2-)-COO-C_6H_4-CH{=}N-C_6H_4-CN$	—	90–98·5			Cholesteric	Reflection of VIS light (frozen in)	20
$CH_3-C(-CH_2-)-COO(CH_2)_2O-C_6H_4-COO-C_6H_4-CH{=}N-\overset{*}{C}H(CH_3)-C_6H_5$	$CH_3-C(-CH_2-)-COO(CH_2)_nO-C_6H_4-COO-C_6H_4-C_6H_4-OCH_3$	6	75–95	247	177	Cholesteric		43

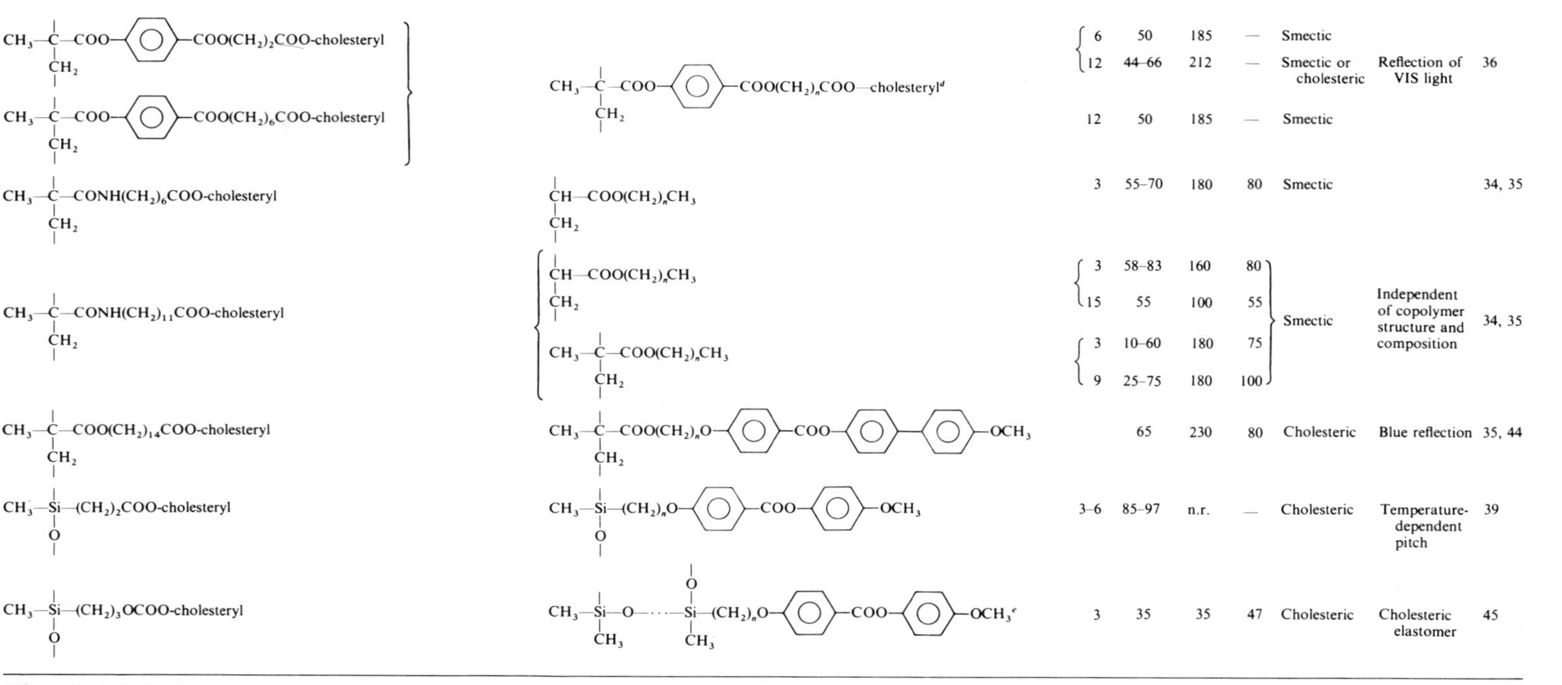

6	50	185	—	Smectic		
12	44–66	212	—	Smectic or cholesteric	Reflection of VIS light	36
12	50	185	—	Smectic		
3	55–70	180	80	Smectic		34, 35
3	58–83	160	80	Smectic	Independent of copolymer structure and composition	34, 35
15	55	100	55			
3	10–60	180	75			
9	25–75	180	100			
	65	230	80	Cholesteric	Blue reflection	35, 44
3–6	85–97	n.r.	—	Cholesteric	Temperature-dependent pitch	39
3	35	35	47	Cholesteric	Cholesteric elastomer	45

[a] If not otherwise indicated.
[b] Maximum value of the isotropization temperature in the series.
[c] Maximum value of the temperature range of mesophase existence.
[d] Optically active residue.
[e] Crosslinked via a α,ω-divinylnona(dimethylsiloxane).
n.r., Not reported.

of the two series of homopolymers based on acrylate and methacrylate monomers are comparable, provided that the length of the spacer between the mesogenic group and the backbone is sufficiently high. There is clear indication, therefore, that the nature of the mesogenic moiety is solely responsible for the anisotropic behaviour. Samples with the mesogenic group relatively close to the polymer backbone do not show clear evidence of liquid crystalline order in the melt phase.[31,32,34,35]

More recently, an elegant route to cholesteric homopolymers based on post-reactions carried out with insertion of inherently chiral mesogens on preformed flexible reactive polymers, such as poly(hydrogen methyl) siloxane, has been introduced.[40] Accordingly, the preparation of the first examples of cholesteric homopolymers has been claimed. However, it is fair to mention that reactions performed even on structurally simple polymers never reach total conversion of functional groups and one must take into account that the presence of any structural irregularity along the polymer backbone might play a role in affecting the ultimate properties of the polymer bulk.

The reported cholesteric polysiloxanes are characterized by typical reflection of incident light in accordance with the establishment of a right-handed helical structure. The pitch of the helix decreases with increasing temperature with a concurrent shift of the maximum of reflection wave-length from near-IR towards the visible range.[40]

B. Copolymers

A more valuable route to realize polymeric materials displaying cholesteric mesophases appears to be that based on processes involving the formation of copolymers[20,33–36,42–45] by copolymerization of:

(i) two optically active, mesogenic monomers;
(ii) an optically active, not mesogenic monomer with an achiral, mesogenic comonomer;
(iii) an optically active, mesogenic monomer with an achiral comonomer mesogenic or not mesogenic.

The procedure based on the copolymerization of a cholesterogenic monomer with a mixture of nematogenic comonomers was, for the first time, introduced by Strzelecki and Liebert[20] in the synthesis of crosslinked polymers that retained frozen cholesteric phases with rather extended helical pitches (1500–4400 nm). Five years later, Ringsdorf and co-workers[43] reported the first preparation of an enantiotropic cholesteric copolymer. This was obtained by copolymerization of appropriate mixtures of mesogenic cholesterol derivatives with the chiral unit spaced from the methacrylate backbone through segments of different length and rigidity.

The formation of copolymer structures tends, as a major effect, to suppress the smectogenic tendency of corresponding homopolymers, and either intrinsic or induced twisted nematic phases can be originated in rather wide temperature intervals. The onset and stability of the cholesteric mesophases are markedly affected by the chemical structure of the two comonomers (nature and length of the spacer and character of the mesogenic core) and chemical composition of the copolymers. Accordingly the larger variety of cholesterogenic materials with copolymer structure with respect to the corresponding homopolymers can be easily justified. It appears evident, therefore, the great value of the copolymerization methods to realize the synthesis of structurally analogous materials with suitable mesomorphic properties. However, by examining the data collected in Table 2, we have to remark that even the indicated procedures lead in several cases to mesophases with positional order of the mesogenic elements typical of smectic structures.[34-36]

Analogous to the described homopolymers, the copolymers are mainly based on stereoirregular methacryloyl or, less frequently, acryloyl derivatives containing cholesterol as the intrinsically chiral component. A series of linear[39] and crosslinked (elastomeric)[45] copolymers based on the highly flexible siloxane backbone has been investigated in great detail and in all cases the copolymers have been found to assume a cholesteric structure.

Most of the reported cholesteric copolymers spontaneously develop planar textures which reflect visible light. As a general remark, it is important to note, from an applicative point of view, that amorphous polymeric materials, that do not possess a smectic phase at low temperatures, can preserve their cholesteric structure in the glassy state.[33,39] This may allow for the preparation of films and coatings provided with intrinsic reflecting properties and bright colours, not imparted by any dye or chromophore in the polymer matrix.

2. Main Chain Chiral Thermotropic Liquid Crystal Polymers

In this chapter chiral thermotropic semisynthetic and synthetic polymers are included. They will be treated in separate sections focused on cellulose and polypeptide systems and on polyesters containing (+)-(*R*)-3-methyladipic acid as the chiral moiety. Mixtures of achiral condensation polymers with optically active low molar mass compounds will be also briefly discussed.

A. Cellulose and Polypeptide Systems

Cellulose derivatives[46] and polypeptides[25] are well known to exhibit lyotropic liquid crystal properties in a large variety of solvents and only recently have been found to possess, additionally, thermotropic characteristics.

A particular position among such chiral liquid crystal polymers is held by the systems based on cellulose derivatives.[47–52] They can be classified as cellulose ethers, namely hydroxypropyl cellulose, with some adjunctive ester functions localized normally on the ether branches. The idealized substituted cellobiose repeating unit is represented as follows:

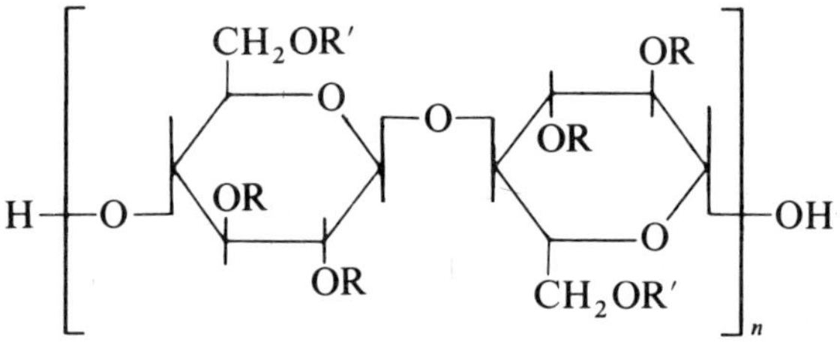

$$R = —CH_2—CH(OX)—CH_3; \quad R' = —CH_2—CH(OCH_2—CH(OX)—CH_3)—CH_3$$

$$X = H, CH_3CO—, CF_3CO—, CH_3CH_2CO—, C_6H_5CO—$$

The molar average degrees of etherification (allowing for the ether functions in R) and esterification are 6–8 and 5–6, respectively.

In Table 3 are summarized the general structures and liquid crystal characteristics of hydroxypropyl cellulose and its ester derivatives.

The cellulose backbone appears to be sufficiently stiff to guarantee the formation of ordered phases with nearly parallel orientation. Preferential chirality of the macromolecular chain imparts a twist to the parallel arrangement, thus inducing the helical structure typical of cholesteric materials.

The average molecular weight of the hydroxypropyl cellulose derivative has been found to affect to a significant extent the thermotropic behaviour, as far as onset and isotropization temperature are concerned. An increase of clearing temperature and breadth of mesophase is observed with increasing molecular weight, a plateau value of T_i being reached at $\bar{M}_w \simeq 5 \times 10^5$ (within the benzoate series), whereas no liquid crystal

TABLE 3
Thermal–Optical Properties of Thermotropic Hydroxypropyl Cellulose Derivatives

Cellulose derivative		*Mesophase*		*Ref.*
Type	$\bar{M}_w \times 10^{-4}$	*Temperature range (°C)*	*Remarks*	
Hydroxypropyl cellulose	6	160–205	Cholesteric reflection of VIS light	47
Hydroxypropyl cellulose acetate	14	room–164	Cholesteric reflection of VIS light	48
Hydroxypropyl cellulose trifluoroacetate	17	115–155	Birefringent	49
Hydroxypropyl cellulose propanoate	22	room–170	Cholesteric reflection of VIS light	50
Hydroxypropyl cellulose benzoate	36	room–164	Birefringent	51, 52

properties were detected for $\bar{M}_w$ values lower than approximately 5×10^4.[52]

The introduction of ester groups, with a consequent depression of intermolecular hydrogen bonding, causes a rather marked decrease of the melting temperature of the ester derivative with respect to the starting unmodified hydroxypropyl cellulose, and materials exhibiting thermotropic mesomorphism even at room temperature are obtained.[48,50–52] Furthermore, side chain substituents control the average spacing among the almost parallel rigid segments with a consequent influence on the pitch of the cholesteric array and, hence, on the optical properties of the corresponding anisotropic melt. Reflection of visible light is observed with samples containing less bulky acyl residues (acetate and propionate esters),[48,50] whereas longer pitches appear to exist in the benzoate derivative.[51,52] The wavelength of the maximum reflected light in hydroxypropyl cellulose acetate increases with temperature corresponding to an extension of the cholesteric helix, in contrast with the most common contraction observed in thermotropic cholesterics.[5]

Recently, the thermotropic behaviour of copoly(α-amino acid) derivatives based on γ-*n*-alkyl-L-glutamates has been reported for the first time.[53]

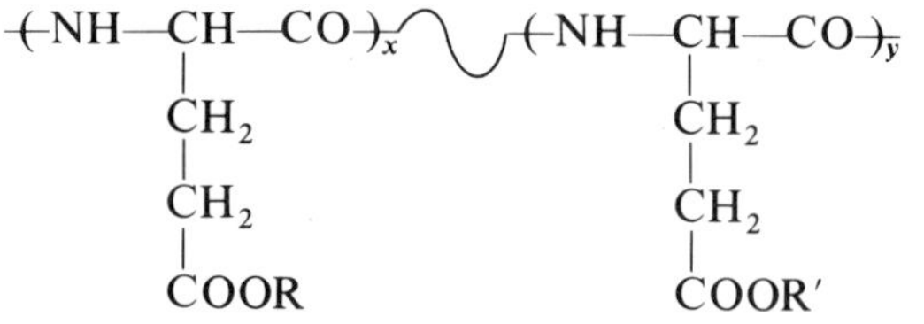

$$\left(\text{NH}-\underset{\substack{|\\ CH_2\\ |\\ CH_2\\ |\\ COOR}}{\text{CH}}-\text{CO}\right)_x \sim \left(\text{NH}-\underset{\substack{|\\ CH_2\\ |\\ CH_2\\ |\\ COOR'}}{\text{CH}}-\text{CO}\right)_y$$

R	R′
CH_3	*n*-C_6H_{13}
CH_3	*n*-C_8H_{17}
n-C_3H_7	*n*-C_8H_{17}

The prepared copolymers give rise to anisotropic melts characterized by selective reflection of visible light in substantial ranges of composition (50–80 % of shorter side chain substituent), consistent with the existence of a right-handed cholesteric helical structure. The temperature intervals of mesophase persistence are relatively narrow ($\simeq$ 30 °C at most) and depend on the chemical structure of the alkyl substituent, minimum values being reached with longer lateral chains. The wavelength of reflected light increases, at any copolymer composition, with increasing temperature, suggesting a significant unwinding of the helical structure. Incidentally, it may be noted that this trend coincides with that shown by lyotropic poly(γ-benzyl-L-glutamate),[4,7] which assumes a right-handed helical conformation in chloroform and dioxane solutions.[4,54] Side chains appear, therefore, to play the role of solvent molecules.[55] In this respect, the thermotropic system should correspond to a highly concentrated solution.[53] By considering the analogy of copolypeptides with cellulose derivatives in their thermotropic behaviour, and in respect also to the relevant structural effects, one may infer that even hydroxypropyl cellulose esters are describable in terms of intrinsic concentrated solutions.

B. Polyesters Based on (R)-3-methyladipoyl Residue

Other than the systems based on cellulose and polypeptide derivatives treated previously, (*R*)-3-methyladipate polyesters containing two-phenyl mesogenic residues are almost the only optically active main chain thermotropic liquid crystal polymers described in the literature.[18,19,56–59] Tables 4 and 5 show the general structures and mesomorphic characteristics of the homo- and co-polycondensate liquid crystal systems, respectively. Chiral thermotropic polyesters have been studied mainly by Blumstein[18,57] and Krigbaum,[19,58,59] almost simultaneously with, or immediately following, the first report by Strzelecki and co-workers.[56] The latter report described also the preparation and properties of a cholesteryl end-capped polyester based on achiral components 4′-hydroxyphenyl 4-hydroxybenzoate and pimelic acid[56] (Table 5).

The bipolycondensates of (*R*)-3-methyladipic acid with mesogenic 4,4′-diphenyloxy or 4,4′-diphenylethylenoxy residues with interconnecting carboxy, azoxy, isopropenyl, or acryloyloxy polarizable groups, exhibited cholesteric Grandjean textures accompanied, in some cases, by reflection of visible light in more or less broad temperature ranges. Depending on the nature of the mesogenic units, the clearing temperatures of polymers varied in the range 150–300 °C. The lowest values were shown by samples incorporating substituted, or less rigid, mesogenic cores.[18,59]

By starting from the same mesogens used in the preparation of bipolycondensates, a series of cholesteric copolymers based on (*R*)-3-methyladipic and linear α,ω-dicarboxylic acids (adipic, pimelic, suberic and dodecanedioic) have been synthesized.[18,19,56–59] Cobipolymers based on 4′-hydroxyphenyl 4-hydroxycinnamate or 4,4′-dihydroxydiphenyl mesogenic units and 1:1 mixtures of aliphatic diacids containing (*R*)-methyladipic acid as the chiral component have been also investigated.[58] The copolymerization procedure offers the possibility of preparing a large variety of mesomorphic polymers displaying improved solubility properties and possessing a much wider interval of mesophase existence in the melt phase. This result, analogous to that already evidenced in the case of copolymers of optically active unsaturated monomers, proves that randomness of different residues in the macromolecular backbone helps to disrupt, even to a greater extent here, the three-dimensional order of polymer segments

TABLE 4
Chiral Thermotropic Bipolymers with Mesogenic Units in the Main Chain

Chiral component residue	*Mesogenic unit*	*Mesophase*			*Ref.*
		Temperature range (°C)	*Structure*	*Remarks*	
(*R*)-3-methyladipoyl	$-O-C_6H_4-CH{=}CH-COO-C_6H_4-O-$	203–210	Cholesteric	Blue reflection	59
	$-O-C_6H_4-CH{=}C(CH_3)-C_6H_4-O-$	110–250	Cholesteric	Blue reflection[a]	58
	$-O-C_6H_4-N{=}N(\rightarrow O)-C_6H_4-O-$	221–295	Cholesteric	Reflection of UV light	18, 57
	$-O-C_6H_3(CH_3)-N{=}N(\rightarrow O)-C_6H_3(CH_3)-O-$	151–201	Cholesteric	Reflection of UV light	18
	$-OCH_2CH_2-C_6H_4-N{=}N(\rightarrow O)-C_6H_4-CH_2CH_2O-$	155–165	Cholesteric	Reflection of UV light	18
	$-O-C_6H_4-COO-C_6H_4-O-$	n.r.	Cholesteric	Reflection of VIS light	56

[a] After mixing with corresponding polymer containing adipic acid residues.
n.r., Not reported.

TABLE 5
Chiral Thermotropic Copolymers with Mesogenic Units in the Main Chain

Chiral component residue	*Achiral component*		*Mesogenic unit*	*Mesophase*			*Ref.*
	Type	*Content (mol %)*		*Temperature range (°C)*[a]	*Structure*	*Remarks*	
(*R*)-3-methyladipoyl	$—CO(CH_2)_4CO—$	50	$—O—C_6H_4—CH{=}C(CH_3)—C_6H_4—O—$	199–300	Cholesteric	Blue reflection	58
(*R*)-3-methyladipoyl	$—CO(CH_2)_{10}CO—$	20–90	$—O—C_6H_4—N(\rightarrow O){=}N—C_6H_4—O—$	162–279	Cholesteric	Reflection of VIS light	18, 57
(*R*)-3-methyladipoyl	$—CO(CH_2)_{10}CO—$	25–75	$—O—C_6H_3(CH_3)—N(\rightarrow O){=}N—C_6H_3(CH_3)—O—$	76–178	Cholesteric	Reflection of VIS light	18
(*R*)-3-methyladipoyl	$—CO(CH_2)_{10}CO—$	25–75	$—OCH_2CH_2—C_6H_4—N(\rightarrow O){=}N—C_6H_4—CH_2CH_2O—$	140–150	Undefined	No iridescence	18
(*R*)-3-methyladipoyl	$—CO(CH_2)_8CO—$	50	$—O—C_6H_4—C_6H_4—O—$	220–275	Cholesteric		19
(*R*)-3-methyladipoyl	$—CO(CH_2)_5CO—$	50–97	$—O—C_6H_4—COO—C_6H_4—O—$	150–260	Cholesteric	Reflection of VIS and IR light	56
Cholesteryl (as end group)	$—CO(CH_2)_5CO—$	100	$—O—C_6H_4—COO—C_6H_4—O—$	n.r.	Cholesteric	Reflection of IR light	

[a] Maximum value in the series.
n.r., Not reported.

and results in a very depressed melting temperature. By contrast, the isotropization temperature appears to be affected in a much less significant way. The maximum value of mesophase breadth usually occurs at an equimolar composition of achiral and chiral diacid residues in the polymer, which corresponds to a maximum in structural randomness.

Interestingly, copolymers containing variable amounts of the chiral component are characterized by a gradual change in the mesophase properties.[18] A lengthening of the helical pitch of the cholesteric mesophase is recorded in going from the corresponding homopolyester of (*R*)-3-methyladipic acid to the homopolyester of the achiral diacid. Correspondingly, different selective reflections are observed at various chemical compositions.[18,57]

C. Chiral Composite Polymeric Systems

It has been established that conventional nematic and tilted smectic phases can be converted, by dissolution of optically active compounds, to mesophases with a preferential chirality (twisted nematics and twisted smectics).[2,60] This technique has been successfully extended to liquid crystal polymers in a diagnostic effort addressed to elucidation of mesomorphic structures of both chiral and achiral polycondensates.

By following the 'contact method',[61] Noel and co-workers[62–64] have reported the isobaric phase diagrams of mixtures of optically active compounds with thermotropic liquid crystal polymers and copolymers, such as those represented by structures **1–5**

1 $-\!\!\!\left(CH{=}N{-}C_6H_3(CH_3){-}N{=}CH{-}C_6H_4\right)\!\!\!-$ $\quad K \xrightarrow{258\,^\circ C} N \xrightarrow{341\,^\circ C} I$

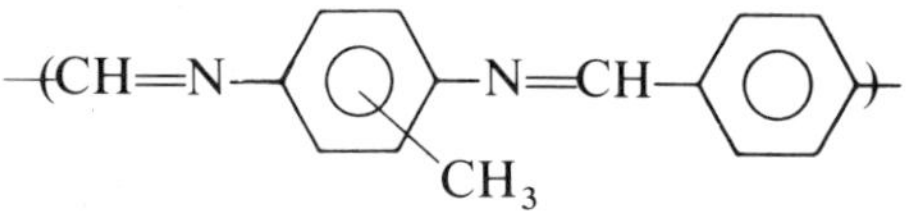

2 $-\!\left(O{-}C_6H_4{-}CO\right)_{0{\cdot}8}\!-\!\left(OCH_2CH_2OCO{-}C_6H_4{-}CO\right)_{0{\cdot}2}$

$K \xrightarrow{298\,^\circ C} N \xrightarrow{350\,^\circ C} I$

3 $-\!\left[CO{-}C_6H_4{-}C_6H_4{-}C_6H_4{-}COO{-}(CH_2CH_2O)_4\right]\!-$

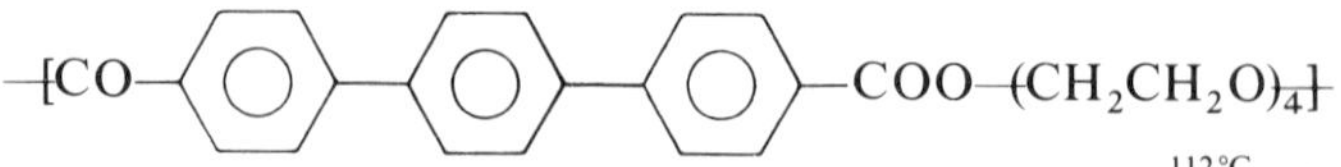

$K \xrightarrow{112\,^\circ C} S_C \xrightarrow{247\,^\circ C} I$

4 $-\!\left(CO{-}C_6H_4{-}CO\right)_{x+y}\!-\!\left(O{-}C_6H_3(CH_3){-}O\right)_x\!-\!\left(O{-}C_6H_4{-}O\right)_y\!-$

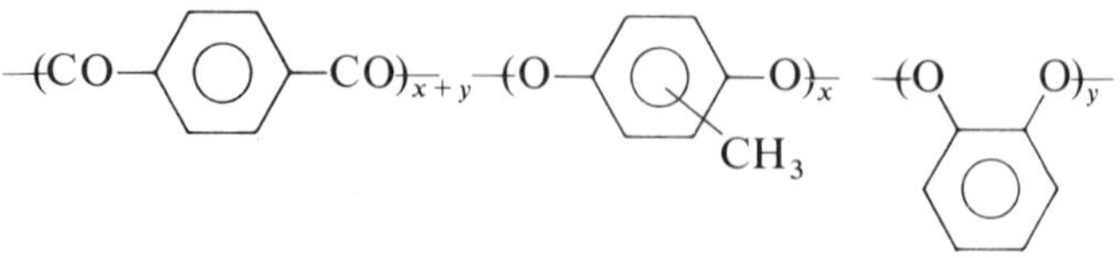

$g \xrightarrow{109\,^\circ C} N \xrightarrow{380\,^\circ C} I$

$$\mathbf{5}\quad -(CO-C_6H_4-O)_{0\cdot6}-(CO-C_6H_4-OCH_2CH_2O-C_6H_4-COOCH_2CH_2O)_{0\cdot4}\quad K \xrightarrow{218\,°C} N \xrightarrow{337\,°C} I$$

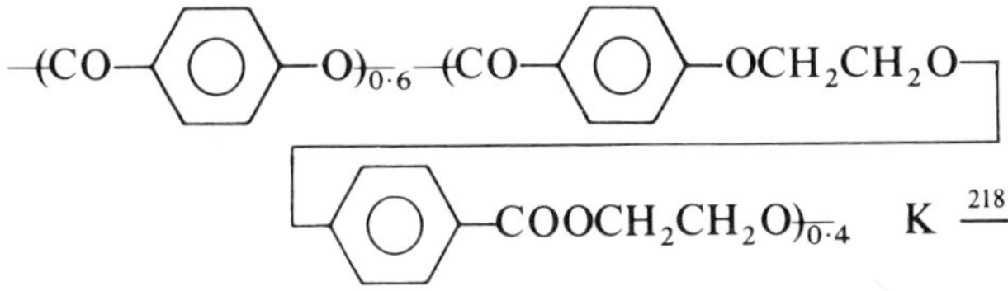

where g = glass; K = semicrystalline; N = nematic; S = smectic; I = isotropic.

The optically active low molecular weight mesogens employed were either 4′-(2-methylhexyloxy)-biphenyl-4-carboxylic acid (**6**) or terephthalylidene-bis[4-(4′-methylhexyloxy)aniline] (**7**).

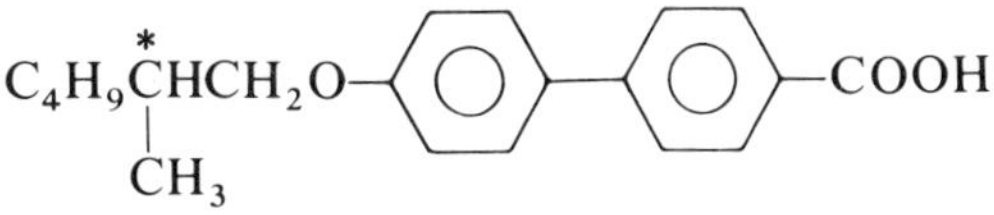

$$C_4H_9\overset{*}{C}H(CH_3)CH_2O-C_6H_4-C_6H_4-COOH\quad \mathbf{6}$$

$$K_I \xrightarrow{103\,°C} K_{II} \xrightarrow{165\,°C} K_{III} \xrightarrow{171\,°C} S_C \xrightarrow{215\,°C} Ch \xrightarrow{229\,°C} I$$

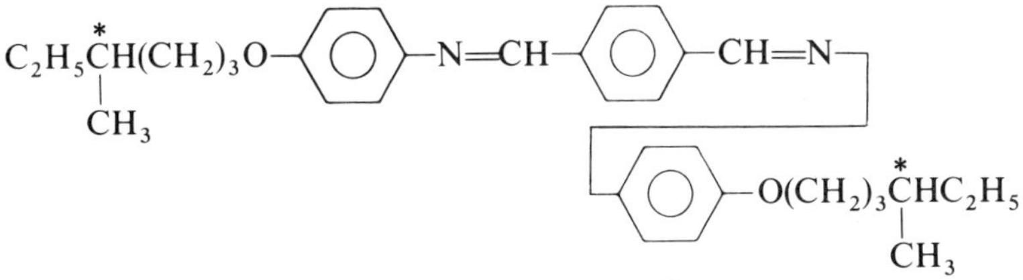

$$C_2H_5\overset{*}{C}H(CH_3)(CH_2)_3O-C_6H_4-N{=}CH-C_6H_4-CH{=}N-C_6H_4-O(CH_2)_3\overset{*}{C}H(CH_3)C_2H_5\quad \mathbf{7}$$

$$K \xrightarrow{144\,°C} S_B \xrightarrow{160\cdot4\,°C} S_C \xrightarrow{270\cdot6\,°C} Ch \xrightarrow{224\cdot5\,°C} I$$

where K = crystalline; S = smectic; Ch = cholesteric; I = isotropic.

The former mesogen was utilized in combination with polyesters which give nematic structures in the melt, with the exception of the *p*-terphenyl-containing polyester that shows a smectic C phase. The latter optically active mesogen was used in combination with the nematogenic azomethine polymer and with the *p*-terphenyl polyester. The last polymer displays, on admixture over a significant range of composition, a twisted smectic C phase,[62] whereas induced cholesteric structures were observed in all other cases.[62–64]

The addition of either a low molecular weight or polymeric liquid crystal, *p*-azoxyanisole and poly(4,4′-dihydroxy-α-methylstilbene adipate) respectively, to cholesteric poly(4,4′-dihydroxy-α-methylstilbene 3-methyladipate) has been also investigated.[58] An extension of the helical pitch was achieved, according to the shift to the visible range of the maximum reflection wavelength that was centred in the UV region in the case of the optically active homopolyester.

One might conclude that the investigations mentioned have utilized a new powerful route to formulate composite polymeric materials with peculiar anisotropic behaviour.

PART II. CHIRAL THERMOTROPIC POLYESTERS BASED ON AROMATIC TRIAD UNITS

The research undertaken in this area stems from our interest, dating back to the late seventies, in the preparation of segmented polycondensation polymers constituted by flexible segments of varying hydrophilic character and rigid segments of either linear or non-linear structure.[65] Among these, the polyesters based on *p*-hydroxybenzoic acid (H) residues built in benzenoid dyads and triads with a 1,4-phenylene [terephthaloyl (T)], 1,2-phenylene [phthaloyl (P)], or 1,3-phenylene [isophthaloyl (I)] diacid moiety interconnected by chiral glycol or glycol ether spacers appeared to us to be of special interest for their potential of displaying thermotropic mesomorphic properties. It was expected, in fact, that the order in the melt phase could be varied on the basis of the structure of both mesogen and flexible diol. The chiral polymeric materials that have been prepared and characterized in the melt state and, whenever possible, in dilute solution have general structures as follows:

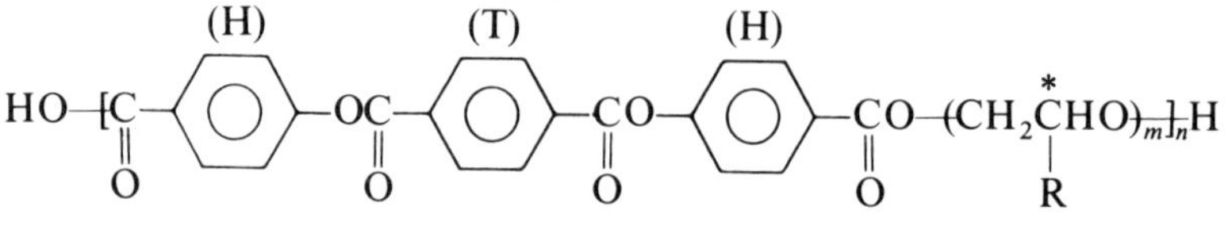

$$HO{+}\underset{\text{(H)}}{C(=O){-}C_6H_4{-}O}C(=O){-}\underset{\text{(T)}}{C_6H_4}{-}C(=O)O{-}\underset{\text{(H)}}{C_6H_4}{-}C(=O)O{+}CH_2\overset{*}{C}H(R)O{+}_m{+}_nH \qquad \mathbf{8}$$

$$HO{+}\underset{\text{(H)}}{C(=O){-}C_6H_4{-}O}C(=O){-}\underset{\text{(T)}}{C_6H_4}{-}C(=O)O{+}CH_2\overset{*}{C}H(R)O{+}_m{+}_nH \qquad \mathbf{9}$$

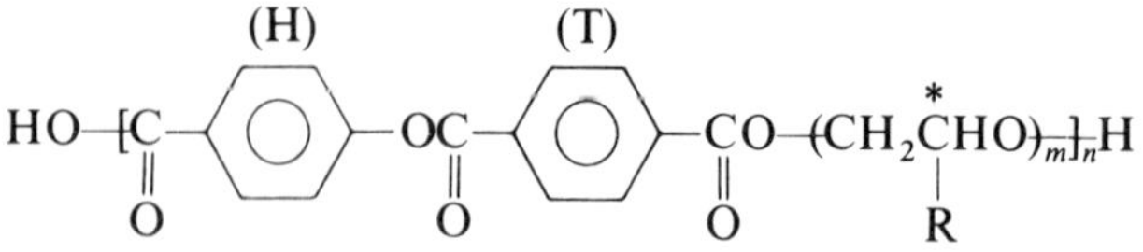

R	*m*
$-CH_3$	1, 2, 3
$-CH_2OR'$	1

Within the two classes, the systems investigated in more detail are those containing aromatic triads of the same structure (bipolycondensates), or triads of linear and non-linear character at the same time (copolycondensates). In this framework, two series (**10**, **11**) of copolymers containing units terephthalate–isophthalate (HTH–HIH) or terephthalate–*ortho*-phthalate (HTH–HPH) were also studied.

(H) (T) (H)

$$-\!\!\left(\text{C(=O)}-\text{C}_6\text{H}_4-\text{OC(=O)}-\text{C}_6\text{H}_4-\text{C(=O)O}-\text{C}_6\text{H}_4-\text{C(=O)OCH}_2\overset{*}{\text{C}}\text{H(CH}_3\text{)O}\right)_{x}\!\!-$$

(H) (I) (H)

$$-\!\!\left(\text{C(=O)}-\text{C}_6\text{H}_4-\text{OC(=O)}-\text{C}_6\text{H}_4-\text{C(=O)O}-\text{C}_6\text{H}_4-\text{C(=O)OCH}_2\overset{*}{\text{C}}\text{H(CH}_3\text{)O}\right)_{y}\!\!- \quad \textbf{10}$$

(H) (T) (H)

$$-\!\!\left(\text{C(=O)}-\text{C}_6\text{H}_4-\text{OC(=O)}-\text{C}_6\text{H}_4-\text{C(=O)O}-\text{C}_6\text{H}_4-\text{C(=O)OCH}_2\overset{*}{\text{C}}\text{H(CH}_3\text{)O}\right)_{x}\!\!-$$

(H) (P) (H)

$$-\!\!\left(\text{C(=O)}-\text{C}_6\text{H}_4-\text{OC(=O)}-\text{C}_6\text{H}_4-\text{C(=O)O}-\text{C}_6\text{H}_4-\text{C(=O)OCH}_2\overset{*}{\text{C}}\text{H(CH}_3\text{)O}\right)_{y}\!\!- \quad \textbf{11}$$

1. Synthesis

Polymer samples were prepared by stepwise condensation of the diacid chloride of aromatic dyad or triad moieties with stoichiometric amounts of glycols or glycol ethers at 60–70 °C in solution of 1,2-dichloroethane/pyridine mixture of pure pyridine. The rigid anisotropic units were preformed prior to polycondensation, according to the procedures reported in Scheme 1. The experimental details have been described elsewhere.[66,67]

Alternative synthetic procedures,[68] based respectively on the preparation of α,ω-bis(4-hydroxybenzoyloxy) alkanes and successive coupling with stoichiometric amounts of terephthaloyl chloride (Series **8**), or of the diacid chloride of α,ω-bis(4-carboxybenzoyloxy) alkanes (Series **9**), were not used; neither was the procedure involving the melt-phase transesterification[69] used. This last synthetic route has been found to work adequately only in the insertion reaction of *p*-oxybenzoyl units in poly(ethylene terephthalate),

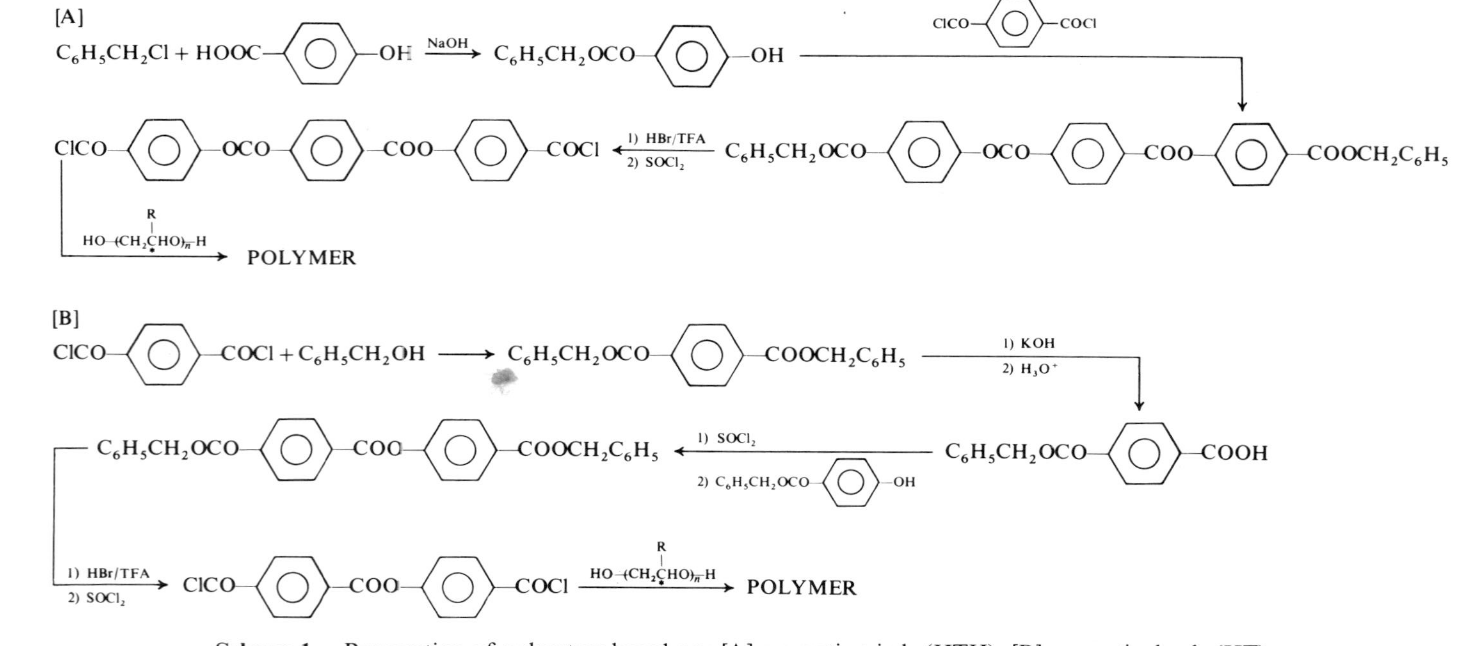

Scheme 1. Preparation of polyesters based on: [A] aromatic triads (HTH), [B] aromatic dyads (HT).

whereas no insertion of practical value was observed with corresponding homologous poly(alkylene terephthalate)s.[70] In any case, it cannot guarantee a predictable length and sequencing of aromatic mesogens along the macromolecular backbone. It should be noted, however, that even by using the stepwise procedures represented in Scheme 1, structurally disordered polycondensates are expected to be obtained in respect to the distribution of the two distinct glycol reactive sites (Series **8**), and of constitutional order of aromatic dyad residues (Series **9**).

If we limit our considerations to a sequence of two repeat polyester units, several structural events can be realized as far as the sequential orientation of glycolic and aromatic segments within the backbone is concerned (Scheme 2). In the case of other polycondensates, such as polyamides derived from diamines with two different reactive amino groups (primary and secondary), it is possible to predetermine the sequential structure of the polyamide by taking into account the known relative reactivity of the two amino groups.[71] Nothing similar seems to be correctly applicable to polyesters of a primary–secondary diol, due to the impossibility of controlling any structural rearrangements due to transesterification phenomena.

The synthesis of the chiral glycerol ethers and glycol ethers was realized by starting from commercially available chiral precursors, such as (D)-mannitol (**12**) and (*S*)-ethyl lactate (**13**), according to Schemes 3 and 4 respectively. The glycerol ethers prepared were (*R*)-methyl glycerol (MG), (*R*)-benzyl glycerol (BzG), (*R*) (*S*)-methoxyethylene oxyglycerol (MEG), (*R*)-methoxydiethylene oxyglycerol (MDEG), and (*R*) (*S*)-methoxytriethylene oxyglycerol (MTEG) (Scheme 3). The glycol ethers synthesized were (*S*)-propylene glycol (PG), (*S*,*S*)-dipropylene glycol (DPG), and

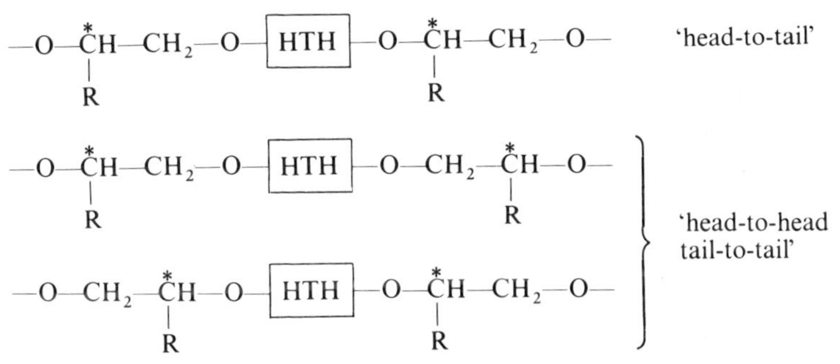

$—O—\overset{*}{C}H(R)—CH_2—O—\boxed{HTH}—O—\overset{*}{C}H(R)—CH_2—O—$ 'head-to-tail'

$—O—\overset{*}{C}H(R)—CH_2—O—\boxed{HTH}—O—CH_2—\overset{*}{C}H(R)—O—$

$—O—CH_2—\overset{*}{C}H(R)—O—\boxed{HTH}—O—\overset{*}{C}H(R)—CH_2—O—$

'head-to-head tail-to-tail'

Scheme 2. Representation of positional isomerism in polyesters of symmetric diacid HO—(HTH)—OH and asymmetric diol.

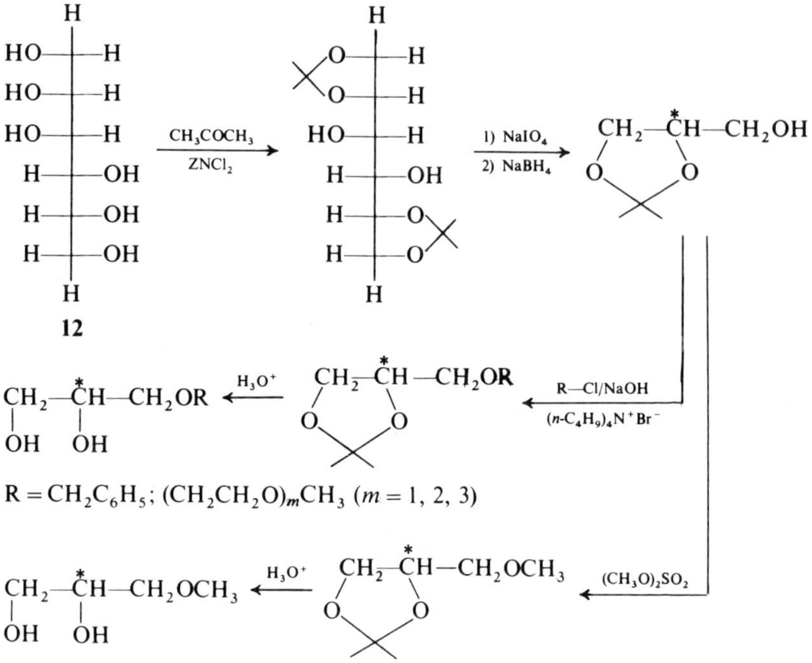

12

CH_3COCH_3 / $ZnCl_2$ → 1) $NaIO_4$ 2) $NaBH_4$ → $\overset{*}{C}H_2$–CH–CH_2OH

R—Cl/NaOH, $(n\text{-}C_4H_9)_4N^+Br^-$ → CH_2–$\overset{*}{C}H$–CH_2OR —H_3O^+→ $CH_2(OH)$–$\overset{*}{C}H(OH)$–CH_2OR

$R = CH_2C_6H_5$; $(CH_2CH_2O)_mCH_3$ ($m = 1, 2, 3$)

$(CH_3O)_2SO_2$ → CH_2–$\overset{*}{C}H$–CH_2OCH_3 —H_3O^+→ $CH_2(OH)$–$\overset{*}{C}H(OH)$–CH_2OCH_3

Scheme 3. Synthesis of chiral glycerol ethers starting from (D)-mannitol (**12**).

(*S*,*S*,*S*)-tripropylene glycol (TPG) (Scheme 4). The experimental details are reported elsewhere.[72–74] No appreciable racemization occurs during the oxidative cleavage of (D)-mannitol diacetonide followed by chemical reduction of isopropylidene glyceraldehyde, whereas a somewhat unexpected racemization was encountered during the alkylation of isopropylidene glycerol and hydrolytic removal of the isopropylidene group under acidic conditions. An almost quantitative racemization was detected in the synthesis of methoxyethylene oxyglycerol and methoxytriethylene oxyglycerol. Research is in progress to establish a better control of the final deblocking stage, which, very likely, is the critical one for the occurrence of a racemization process. In fact, isopropylidene glycerol is known to lose enantiomeric purity during its storage in glass[75] as promoted by residual glass acidity.

On the contrary, the reactions leading to (*S*)-1,2-propanediol and its head-to-tail dimer and trimer glycol ethers occur without any appreciable racemization.[73] The preparation of glycol ethers by catalytic oligomerization of the corresponding epoxides is under investigation.[73,76] It must

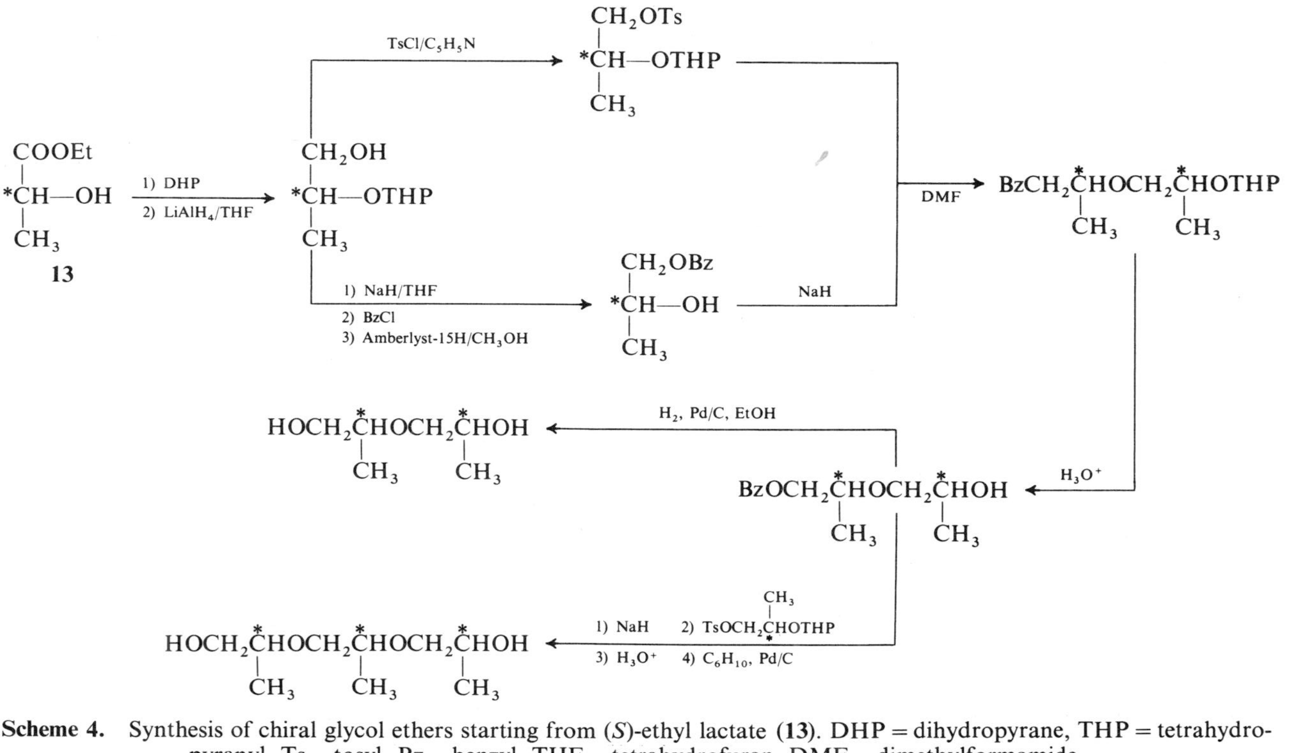

Scheme 4. Synthesis of chiral glycol ethers starting from (*S*)-ethyl lactate (**13**). DHP = dihydropyrane, THP = tetrahydropyranyl, Ts = tosyl, Bz = benzyl, THF = tetrahydrofuran, DMF = dimethylformamide.

be emphasized, however, that the selected routes to chiral hydroxylated derivatives usable in the synthesis of a great variety of polycondensation polymers open the way, for the rather simple chemistry involved, to reasonable exploitations of any polymeric material of interest derived therefrom.

2. Properties in Bulk

A. Bipolyesters of Chiral Glycol Ethers and Glycerol Ethers

In Tables 6 and 7 are collected data relevant to the solubility behaviour, optical rotation in solution, and thermal properties in bulk of the two series of bipolyesters based on linear diacid mesogens (HT and HTH) and head-to-tail oligomers of (*S*)-1,2-propanediol and (*R*)-1-alkyl (or arylalkyl) glycerols, respectively. Optically active components with high optical purity or racemic samples were used in both series. The polymerization products obtained in variable yields (20–80 %) had intrinsic viscosity values, determined whenever possible in organic solvents at 25–35 °C, that ranged from 0·1 to 0·2 dl/g, thus indicating a fairly low molecular weight, but sufficiently high[63] to retain the typical behaviour of a polycondensate material in both solution and melt phase. The two polymeric samples containing the aromatic dyad as the mesogenic core are soluble in most common organic solvents. The polymers containing the aromatic triads are comparatively less soluble, the minimum tendency to dissolve being presented by the sample based on (*S*)-1,2-propanediol, which contains the highest weight per cent of aromatic units. The corresponding sample based on racemic propanediol displays higher solubility in agreement with a random distribution of enantiomeric residues.

All the polymeric products, with the exception of the two waxy samples based on HT units, are solid at room temperature and semicrystalline at the x-ray analysis. The melting temperature (T_m) values decrease on increasing the length of the spacer in the main chain, as observed for an analogous series of achiral polyesters,[66,67] or on increasing the length, or bulkiness, of the side chain substituent,[77,78] and according to a trend denoting an odd–even effect.[78,79] By contrast, the isotropization temperature (T_i) does not appreciably follow that tendency and is fairly high (≥ 320 °C) for the series HTH/propanediol oligomers. A drop of 50–100 °C in T_i is observed in the series HTH/alkyl glycerols. Configurational irregularities introduced into the main chain by the polymerization of diols, either racemic or with low optical purity, cause an appreciable depression of the melting temperature, whereas T_i is not affected that much. No information can be

TABLE 6

Properties[a] of Thermotropic Polyesters Based on Mono(4-carboxyphenyl) Terephthalate (HT) or Bis(4-carboxyphenyl) Terephthalate (HTH) and Glycol Ethers $HO{-}(CH_2\overset{*}{C}H(CH_3)O)_n{-}H$

Sample	Glycol ether		Solubility[b]	$[\alpha]_{578}^{25}$	Mesophase		
	n	Absolute configuration			T_m (°C)	T_i (°C)	Remarks
HT/PG	1	(S)	A, B, D	+26·7[c]	47	110	Undefined
HT/DPG	2	(S; S)	A, B, D	+24·0[c]	<20	40	Undefined
HTH/PG	1	(S)	C	+9·5[d]	334	360	Cholesteric?
HTH/PG	1	(R)(S)	D	—	262	350	Nematic
HTH/DPG	2	(S; S)	A, B, C	+17·2[d]	130	320	Smectic and cholesteric[e]
HTH/TPG	3	(S; S; S)	C	+20·0[d]	275	321	Smectic and cholesteric[f]

[a] Intrinsic viscosity values 0·02–0·15 dl/g in fuming H_2SO_4.
[b] Apparent solubility in: (A) chloroform; (B) dioxane; (C) fuming sulphuric acid; (D) trifluoroacetic acid.
[c] In trifluoroacetic acid.
[d] In fuming sulphuric acid.
[e] Smectic–cholesteric transition at 286 °C.
[f] Smectic–cholesteric transition at 305 °C.

TABLE 7

Properties[a] of Thermotropic Polyesters Based on Bis(4-carboxyphenyl) Terephthalate (HTH) and Glycerol Ethers

$$\mathrm{HOCH_2\overset{*}{C}HOH}$$
$$\mathrm{|\ CH_2OR}$$

Sample	Glycerol ether		$[\alpha]_D^{25}$[b]	Mesophase		
	R	*Absolute configuration*		T_m (°C)	T_i (°C)	*Remarks*
HTH/MG	CH_3	(*R*)	+36·6	135	287	Smectic and cholesteric[c]
HTH/BzG	$CH_2C_6H_5$	(*R*)	+23·1	145	202	Cholesteric
HTH/MEG	$CH_2CH_2OCH_3$	(*R*)(*S*)	—	151	270	Nematic
HTH/MDEG	$(CH_2CH_2O)_2CH_3$	(*R*)	+34·7	103	205	Cholesteric
HTH/MTEG	$(CH_2CH_2O)_3CH_3$	(*R*)(*S*)	—	115	160	Nematic

[a] Samples soluble in chloroform and dioxane; intrinsic viscosity values 0·1–0·2 dl/g in dioxane.
[b] In dioxane.
[c] Smectic–cholesteric transition at 143 °C.

given at present of any influence of the orientational order of the glycol moieties due to the two different functionalities of the hydroxyl groups of the diols (Scheme 2). Correspondingly, the structure and stability of the mesophase are affected, and in principle can be predetermined, by the configurational homogeneity of the chiral centres present in the main chain.[80,81] The optically active samples show, quite unusually, on cooling from the isotropic melt, a quite high degree of supercooling (50 °C) for the mesophase transition, similar to that observed for the crystallization.

Polymorphism from solid to smectic and then to twisted nematic phase was detected in samples of the homologous series with longer spacers. In particular, the smectic phase observed in the HTH/DPG sample by x-ray analysis possesses a highly ordered structure (S_E or S_H)[82] characterized by an unusually high enthalpy of transition to a less ordered mesophase. After the transition from the smectic phase, red–orange and green–blue reflections were observed for HTH/DPG and HTH/TPG samples, respectively, over a rather broad temperature range. Quite interestingly, HTH/DPG on cooling from the anisotropic state retains the iridescent texture down to room temperature. Moreover, on cooling from the homeotropic state established at 230 °C, a mesophase with tendency to nucleate in spheres and to coalesce in extended domains with a striated pattern resembling a myeline structure was also observed.[80] Two pictures of typical textures shown by HTH/DPG between glass slides without any orienting pretreatment are reported in Figs 1 and 2.

Twisted nematic structures were encountered in all the optically active polymers containing glycerol derivatives (Fig. 3). In the case of HTH/MG polymer, a disordered smectic phase (S_A) preceded at low temperatures the onset of the cholesteric phase. As expected, configurational irregularities in the repeating unit of HTH/MEG and HTH/MTEG samples lead to nematic structures in the melt (Fig. 4).

It is clear that the substitution of a hydrogen atom of the α-methyl group present in the side chain of HTH/PG polymer samples with the same structural features in the backbone has a strong effect on the morphological properties of the thermotropic mesophase.

B. Copolyesters of Optically Active (S)-1,2-propanediol

Data relevant to the chemical constitution, properties in solution, and thermal behaviour of the two copolymer systems based on (*S*)-1,2-propanediol and different amounts of linear HTH mesogen and non-linear HIH and HPH isomers are reported in Tables 8 and 9 respectively. For the first series, data for the homopolymer HIH/PG are reported for

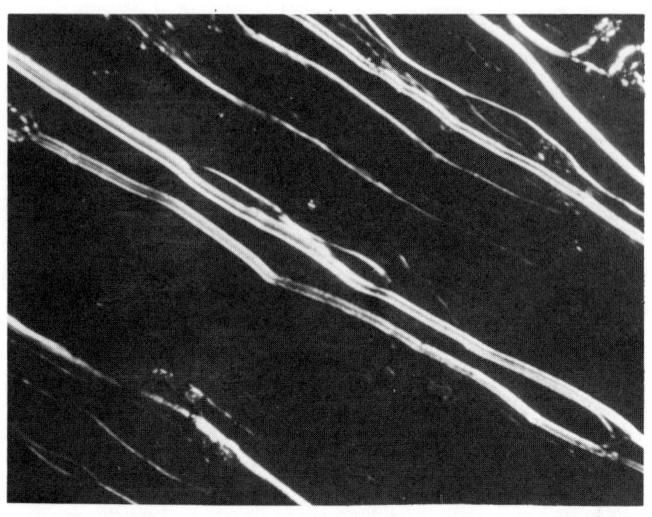

Fig. 1. Cholesteric texture with oily streaks of polyester HTH/DPG sample at 290 °C (original magnification 300 ×).

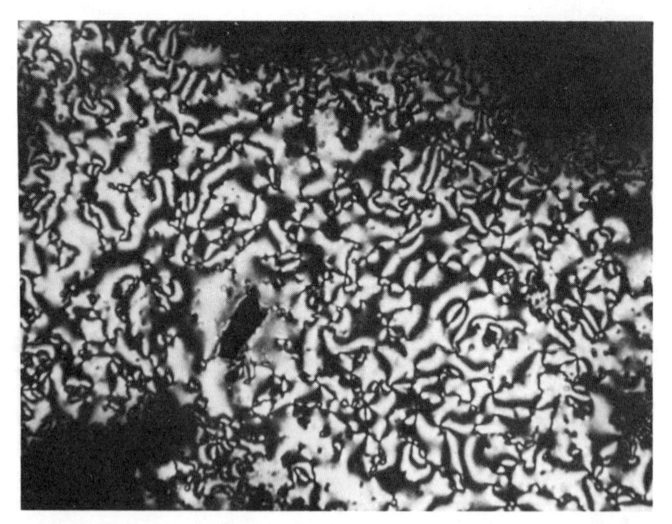

Fig. 2. Schlieren texture of polyester HTH/DPG sample at 298 °C (original magnification 300 ×).

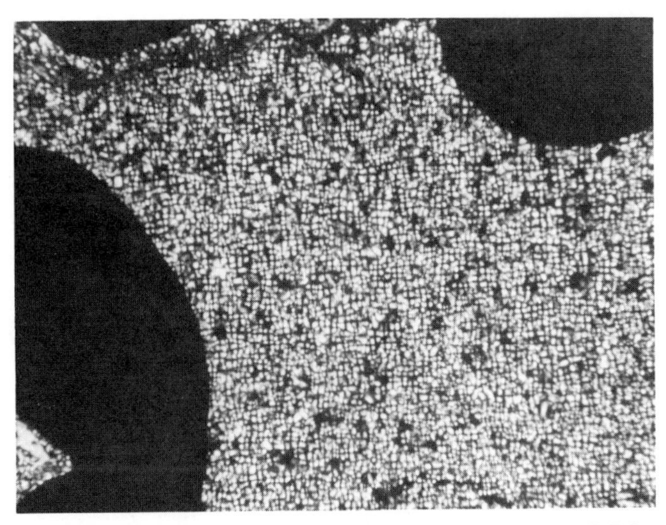

Fig. 3. Cholesteric texture with focal conics of polyester HTH/MDEG sample at 110 °C (original magnification 300 ×).

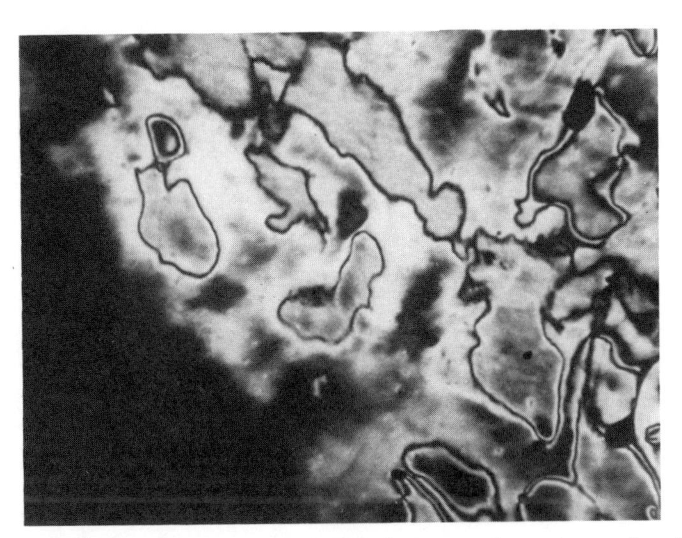

Fig. 4. Threaded nematic texture with homeotropic regions of polyester HTH/MEG sample at 234 °C (original magnification 300 ×).

TABLE 8

Properties[a] of Thermotropic Copolyesters Based on Bis(4-carboxyphenyl) Terephthalate (HTH) and Bis(4-carboxyphenyl) Isophthalate (HIH) and (*S*)-1,2-propanediol (PG)

Sample	*Composition (% HIH)*	*Solubility*[b]	$[\alpha]_D^{25}$[c]	*Mesophase*		
				T_m (°C)	T_i (°C)	*Remarks*
HTH/PG	0	A	+9·5[d]	334	360	Cholesteric?
HTH–HIH/PG (9:1)	10	B	+91·7	222	325	Smectic and cholesteric[e]
HTH–HIH/PG (7:3)	30	C	+81·5	145	285	Cholesteric
HTH–HIH/PG (1:1)	50	C, D	+82·2	139	220	Cholesteric
HTH–HIH/PG (3:7)	70	E, B	+66·5	112	115	Undefined
HIH/PG	100	E, B	+66·5	130	—	Not liquid crystalline
HIH/PG[f]	100	E, B	—	100	—	Not liquid crystalline

[a] Intrinsic viscosity values 0·18–0·25 dl/g in trifluoroacetic acid.
[b] Apparent solubility in: (A) fuming sulphuric acid; (B) trifluoroacetic acid; (C) hot trifluoroacetic acid; (D) hot chloroform; (E) chloroform.
[c] In trifluoroacetic acid.
[d] In fuming sulphuric acid.
[e] Smectic–cholesteric transition at 240 °C.
[f] Racemic.

TABLE 9

Properties[a] of Thermotropic Copolyesters Based on Bis(4-carboxyphenyl) Terephthalate (HTH) and Bis(4-carboxyphenyl) Phthalate (HPH) and (*S*)-1,2-propanediol (PG)

Sample	*Composition (% HPH)*	*Solubility*[b]	$[\alpha]_D^{25\,c}$	*Mesophase*		
				T_m (°C)	T_i (°C)	*Remarks*
HTH/PG	0	A	+9·5[d]	334	360	Cholesteric?
HTH–HPH/PG (9:1)	10	B	+75·5	208	290	Smectic and cholesteric[e]
HTH–HPH/PG (7:3)	30	B	+63·4	182	271	Cholesteric
HTH–HPH/PG (1:1)	50	C	+58·0	130	277	Cholesteric

[a] Intrinsic viscosity values 0·14–0·16 dl/g in trifluoroacetic acid.
[b] Apparent solubility in: (A) fuming sulphuric acid; (B) trifluoroacetic acid; (C) chloroform.
[c] In trifluoroacetic acid.
[d] In fuming sulphuric acid.
[e] Smectic–cholesteric transition at 251 °C.

comparison. ^{1}H- and ^{13}C-NMR spectroscopic measurements are consistent for both copolymer classes with a chemical composition identical to that of the starting diacid mixtures. This fact, given the 20–80 % conversion, is taken as a proof of the anticipated equivalence in reactivity of the triphenyl-containing diacid isomers. A random distribution of linear and non-linear aromatic triads along the macromolecular chain can be reasonably assumed therefore. Correspondingly, a more pronounced solubility, associated also with an abatement of the linearity of the polymer backbone, with respect to HTH homopolymer, is observed.

In both series of copolymers the melting temperature, and the onset of mesomorphic behaviour, and the clearing temperature are markedly affected by the chemical composition. In particular, with increasing content of non-linear aromatic units, melting or softening temperature and isotropization temperature decrease. The trend of this decrease is consistent with larger effects on T_m that reaches a minimum value at about 70 % non-linear units. Mesophases were observed by polarizing microscopy in copolymers containing up to 70 % HIH units and 50 % HPH units. First-order melting transitions were detected by DSC analysis in samples containing up to 30 % non-linear triads. The phase diagram (determined by combined DSC and polarizing microscopy techniques) of HTH–HIH/PG copolymers is represented in Fig. 5.

It is of interest to note that the introduction of only 10 % non-linear triads, while decreasing the melting temperature by more than 100 °C,

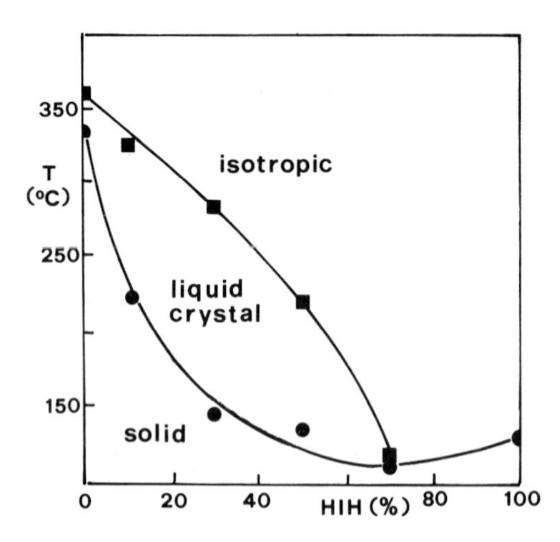

Fig. 5. Phase diagram of copolyester HTH–HIH/PG samples: ●, melting temperature (T_m); ■, isotropization temperature (T_i).

depresses the isotropization temperature by only about 30 °C. As a consequence the existence of a mesomorphic state over a rather broad temperature range (100–140 °C) can be realized at intermediate copolymer compositions. These effects are of the same order of magnitude as those observed in copolymers of mesogenic residues having different chemical and stereochemical structures.[18,83,84] A concentration of about 60 % of distorted aromatic units in copolymers with linear aromatic mesogens has been already claimed in the case of fully aromatic systems as the upper limit to allow for the existence of thermotropic properties.[85]

Starting from anisotropic components, whose homopolymers do not show mesomorphic behaviour, it is therefore possible to build up copolymers, that, on the contrary, display liquid crystalline properties in the melt phase. It must be stressed again the importance and versatility of the copolymerization process in producing materials with fairly low three-dimensional crystal order and still rather marked tendency to originate stable mesophases.

In samples with contents in HIH units higher than 50 %, rather high glass transition temperature values were detected (T_g = 90–100 °C). These are very close to the softening temperature values, suggesting that a certain degree of molecular order can be retained by the mesophase in the amorphous state.[63,86,87] As anticipated, the persistence of anisotropy in the melt was clearly demonstrated, up to a content of 60–70 % non-linear triads, by optical microscopy observations. The texture of the copolymer mesophases changed with chemical composition and temperature. The presence of a planar texture with oily streaks and selective reflection variable with temperature was revealed in samples containing up to 30 % distorted aromatic units. At higher contents, the occurrence of a cholesteric structure was not unambiguously determined. A typical cholesteric texture relevant to copolyester HTH–HIH/PG (9:1) is represented in Fig. 6.

It is worth mentioning that the observed changes in the iridescence properties of these copolymers with increasing temperature are associated with an extension of the cholesteric pitch, in contrast to the common trend shown by low molecular weight cholesterics[5] and HTH/MG homopolyester. At present we do not have a reasonable explanation for this behaviour, but it may be recalled that analogous phenomena have been previously observed in thermotropic hydroxypropyl cellulose acetate[48] and copolypeptides[53] and lyotropic poly(γ-benzyl-L-glutamate) solutions.[4,7] In these last systems, the extension of the cholesteric pitch with increasing temperature is usually attributed to the free rotation of the polymer chain around its long molecular axis, which would reduce the anisotropy in the intermolecular potential.[88]

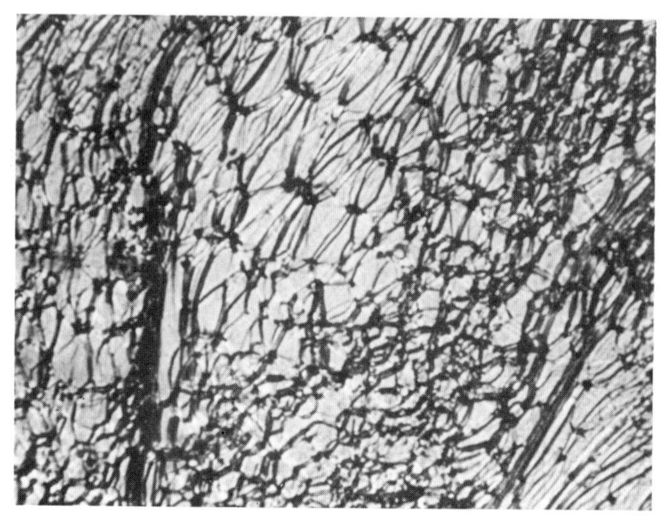

Fig. 6. Cholesteric texture with oily streaks of copolyester HTH–HIH/PG (9:1) sample at 260 °C (original magnification 300 ×).

3. Chiroptical Properties in Dilute Solution

All the polymeric samples prepared are characterized by a fairly high molar optical rotation in various solvents (Tables 6–9); that is almost four–five times higher than that of the starting optically active precursors and low molecular weight model compounds **14** and **15** in the same solvent.

$$C_2H_5\overset{*}{C}H(CH_3)CH_2OCO\text{–}C_6H_4\text{–}OCO\text{–}C_6H_4\text{–}COO\text{–}C_6H_4\text{–}COOCH_2\overset{*}{C}H(CH_3)C_2H_5$$

14

$$C_2H_5\overset{*}{C}H(CH_3)CH_2OCO\text{–}C_6H_4\text{–}OCO\text{–}C_6H_4\text{–}COO\text{–}C_6H_4\text{–}COOCH_2\overset{*}{C}H(CH_3)C_2H_5$$

15

This finding supports the existence, even in dilute solution, of a somewhat homogeneous conformational order of the polymer segments.[89] Further evidence substantiating this aspect will be better produced by measurements of circular dichroism absorption. Moreover, in the prepared copolyesters the optical rotation changes with chemical composition in a nearly linear fashion, thus indicating that no strong variation of the local conformational environment is generated by the proximity of structurally identical or different residues.

The conformational properties of mesogenic macromolecules, which tend to establish a rather pronounced intramolecular organization in the melt phase, are only sparingly investigated in dilute solution and generally related to side chain liquid crystal polymers.[90,91] Rather, extensive studies have been addressed to comb-like polymers and aromatic polyamides, that are known to give lyotropic order in fairly concentrated solutions.[92] Optically active thermotropic polyesters have been more recently characterized by measurements of magnetic birefringence and found to assume in dilute solution a random coil conformation.[93]

In a previous paper[94] we have reported on the chiroptical properties in dilute solution of polyesters based on HTH and (*S*)-1,2-propanediol and its oligomers. Two rather strong dichroic absorptions of opposite sign were detected for all the samples around 260–265 nm and 235–240 nm, in close correspondence to the maximum UV absorption at 245–250 nm. The positive dichroic signal extends down to 320 nm with a marked shoulder around 300 nm (Fig. 7).[94] In both absorption regions, related to the $\pi \rightarrow \pi^*$

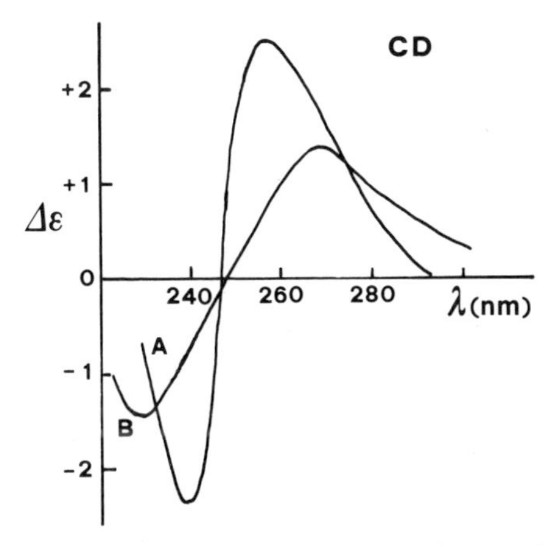

Fig. 7. Circular dichroism spectra (CD) in chloroform solution of polyester samples: (A) HTH/DPG; (B) HTH/TPG.

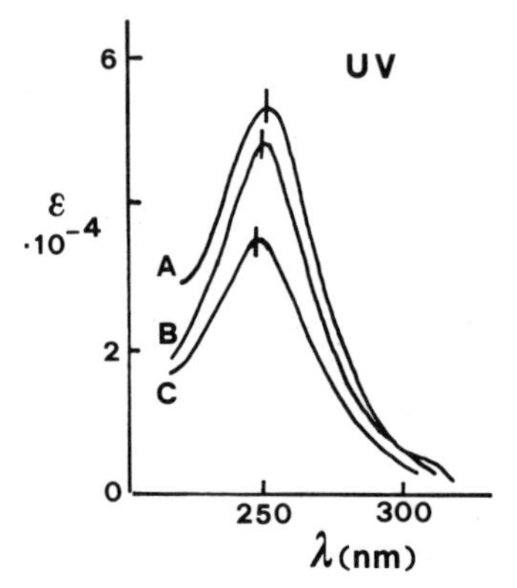

Fig. 8. Ultraviolet absorption spectra (UV) in dioxane solution of polyester samples: (A) HTH/PG; (B) HTH/DPG; (C) HTH/TPG.

electronic transitions of the rather complex aromatic chromophore,[95,96] the model compound bis{4-[(*S*)-2-methylbutoxycarbonylphenyl]}terephthalate (**14**) does not show any appreciable dichroic absorption. The UV absorption spectra in the shorter wavelength region are characterized in the homologous series by a hypochromic effect, with a concurrent hypsochromism, with lengthening of the flexible segment (Fig. 8).

Comparable phenomena are observed in the UV absorption of copolyesters of (*S*)-1,2-propanediol and linear and non-linear aromatic

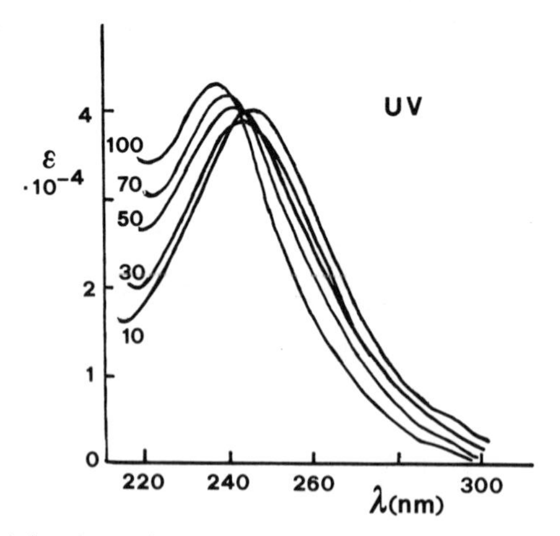

Fig. 9. Ultraviolet absorption spectra (UV) in dioxane solution of copolyester HTH–HIH/PG samples at different HIH per cent content.

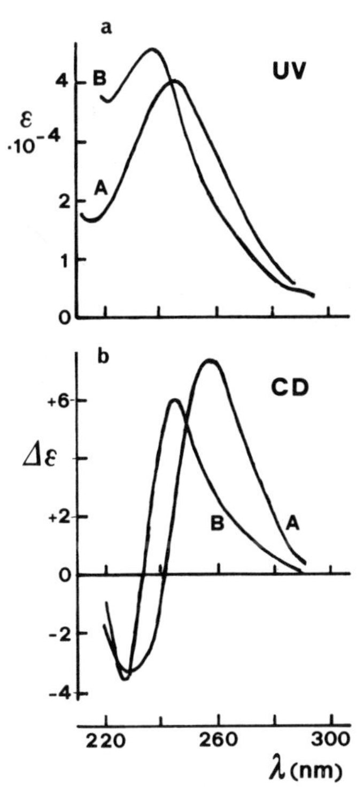

Fig. 10. (a) Ultraviolet absorption (UV) and (b) circular dichroism (CD) spectra in dioxane solution of: (A) copolyester HTH–HIH/PG (9:1) sample; (B) homopolyester HIH/PG sample.

diacids as a function of copolymer composition. In Fig. 9 there is represented the trend observed in the series of copolyesters based on HTH–HIH units. In this system, the higher the content of non-linear units, the higher the blue shift of the wavelength related to the maximum absorption band. Such behaviour has been described for a wide variety of vinyl aromatic polymers and attributed to the presence of electronically interacting neighbouring chromophores.[97–99] It is reasonable, therefore, to conclude that, even in the investigated polyesters, the aromatic diacid residues, either of the same or of different structure, experience intra-molecular electronic interactions.

The profile and the intensity of the dichroic absorption observed in the copolyesters closely resemble that recorded in the case of homopolyesters and are in strict correspondence with the UV absorption band. In Fig. 10 there are reported, as typical examples, the circular dichroism spectra, with corresponding UV bands, of the copolymer containing 10% HIH units and of the homopolymer HIH/PG.

It is noteworthy that in all cases (homopolymers and copolymers):

(i) the fairly high intensity of the induced circular dichroism is similar to that reported for poly(α-amino acids)[100] and synthetic optically active polyamides[101,102] that are known to assume a helical conformation with a preferential screw sense in solution;

(ii) the sigmoid profile, even though partially distorted, with positive and negative dichroic bands is almost symmetrically placed with respect to the maximum of the corresponding UV absorption band. This behaviour is very likely due to an exciton splitting phenomenon[103] involving the electronic transitions of the aromatic chromophores.

As a conclusion, we may stress that UV and CD characteristics are consistent with the existence in dilute solution of a conformationally homogeneous placement of aromatic segments in a helical array. The preferential chirality of the polymer backbone is not appreciably disrupted by incorporation of non-linear aromatic residues.

ACKNOWLEDGEMENTS

The authors wish to thank Professor T. L. St Pierre, University of Alabama in Birmingham, for a critical reading of the manuscript. This work was performed with the financial support of Fondi Ministeriali Ricerca Scientifica, Quota 40%.

REFERENCES

1. Mauguin, C., *Compt. Rend.*, 1910, **151**, 1141.
2. Vorlander, D. and Janecke, F., *Z. Phys. Chem.*, 1913, **85**, 691.
3. Giesel, F., *Phys. Z.*, 1910, **11**, 192.
4. DuPré, D. B. and Duke, R. W., *J. Chem. Phys.*, 1975, **63**, 143.
5. Chandrasekhar, S., *Liquid Crystals*, Cambridge University Press, Cambridge, 1977.
6. Patel, D. L. and DuPré, D. B., *J. Polym. Sci., Polym. Lett. Ed.*, 1979, **17**, 299.
7. Toriumi, H., Kusumi, Y., Uematsu, I. and Uematsu, Y., *Polymer J.*, 1979, **11**, 863.

8. Baessler, H. and Labes, M. M., *J. Chem. Phys.*, 1970, **52**, 631.
9. Dolphin, D., Muljani, Z., Cheng, J. and Meyer, R. B., *J. Chem. Phys.*, 1973, **58**, 413.
10. Adams, J. E. and Hass, W. E., *Mol. Cryst. Liq. Cryst. Lett.*, 1971, **15**, 27.
11. Stegemeyer, H. and Finkelmann, H., *Chem. Phys. Lett.*, 1971, **23**, 227.
12. Hauson, H., Dekker, A. J. and van der Wonde, F., *Mol. Cryst. Liq. Cryst.*, 1977, **42**, 15.
13. DeVries, H. L., *Acta Cryst.*, 1951, **4**, 219.
14. Elser, W. and Ennulat, R. D., in *Advances in Liquid Crystals*, Vol. 2, Brown, G. H. (Ed.), Academic Press, New York, 1976, p. 73.
15. Kahn, F., *J. Appl. Phys. Lett.*, 1971, **18**, 231.
16. Adams, J. E., Haas, W. E. and Daily, J., *J. Appl. Phys.*, 1971, **42**, 4096.
17. Scheffer, T. J., *J. Phys. D*, 1975, **8**, 1441.
18. Blumstein, A., Vilasagar, S., Ponrathnam, S., Clough, S. B., Blumstein, R. B. and Maret, G., *J. Polym. Sci., Polym. Phys. Ed.*, 1982, **20**, 877.
19. Asrar, J., Toriumi, H., Watanabe, J., Krigbaum, W. R., Ciferri, A. and Preston, J., *J. Polym. Sci., Polym. Phys. Ed.*, 1983, **21**, 1119.
20. Strzelecki, L. and Liebert, L., *Bull. Soc. Chim. France*, 1973, 597.
21. Tsutsui, T. and Tanaka, L., *J. Polym. Sci., Polym. Lett. Ed.*, 1980, **18**, 17.
22. Tsutsui, T. and Tanaka, L., *Polymer*, 1980, **21**, 1351.
23. Brown, G. H. and Crooker, P. P., *Chem. Eng. News*, 1983, 24.
24. Brown, G. H. and Wolken, J. J. (Eds), *Liquid Crystals and Biological Structures*, Academic Press, New York, 1979.
25. DuPré, D. B. and Samulski, E. T., in *Liquid Crystals*, Saeva, F. D. (Ed.), Marcel Dekker, New York, 1979, p. 203.
26. Gallot, B., in *Liquid Crystalline Order in Polymers*, Blumstein, A. (Ed.), Academic Press, New York, 1978, p. 191.
27. Iizuka, E. and Yang, J. T., in *Liquid Crystals and Ordered Fluids*, Johnson, J. F. and Porter, R. S. (Eds), Vol. 3, Plenum Press, New York, 1978.
28. Gros, L., Ringsdorf, H. and Skura, J., *Angew. Chem. Int. Ed. Engl.*, 1981, **20**, 305.
29. Hsu, E. C., Clough, R. B. and Blumstein, A., *J. Polym. Sci., Polym. Lett. Ed.*, 1977, **15**, 545.
30. Blumstein, A., *Macromolecules*, 1977, **10**, 872.
31. Shibaev, V. P., Kharitonov, A. V., Freidzon, Ya. S. and Platè, N. A., *Vysokomol. Soed. Ser. A*, 1979, **21**, 1849.
32. Shibaev, V. P., in *Advances in Liquid Crystal Research and Application*, Bata, L. (Ed.), Pergamon Press, Oxford, 1980, p. 869.
33. Freidzon, Ya, S., Kostromin, S. G., Boiko, N. I., Shibaev, V. P. and Platè, N. A., *ACS Polym. Prep.*, 1983, **24**(2), 279.
34. Shibaev, V. P., Platè, N. A. and Freidzon, Ya. S., *J. Polym. Sci., Polym. Chem. Ed.*, 1979, **17**, 1655.
35. Platè, N. A. and Shibaev, V. P., *J. Polym. Sci., Polym. Symp.*, 1980, **67**, 1.
36. Finkelmann, H., Ringsdorf, H., Siol. W. and Wendorff, J. H., *Makromol. Chem.*, 1978, **179**, 829.
37. Shibaev, V. P., Moisenko, V. M., Lukin, N. Yu. and Platè, N. A., *Dokl. Akad. Nauk, SSSR*, 1977, **237**, 401.

38. Finkelmann, H. and Rehage, G., *Makromol. Chem., Rapid Commun.*, 1980, **1**, 31.
39. Finkelmann, H. and Rehage, G., *Makromol. Chem., Rapid Commun.*, 1980, **1**, 733.
40. Finkelmann, H. and Rehage, G., *Makromol. Chem., Rapid Commun.*, 1982, **3**, 859.
41. Finkelmann, H. and Rehage, G., *ACS Polym. Prep.*, 1983, **24**(2), 277.
42. Mousa, A. M., Freidzon, Ya. S., Shibaev, V. P. and Platè, N. A., *Polym. Bull.*, 1982, **6**, 485.
43. Finkelmann, H., Koldehoff, J. and Ringsdorf, H., *Angew. Chem. Int. Ed. Engl.*, 1978, **17**, 935.
44. Shibaev, V. P., Finkelmann, H., Kharitonov, A. V., Portugall, M., Ringsdorf, H. and Platè, N. A., *Vysokomol. Soed. Ser. A*, 1981, **23**, 919.
45. Finkelmann, H., Kock, H. J. and Rehage, G., *Makromol. Chem., Rapid Commun.*, 1981, **2**, 317.
46. Gray, D. G., *Appl. Polym. Symp.*, 1983, **37**, 179.
47. Shimamura, K., White, J. L. and Fellers, J. F., *J. Appl. Polym. Sci.*, 1981, **26**, 2165.
48. Tseng, S. L., Valente, A. and Gray, D. G., *Macromolecules*, 1981, **14**, 715.
49. Aharoni, S. M., *J. Polym. Sci., Polym. Lett. Ed.*, 1981, **19**, 495.
50. Tseng, S. L., Laivins, G. V. and Gray, D. G., *Macromolecules*, 1982, **15**, 1262.
51. Bhadani, S. N. and Gray, D. G., *Makromol. Chem., Rapid Commun.*, 1982, **3**, 449.
52. Bhadani, S. N., Tseng, S. L. and Gray, D. G., *Makromol. Chem.*, 1983, **184**, 1727.
53. Kasuya, S., Sasaki, S., Watanabe, J., Fukuda, Y. and Uematsu, I., *Polym. Bull.*, 1982, **7**, 241.
54. Robinson, C., *Tetrahedron*, 1961, **13**, 219.
55. DiMarzio, E. A., *J. Chem. Phys.*, 1961, **35**, 658.
56. Van Luyen, D., Liebert, L. and Strzelecki, L., *Eur. Polym. J.*, 1980, **16**, 307.
57. Vilasagar, S. and Blumstein, A., *Mol. Cryst. Liq. Cryst. Lett.*, 1980, **56**, 203.
58. Krigbaum, W. R., Ciferri, A., Asrar, J., Toriumi, H, and Preston, J., *Mol. Cryst. Liq. Cryst.*, 1981, **76**, 79.
59. Krigbaum, W. R., Ishikawa, T., Watanabe, J., Toriumi, H. and Kubota, K., *J. Polym. Sci., Polym. Phys. Ed.*, 1983, **21**, 1851.
60. Saupe, A., in *Liquid Crystals & Plastic Crystals*, Vol. 1, Gray, G. W. and Winsor, P. A. (Eds), Ellis Horwood, New York, 1978, p. 19.
61. Kofler, L. and Kofler, A., *Thermomikromethoden*, Verlag Chemie, Weinheim, 1954.
62. Fayolle, B., Noel, C. and Billard, J., *J. Phys. C3*, 1979, **40**, 485.
63. Noel, C., Billard, J., Bosio, L., Friedrich, C., Laupetre, F. and Strazielle, C., *Polymer*, 1984, **25**, 263.
64. Noel, C., Laupetre, F., Friedrich, C., Fayolle, B. and Bosio, L., *Polymer*, in press.
65. Chiellini, E., Galli, G., Ciardelli, F., Palla, R. and Carmassi, F., *Inf. Chim.*, 1978, **176**, 221.

66. Galli, G., Chiellini, E., Ober, C. and Lenz, R. W., *Makromol. Chem.*, 1982, **183**, 2693.
67. Ober, C., Lenz, R. W., Galli, G. and Chiellini, E., *Macromolecules*, 1983, **16**, 1034.
68. Ober, C., Jin, J. I. and Lenz, R. W., *Polymer J.*, 1982, **14**, 9.
69. Jackson, W. J., Jr, and Kuhfuss, H. F., *J. Polym. Sci., Polym. Chem. Ed.*, 1976, **14**, 2043.
70. Chiellini, E., Lenz, R. W. and Ober, C., in *Polymer Blends*, Martuscelli, E., Palumbo, R. and Kryszewski, M. (Eds), Plenum Press, New York, 1980, p. 373.
71. (a) Pino, P., *28th International IUPAC Macromolecular Symposium*, Amherst, 1982, p. 121; (b) Suter, U. W. and Pino, P., *Macromolecules*, in press.
72. Malanga, C., Spassky, N., Menicagli, R. and Chiellini, E., *Synth. Commun.*, 1982, **12**, 67.
73. Malanga, C., Spassky, N., Menicagli, R. and Chiellini, E., *Polym. Bull.*, 1983, **9**, 328.
74. Chiellini, E., Nieri, P. and Galli, G., *Mol. Cryst. Liq. Cryst.*, 1984, in press.
75. Baer, E. and Fischer, H. O. L., *J. Am. Chem. Soc.*, 1945, **67**, 2031.
76. Schleier, G., Galli, G. and Chiellini, E., *Polym. Bull.*, 1982, **6**, 529.
77. Aharoni, S. M., *Macromolecules*, 1979, **12**, 94.
78. Zhou, Q. F. and Lenz, R. W., *J. Polym. Sci., Polym. Chem. Ed.*, 1983, **21**, 3313.
79. Roviello, A. and Sirigu, A., *Makromol. Chem.*, 1982, **183**, 895.
80. Chiellini, E., Galli, G., Malanga, C. and Spassky, N., *Polym. Bull.*, 1983, **9**, 336.
81. Chiellini, E., Galli, G., Lenz, R. W. and Ober, C., in *Polymer Blends*, Vol. 2, Kryszewski, M. and Martuscelli, E. (Eds), Plenum Press, New York, 1984, p. 267.
82. Gallot, B., Galli, G. and Chiellini, E., in preparation.
83. Iannelli, P., Roviello, A. and Sirigu, A., *Eur. Polym. J.*, 1982, **18**, 753.
84. Roviello, A., Santagata, S. and Sirigu, A., *Makromol. Chem., Rapid Commun.*, 1983, **4**, 281.
85. Lenz, R. W. and Jin, J. I., *Macromolecules*, 1981, **14**, 1405.
86. Menczel, J. and Wunderlich, B., *Polymer*, 1981, **22**, 778.
87. Frosini, V., de Petris, S., Chiellini, E., Galli, G. and Lenz, R. W., *Mol. Cryst. Liq. Cryst.*, 1983, **98**, 223.
88. Samulski, T. V. and Samulski, E. T., *J. Chem. Phys.*, 1977, **67**, 824.
89. Pino, P., *Adv. Polym. Sci.*, 1965, **4**, 393.
90. Springer, J. and Weigelt, F. W., *Makromol. Chem.*, 1983, **184**, 1489.
91. Springer, J. and Weigelt, F. W., *Makromol. Chem.*, 1983, **184**, 2635.
92. Tsvetkov, V. N. in *Advances in Liquid Crystal Research and Applications*, Bata, L. (Ed.), Pergamon Press, Oxford, 1980, p. 813.
93. Blumstein, A., Maret, G. and Vilasagar, S., *Macromolecules*, 1981, **14**, 1543.
94. Chiellini, E. and Galli, G., *Makromol. Chem., Rapid Commun.*, 1983, **4**, 285.
95. Cilento, G., *J. Am. Chem. Soc.*, 1953, **75**, 3748.
96. Martin, R. and Coton, G., *Bull. Soc. Chim. France*, 1973, 1442.
97. Okamoto, K., Itaya, A. and Kusabayashi, S., *Chem. Lett.*, 1974, 1167.

98. Chiellini, E., Solaro, R., Ledwith, A. and Galli, G., *Eur. Polym. J.*, 1980, **16**, 875.
99. Galli, G., Solaro, R., Chiellini, E. and Ledwith, A., *Macromolecules*, 1983, **16**, 497.
100. Woody, R. W., *Macromol. Rev.*, 1977, **12**, 181.
101. Overberger, C. G. and Kozlowski, J. H., *J. Polym. Sci., A1*, 1972, **10**, 2291.
102. Montaudo, G. and Overberger, C. G., *J. Polym. Sci., Polym. Chem. Ed.*, 1973, **11**, 2739.
103. Hug, W., Ciardelli, F. and Tinoco, I., Jr, *J. Am. Chem,. Soc.*, 1974, **96**, 3407.

3

SOLID STATE PHYSICS OF THERMOTROPIC POLYESTERS: INTERNAL FRICTION OF MESOMORPHIC STRUCTURES

V. FROSINI and S. DE PETRIS

Dipartimento di Ingegneria Chimica, Centro CNR Processi Ionici, Università di Pisa, Italy

G. GALLI

Istituto Chimica Organica Industriale, Università di Pisa, Italy

and

E. CHIELLINI

Istituto Chimica Generale (Facoltà di Ingegneria), Università di Pisa, Italy

INTRODUCTION

There is currently considerable interest in the solid state physics of thermotropic liquid crystalline (LC) polymers and much emphasis is placed on studies regarding the solid–solid transitions and the chain dynamics of these systems. Recently, evidence for the occurrence of supercooled mesomorphic structures in the glassy state of polymers quenched from the anisotropic melt has been presented.[1-3]

In this work we wish to report some new results on the dynamic–mechanical behaviour *below* the glass transition of these supercooled mesophases in thermotropic liquid crystalline polyesters based on bis(4-carbonylphenyl) terephthalate units (HTH) and alkylene diols (C_3, C_{10}) or

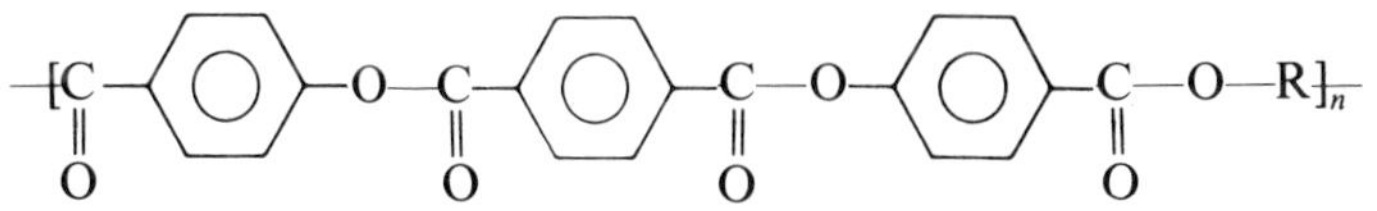

$$-\!\!\left[\underset{\underset{O}{\|}}{C}-\langle\bigcirc\rangle-O-\underset{\underset{O}{\|}}{C}-\langle\bigcirc\rangle-\underset{\underset{O}{\|}}{C}-O-\langle\bigcirc\rangle-\underset{\underset{O}{\|}}{C}-O-R\right]_n\!\!-$$

$$R = -(CH_2)_x-O- \quad x = 3, 5, 10$$
$$R = -(CH_2CH_2O)_y- \quad y = 4$$

tetraethylene glycol (TEG). Copolyesters containing the same mesogenic unit and diols of different length (C_3/C_5 or C_5/C_{10}) were also investigated.

EXPERIMENTAL

Polymer samples HTH–C_3 and HTH–C_{10} were prepared from the corresponding α,ω-bis(4-hydroxybenzoyloxy)alkanes and terephthaloyl chloride according to a previous procedure.[4] All the other polymers were obtained from bis(4-carboxyphenyl)terephthalate and the appropriate diol as already reported.[5]

Calorimetric measurements were performed with a Perkin–Elmer DSC-2 instrument under dry nitrogen flow. The temperature of the maximum in the enthalpic peak was taken as the transition temperature.

Dynamic elastic modulus and mechanical loss were determined using a resonance electrostatic apparatus at frequencies of 10^3–10^4 Hz, with specimens of the shape of circular plates of 36 mm diameter and 2 mm thickness.

RESULTS

Transition temperatures of the LC polyesters and copolyesters are reported in Table 1. Original and melt crystallized samples of the polyesters HTH–C_3, HTH–C_{10}, and HTH–TEG did not show any transition below room temperature, whereas the polymers quenched from the LC state were characterized by the existence of two or more phase modifications which

TABLE 1

Phase Transition Temperatures of Thermotropic Liquid Crystalline Polyesters and Copolyesters

Sample	$T_g{}^a$ (K)	T_m (K)	$T_{LC}{}^b$ (K)	T_i (K)
HTH–C_3	400	513	538	585
HTH–C_{10}	310	494	—	540
HTH–TEG	300	423	455	490
HTH–C_3/C_5	330	460	—	523
HTH–C_5/C_{10}	320	435	486	535

[a] Glass transition of the amorphous phase of virgin samples.
[b] Liquid crystal–liquid crystal transition.

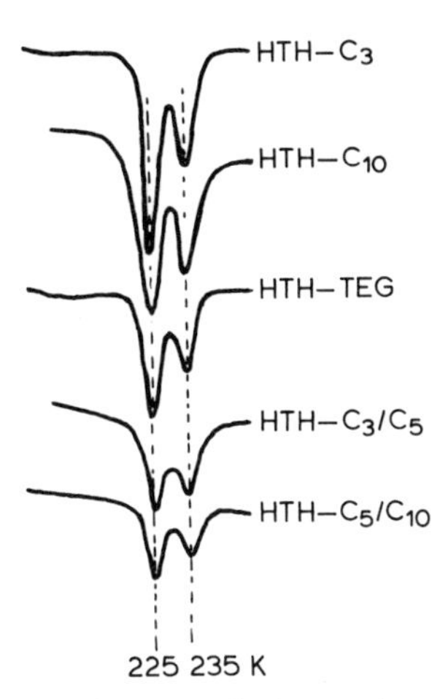

Fig. 1. DSC heating curves for quenched samples of HTH–C_3, HTH–C_{10}, HTH–TEG, HTH–C_3/C_5 and HTH–C_5/C_{10} (heating rate $10\,K\,min^{-1}$).

differed in terms of energy. In fact, by heating the samples from very low temperatures, two endothermic peaks were found to occur at 225 and 235 K (Fig. 1). The quenched samples of the three polyesters still retained a certain degree of crystallinity which was estimated by x-ray analysis to be of the order of 30–40 % for HTH–C_{10} and HTH–TEG, and less than 30 % for HTH–C_3. It should be pointed out that the crystal form of the quenched samples is very nearly the same as that found in samples slowly crystallized from the anisotropic melt. However, this crystal form (form I) is different from the crystal modification (form II) which was detected in the original untreated samples of these polyesters at room temperature.[3] The low temperature transition was invariably detected by DSC analysis in all the polymers quenched from the LC state, whereas it tended to disappear in samples rapidly quenched from the isotropic melt. Therefore, it may be concluded that the crystal phase (form I) in quenched samples is not involved in the observed phase transition, which on the contrary must be connected to the existence of a supercooled mesophase.

Further evidence of the presence of a glass-like mesophase was provided by the DSC data obtained at high temperatures for the quenched polymers. In general a change in heat capacity, characteristic of a glass transition, was found to occur at ~100 K above the T_g of the amorphous material in the semicrystalline samples. A typical example of this behaviour is represented in Fig. 2. Experimental results thus supported the conclusion that a metastable glass-like modification is formed by quenching from the LC

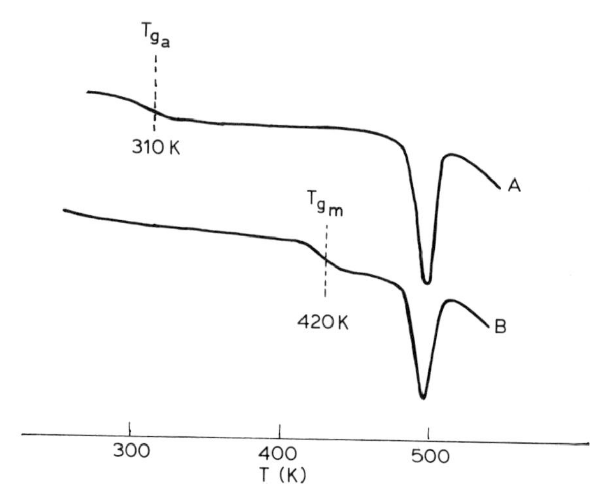

Fig. 2. DSC heating curves at high temperatures for HTH–C_{10} samples: (A) melt crystallized; (B) quenched from LC state (heating rate $10\,K\,min^{-1}$).

state and such supercooled mesophase is transformed below 235 K into a more stable phase. However, in view of the low value of the enthalpy of the transition ($\Delta H = 0{\cdot}196$ kcal mole^{-1} HTH for HTH–C_{10} sample), which is rather of the order of magnitude of the enthalpy changes for the transitions between smectic polymorphs,[6] it may be speculated that a more ordered smectic mesophase, instead of a true crystalline phase, is formed.

Calorimetric data also demonstrated that the low temperature first-order transition occurs in quenched samples independent of the particular chemical nature and length of the flexible spacers. Since the structure of the rigid core of the mesogenic groups is the same for all the polyesters examined, it can be reasonably assumed that such a transition involves exclusively the aromatic moiety of the repeating unit. This conclusion is also supported by the calorimetric data obtained for copolyesters HTH–C_3/C_5 and HTH–C_5/C_{10}, which still exhibit the double-peaked transition in their DSC curves at the same temperature as that of the corresponding homopolymers (Fig. 1). It must be noted, however, that copolymer samples quenched from the LC state had very poor crystallinity and attempts to clarify the nature and state of order of the supercooled mesophase by x-ray analysis have been unsuccessful to date.

All the polymers were subjected to a dynamic mechanical investigation in order to study their relaxation properties in the range of linear viscoelasticity. Comparative plots of the dynamic modulus E' and internal

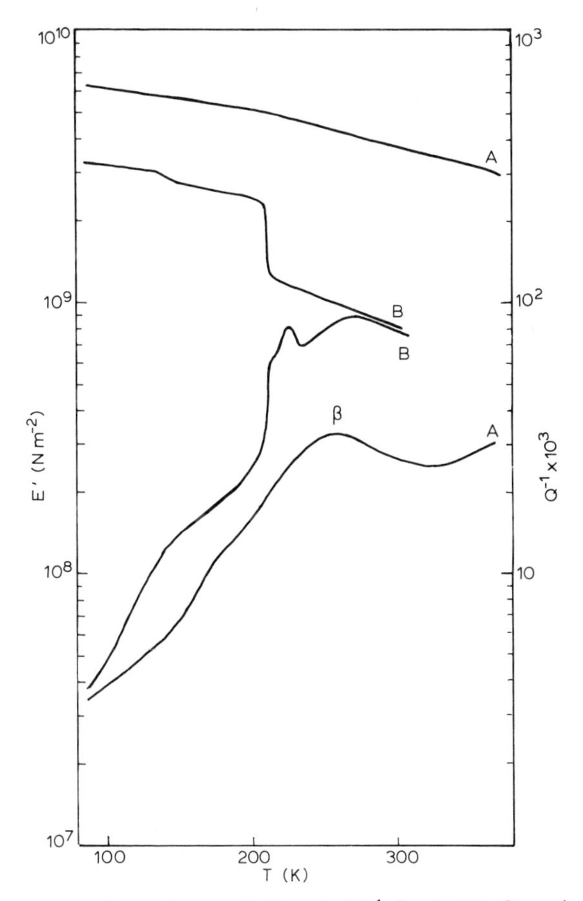

Fig. 3. Temperature dependence of E' and Q^{-1} for HTH–C_3 polyester: (A) original; (B) quenched from LC state.

friction Q^{-1} vs. temperature for the original and quenched samples are reported in Figs 3–7.

Original samples of polyesters and copolyesters showed the usual dynamic-mechanical behaviour which is expected for semi-rigid macromolecules. Polyester HTH–C_3 (Fig. 3) exhibited a β-loss peak at about 260 K, typically characterized by a broad distribution of relaxation times. The dynamic elastic modulus decreased steadily with increasing temperature, showing a little dispersion in the temperature interval of the β-relaxation. Polymer HTH–TEG (Fig. 4) presented a complex β-relaxation in the range 200–260 K, in which well-shaped loss peaks were not observed;

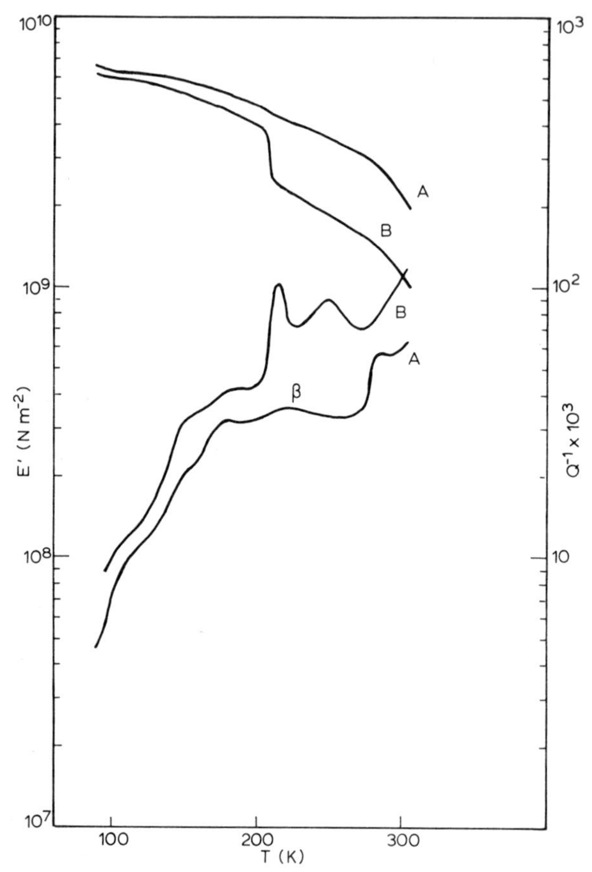

Fig. 4. Temperature dependence of E' and Q^{-1} for HTH–TEG polyester: (A) original; (B) quenched from LC state.

rather, a continuous spectrum of several relaxation effects resulted. In the E' vs. T curve, two or more dispersion phenomena occurred, that were correlated to the complex loss behaviour of this polymer. Polyester HTH–C_{10} (Fig. 5) showed two distinct relaxation effects: a weak β-peak at about 260 K and a well pronounced γ-peak at much lower temperatures. In the original sample, as well as in that slowly crystallized from the anisotropic melt, the γ-relaxation was always present as a single loss peak centred at 150 K. For these samples no particular dispersion in the modulus vs. temperature curves was observed, except for a little change at about 210 K.

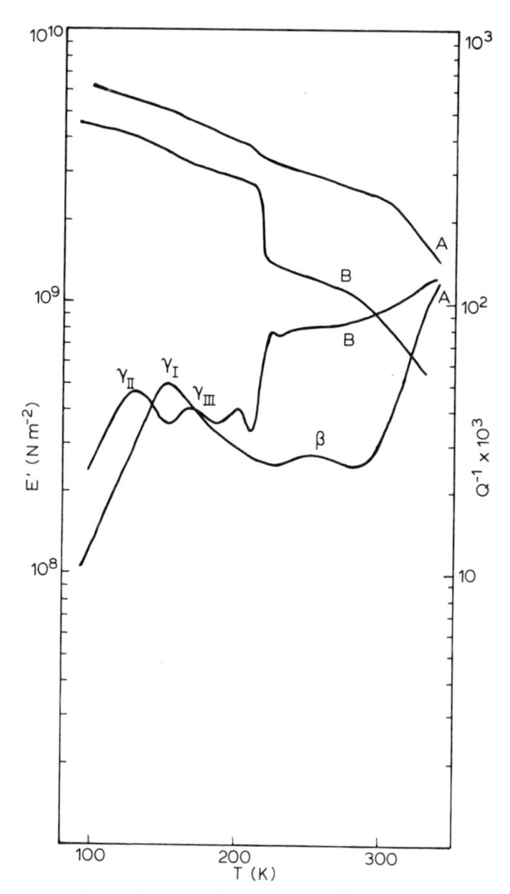

Fig. 5. Temperature dependence of E' and Q^{-1} for HTH–C_{10} polyester: (A) original; (B) quenched from LC state.

The internal friction of copolyester HTH–C_3/C_5 (Fig. 6) showed a maximum at about 250 K, with a large shoulder on the low temperature side. Corresponding with the β-relaxation a noticeable dispersion in the E' vs. T curve was also noted. Copolyester HTH–C_5/C_{10} (Fig. 7) presented two separate loss peaks: the former at 260 K (β-relaxation) and the latter at 150 K (γ-relaxation) with a smaller shoulder on the low temperature side. In this polymer only a little change in the modulus was observed at 210 K, in addition to the normal decrease of E' with increasing temperature.

The dynamic-mechanical behaviour of samples of polyesters and copolyesters quenched from the LC state presented dramatic differences

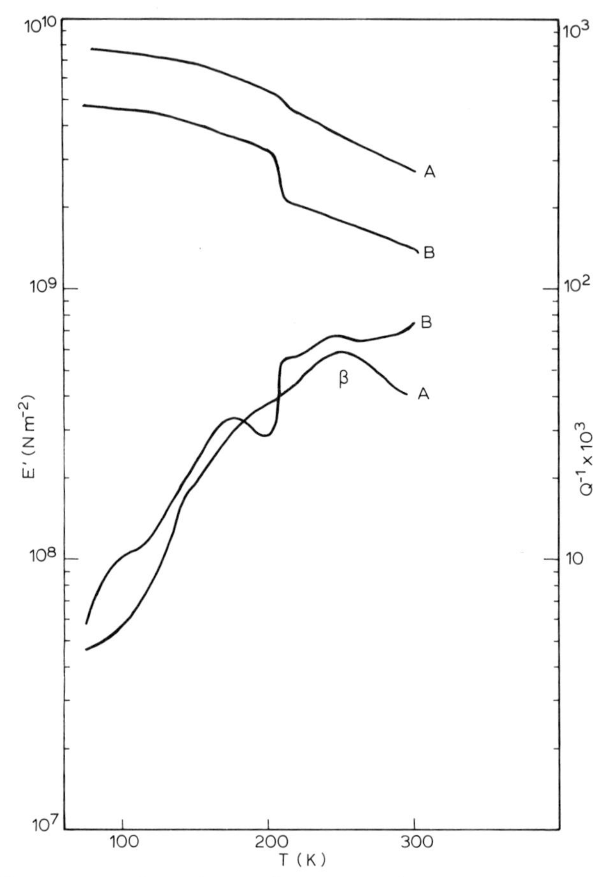

Fig. 6. Temperature dependence of E' and Q^{-1} for HTH–C_3/C_5 copolyester: (A) original; (B) quenched from LC state.

from that of original and melt-crystallized samples. Some unusual features shown in their dynamic viscoelastic patterns (Figs 3–7) deserve a particular comment. In all the polymers there was a strong and sharp change in modulus at 210 K. The change in modulus was always accompanied by an abrupt change of internal friction. In all cases such mechanical transition was enantiotropic in character and reproducible in cooling and heating cycles throughout the transformation range. The strength of the transition, on the contrary, was found to change either by varying the quench-cooling conditions or upon annealing the sample after quenching. In Fig. 8 the E' vs. T and Q^{-1} vs. T plots of quenched samples of polyester HTH–C_{10}

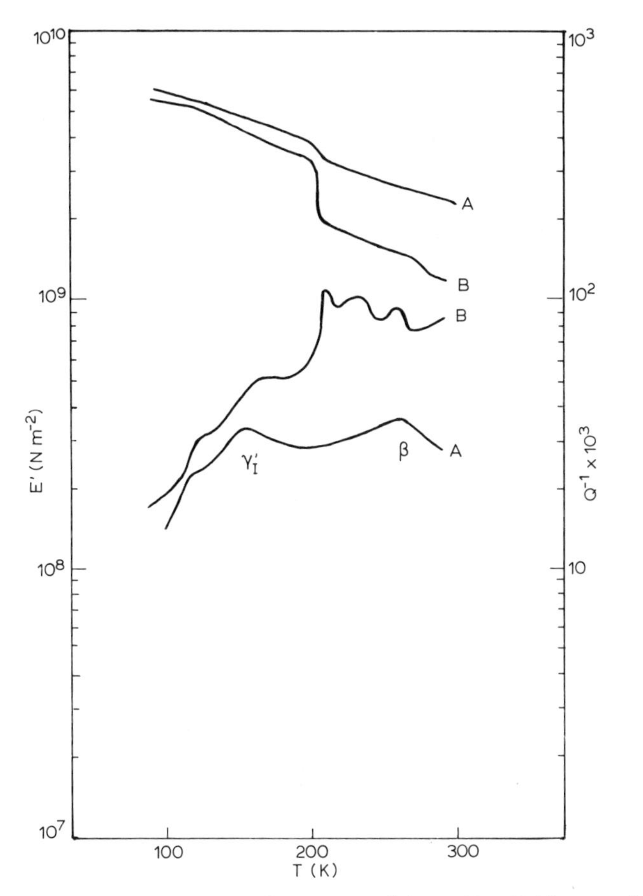

Fig. 7. Temperature dependence of E' and Q^{-1} for HTH–C_5/C_{10} copolyester: (A) original; (B) quenched from LC state.

obtained by different thermal treatments are reported. Some data relevant to the mechanical transition are reported in Table 2. Figure 9 shows the close correlation existing between the transition strength, $\Delta E'/E'$, and the change of internal friction, ΔQ^{-1}, at the transformation temperature for the HTH–C_{10} sample.

Above the transition temperature the mechanical losses were maintained at a quite high level, whereas below the transition point the losses fell down to values normally found in polymers below T_g. For the HTH–C_{10} sample quenched from 513 K, that is from the smectic state, the level of dissipation between 210 and 300 K was so high as to submerge completely

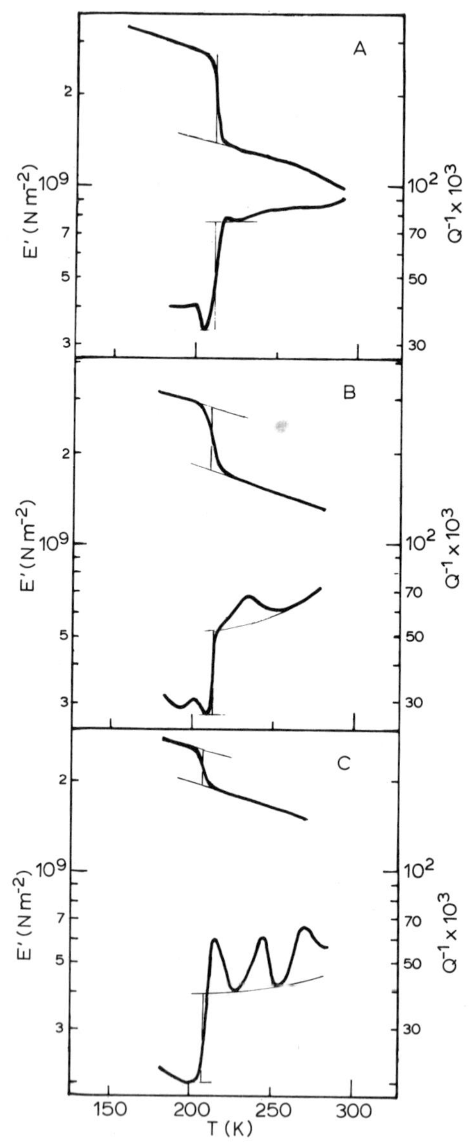

Fig. 8. E' vs. T and Q^{-1} vs. T plots of quenched samples of HTH–C_{10} polyester with different thermal histories: (A) quenched from LC melt; (B) quenched from isotropic melt; (C) quenched from LC melt and then annealed at 540 K.

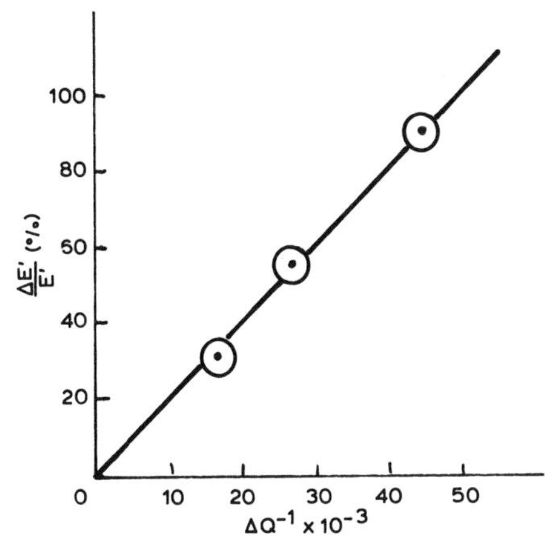

Fig. 9. Relationship between transition strength, $\Delta E'/E'$ and change of internal friction, ΔQ^{-1}, for HTH–C_{10} polyester at the transformation temperature.

the secondary relaxation effects existing in this temperature range. For the other two HTH–C_{10} samples with a different thermal history, the progressive reduction of dissipation, which was seen to parallel the decrease of the transition strength, allowed the secondary loss peaks to emerge above the background dissipation and to become more and more evident.

In the lower temperature region other significant changes in the mechanical spectra of the quenched samples must be pointed out. Apart from a general increase of losses in the quenched polymers with respect to the unquenched samples, and some other minor changes occurring

TABLE 2
Modulus Change, $\Delta E'$, and Transition Strength, $\Delta E'/E'$,[a] at 210 K for Polyesters and Copolyesters Quenched from the LC State

Sample	$\Delta E' \times 10^{-9}$ $(N m^{-2})$	$\Delta E'/E'$
HTH–C_3	1·10	0·92
HTH–C_{10}	1·23	0·89
HTH–TEG	1·40	0·56
HTH–C_3/C_5	1·00	0·46
HTH–C_5/C_{10}	1·30	0·68

[a] E' is the value of the modulus at temperatures just above the transformation temperature T_{tr}.

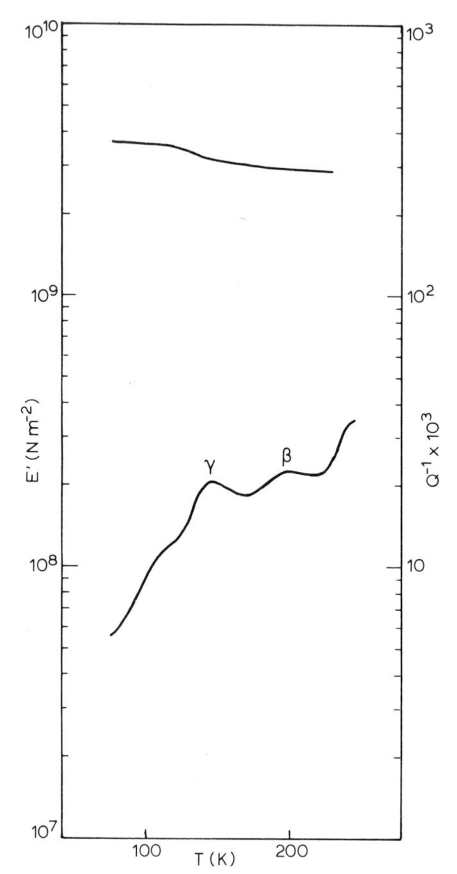

Fig. 10. Temperature dependence of E' and Q^{-1} for the melt crystallized sample of model compound bis(4-decyloxycarbonylphenyl) terephthalate.

in HTH–C_3, HTH–TEG, and HTH–C_3/C_5 polyesters, an interesting phenomenon was observed in the region of the γ-relaxation in the HTH–C_{10} and HTH–C_5/C_{10} samples. As regards HTH–C_{10}, it can be noted that the γ-peak at 150 K (henceforth labelled as γ_I) characteristic for the unquenched sample, was replaced in the quenched polymer by two distinct loss peaks, γ_{II} and γ_{III}, at 130 and 170 K respectively. A partial overlap of these two relaxation effects was always observed, owing to the relative vicinity of the two peaks on the temperature scale and the rather broad distribution of relaxation times of the two processes. In the quenched samples of HTH–C_5/C_{10} copolyester, the γ_I-peak was also missing, but contrary to HTH–C_{10}, only a small shoulder at about 130 K and a barely resolved peak at 170 K were observed.

In addition to the polymers mentioned above, the model compound bis(4-decyloxycarbonylphenyl) terephthalate was also examined (Fig. 10).

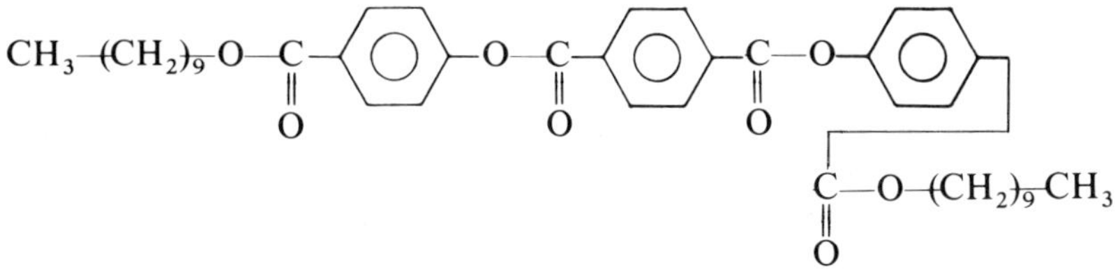

$$CH_3\text{—}(CH_2)_9\text{—}O\text{—}\underset{\underset{O}{\|}}{C}\text{—}\langle\bigcirc\rangle\text{—}O\text{—}\underset{\underset{O}{\|}}{C}\text{—}\langle\bigcirc\rangle\text{—}\underset{\underset{O}{\|}}{C}\text{—}O\text{—}\langle\bigcirc\rangle\text{—}\underset{\underset{O}{\|}}{C}\text{—}O\text{—}(CH_2)_9\text{—}CH_3$$

The loss pattern of the melt-crystallized sample was characterized by two relaxation processes, β and γ, occurring at 200 and 150 K respectively. It is interesting to note that in the model compound the γ-peak manifests at the same temperature as the γ_I-process of unquenched polyesters containing $—(CH_2)_{10}$ sequences, although in the former the intensity of the peak is somewhat lower.

DISCUSSION

The Sólid–Solid Transition at Low Temperature

Calorimetric data supported the conclusion that, during the quenching of these materials from the anisotropic melt, the LC phase is prevented from crystallizing and remains as a highly supercooled mesophase. This metastable glass-like modification is subsequently transformed below 235 K into a more ordered, stable phase.

The starting point for the interpretation on a molecular scale is represented by the finding that such a transition is apparently independent of the chemical structure of the flexible spacers. Therefore, the rigid core of the repeating unit must be involved in the transformation. For polyester HTH–C_{10}, the entropy of transition, extrapolated for the 100% smectic glassy polymer, is 1·29 cal (K mole)$^{-1}$ per rigid core unit. Many molecular crystals undergo a solid–solid phase transition at a certain temperature T_{tr} below the melting point T_m. At $T > T_{tr}$ the crystal is hard and the molecules comparatively immobile, whereas at $T < T_{tr}$ (in the 'plastic' range) the crystal is soft and the molecules can rotate almost as rapidly as in the liquid phase. For solids formed by molecules with certain restricted symmetry,

ΔS_{tr} can be estimated if the transition is considered as an order–disorder transition.[7] In this case $\Delta S_{tr}/R = \ln(N_2/N_1)$, where N_1 and N_2 are the number of states of disorder statistically occupied below and above T_{tr}, respectively. In the present case the entropy change ($\Delta S \simeq 1{\cdot}29$ cal (K mole)$^{-1}$) is close to $R \ln 2 = 1{\cdot}38$ cal (K mole)$^{-1}$. This value can be accounted for, by admitting that at $T < T_{tr}$ there is only one conformational state allowed for the polymer segment ($N_1 = 1$), whereas at $T > T_{tr}$ a random distribution of segments at lattice sites in two equivalent conformations ($N_2 = 2$) occurs.

Even though this interpretation appears consistent with the experimental data, the possibility that the entire repeating unit, including the flexible spacer, is cooperatively involved in the rotational transition cannot be completely ruled out. In fact, whereas the transition involving the sole aromatic triads can occur only by a crankshaft motion of the aromatic moiety in a *cisoid* conformation, the rotation of the entire portion of the chain in the smectic layers would allow the aromatic triads to exist in the *transoid* conformation. This conformation has been found to be energetically favoured in most aromatic polyesters.[8–10] Additionally, it may be observed that in several series of homologous compounds the transition temperatures for the order–disorder phenomena were found to be approximately a constant fraction of the absolute melting temperature.[11] Analogously, for all the polyesters and copolyesters examined the ratio T_{tr}/T_m does not change substantially, being equal to $0{\cdot}45 \pm 0{\cdot}05$ ($T_{tr} = 210$ K as determined by dynamic mechanical data).

The dynamic viscoelastic properties of the quenched polymers demonstrate the internal freedom and the high internal mobility of the mesomorphic structures at $T > T_{tr}$. In fact, in the temperature interval between 210 K and T_g, the internal friction of the quenched samples is as high as that of many linear semicrystalline polymers (see for instance polypropylene, polyoxymethylene, polyethylene terephthalate)[12] at temperatures much higher than the respective T_g. A comparison between the dissipation properties of these polymers and two HTH–polyesters is given in Table 3, where reduced temperatures at which the loss modulus, $E'' = E' \times Q^{-1}$, rises to a value of 10^8 N m^{-2} are reported together with the T_g values and the apparent degree of crystallinity, χ_c. It should be noted that in the semicrystalline polymers reported above, the background losses approach such high values only at temperatures where cooperative segmental motions of amorphous chains are already fully activated and mobility even in the crystalline phase starts to occur.

In conclusion we assume that, basically, the first order transition in the supercooled mesophase is an order–disorder transition and that segments

TABLE 3
Glass Transition Temperature and Reduced Temperature, T/T_g, for Different Polymers

Polymer[a]	χ_c[b]	T_g (K)	T/T_g[c]
PP	0·7	263	1·22
PET	0·4	343	1·43
POM	0·8	193	2·03
HTH–C_{10}	0·3–0·4	310[d]	0·68
HTH–C_3	$\leq$0·3	400[d]	0·52

[a] See text.
[b] Degree of crystallinity.
[c] T is the temperature at which the measured loss modulus is close to 10^8 N m^{-2}.
[d] Glass transition temperature of the amorphous phase of the original polyester sample.

of the polymer chain that possess some degree of symmetry around the long axis, at $T > T_{tr}$ are in a state of definable disorder and display a high rotational mobility around this axis. The occurrence of cooperative motions of mesogenic side-groups, which originate exclusively in a solid smectic phase, has been already reported for poly(p-biphenyl acrylate).[13]

The Relaxation Behaviour at $T < T_{tr}$

There are two general theories[12] for mechanical relaxations in polymers below T_g: *site theory*, which postulates that relaxation is due to the movement of the polymer segments passing over a potential energy barrier from one stable conformation to another; *local mode theory*, which involves the damped oscillation of chain segments about their equilibrium position. The molecular weight independence of secondary relaxations and the observation that the magnitude of their activation energy is often of the same order as the barrier height indicate that such relaxation phenomena are associated with short-range conformational changes. Moreover, there is some parallelism between the loss patterns of small molecules and the analogous groups incorporated into the repeating unit of macromolecules. This is particularly true for the γ-relaxation in polyethylene and in polymers containing $-(CH_2)_n-$ sequences with $n > 4$–6. This relaxation is commonly interpreted as a restricted rotational motion of sub-chain units of the paraffinic segments. The kinetic process has been rationalized in terms of coordinated rotational transitions* of the two C—C bonds adjacent to the central C—C bond which undergoes the transition over the potential energy barrier. The two rotating bonds display a crank-like counter-rotation with respect to each other of an angle $+\Phi$ and $-\Phi$, respectively, in order to minimize the swinging of the chain for the rotation of the central bond.[14] This localized mode of motion of the $-(CH_2)_n-$ sequences is thus responsible for the γ-relaxation and it has been shown to occur in both crystalline and amorphous regions of polymer.

Phenomenological Aspects of the γ-Relaxation of Mesomorphic Polyesters

The marked change in the loss pattern observed in the region of γ-relaxation for HTH–C_{10} samples with different thermal histories (Fig. 5) can be described in two alternative ways. On the one hand, the change can be simply considered as due to a shift of the γ_I-peak towards lower temperatures on passing from untreated samples or samples crystallized from the anisotropic melt to samples rapidly quenched from the LC state. As a consequence of the 20 K shift of the main relaxation peak, another loss peak (γ_{III}), previously submerged in the prominent γ_I-peak, appears at higher temperatures. Accordingly, the shift to lower temperatures would indicate a decrease in the hindrance to rotation for $-(CH_2)_n-$ sequences in the smectic layers. This might be a consequence either of looser chain packing with respect to the crystalline order, or of reduced strain imposed on the molecules in the mesomorphic regions of the polymer. However, since the quenched samples can be regarded as two-phase systems, in which a crystalline phase is dispersed in a continuous supercooled smectic phase, the attribution of the γ_{III}-peak at higher temperatures to a relaxation process within an amorphous phase is doubtful. On this line, the origin of this loss peak remains unknown.

On the other hand, the modification of the loss pattern can be caused by a splitting of the γ_I-peak at 150 K into two new peaks, γ_{II} and γ_{III}, centred at 130 and 170 K, respectively. This second interpretation is attractive because it entails the resolution of one relaxation into two different relaxation processes. This assumption is founded on the observation that

* The basic difference between such a relaxation process and a phase transition such as that discussed in this chapter consists in the frequency dependence of the temperature at which the relaxation processes occur, and in the temperature invariance of phase transition. Moreover, contrary to a relaxation process, which gives rise to a maximum in the mechanical energy absorption at a particular combination of frequency and temperature, a phase transition of the order–disorder type causes just a step change in the energy absorption at T_{tr}.

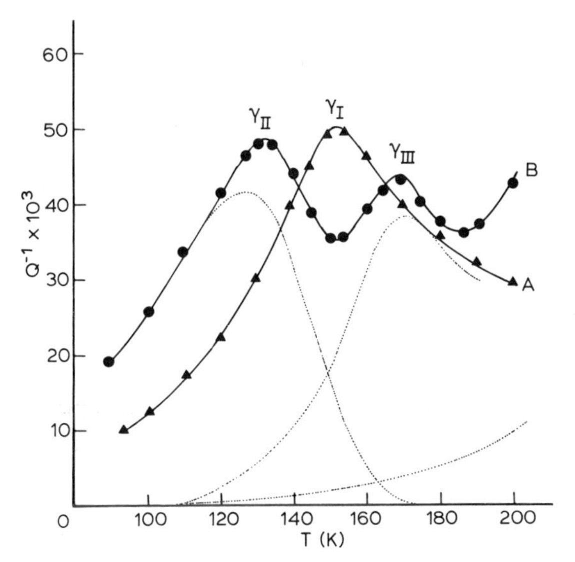

Fig. 11. The γ-loss process in HTH–C_{10} polyester: (A) original; (B) quenched from LC state. Dotted line represents the graphical resolution of γ_{II} and γ_{III} loss peaks.

the two loss peaks γ_{II} and γ_{III} are symmetrically placed with respect to the γ_I-peak, and that their intensity is not too different (Fig. 5). A deconvolution of the portion of the spectrum in the γ-relaxation region of the quenched polymer in terms of background contribution and partial overlap of γ_{II} and γ_{III} processes is presented in Fig. 11.

This also implies that the average relaxation time, τ, of the γ_I-process at a given temperature lies intermediate between those of the γ_{II}- and γ_{III}-processes. For the sub-chain relaxations the temperature dependence of τ ($\tau = \omega^{-1}$, where $\omega = 2\pi\nu$ and ν is the measured frequency in cps) is given by $\tau = \tau_0 \exp(\Delta E/RT)$, in which the activation energy ΔE is independent of T. Assuming for the γ-type relaxation process an activation energy of 12 kcal mole^{-1},[12] we obtain at the temperature of 170 K the values of $\tau_{II} \simeq 7 \times 10^{-10}$ s, $\tau_I \simeq 3 \times 10^{-7}$ s, and $\tau_{III} \simeq 4 \times 10^{-5}$ s. Therefore, the splitting of the γ_I-peak into two almost symmetrical loss peaks would also result in a plot of Q^{-1} vs. $\ln \omega$ at constant temperature.

The other polymer of the investigated series which presents the γ_I-peak at 150 K is the copolyester HTH–C_5/C_{10} (Fig. 7). Also in this case the γ_I-peak disappears on passing from unquenched to quenched samples. In the latter, however, the appearance of the γ_{II}-peak at 130 K is not observed, only a small shoulder in the Q^{-1} vs. T curve being present at this temperature. On the other hand, the γ_{III}-peak at 170 K is very poorly defined but is still detectable above the background dissipation line. Thus, for the HTH–C_5/C_{10} copolymer the γ_{III}-process at higher temperature appears to be the dominant effect in the γ-relaxation. This is inconsistent with the assumption that the modification of the loss pattern is caused by a shift of the γ_I-peak towards lower temperatures. A splitting of the original γ_I-peak into two discrete processes seems, at present, the most convincing hypothesis, although the results so far obtained are not conclusive.

A possible explanation of the bimodal character of the γ-relaxation in the quenched HTH–C_{10} polymer is given below.

Mechanistic Model for the γ-Relaxation in Quenched HTH–C_{10} Polyester

The smectic mesophase of these macromolecules can be considered to consist of two separate sub-phases of different chemical nature and molecular structure, linked by primary chemical bonds. These two sub-phases, one composed of the aromatic portion of the polymer repeating unit, the other formed by the paraffinic sequences, are arranged in a regularly alternating fashion to form a multilayer sandwich-like structure. The overall properties of such smectic polymer phase will result from a combination of the specific properties of the two different sub-phases. In particular, the mechanical response of the bulk mesophase will depend on the elastic constants and compliance of the aromatic and aliphatic subphases. The system can thus be simulated as a molecular composite formed by two continuous phases of different stiffness. It can be postulated that these structures are amenable to a composite analogy for prediction of their mechanical and electrical properties. From geometric considerations on the arrangement of the sub-phases in the smectic structure, a series or a parallel model can be derived depending on the way the stress is applied to the composite (Fig. 12).

If the stress is applied normal to the layers, the stress is constant and the strain on the individual sub-phase is different. Applied stress will bias segmental motion of paraffinic chains in the direction parallel to the chain axis. Alternatively, if the stress is applied parallel to the layers, a constant strain is induced in each sub-phase. Applied strain will deform paraffinic segments in the direction normal to the chain axis. The splitting of the γ_I-peak in the quenched samples of HTH–C_{10} polyester into two separate loss peaks could then reflect the different response to the series (γ_{II}) and parallel (γ_{III}) models to an applied external stress.

The reason why the splitting of the γ_I-peak in quenched HTH–C_5/C_{10}

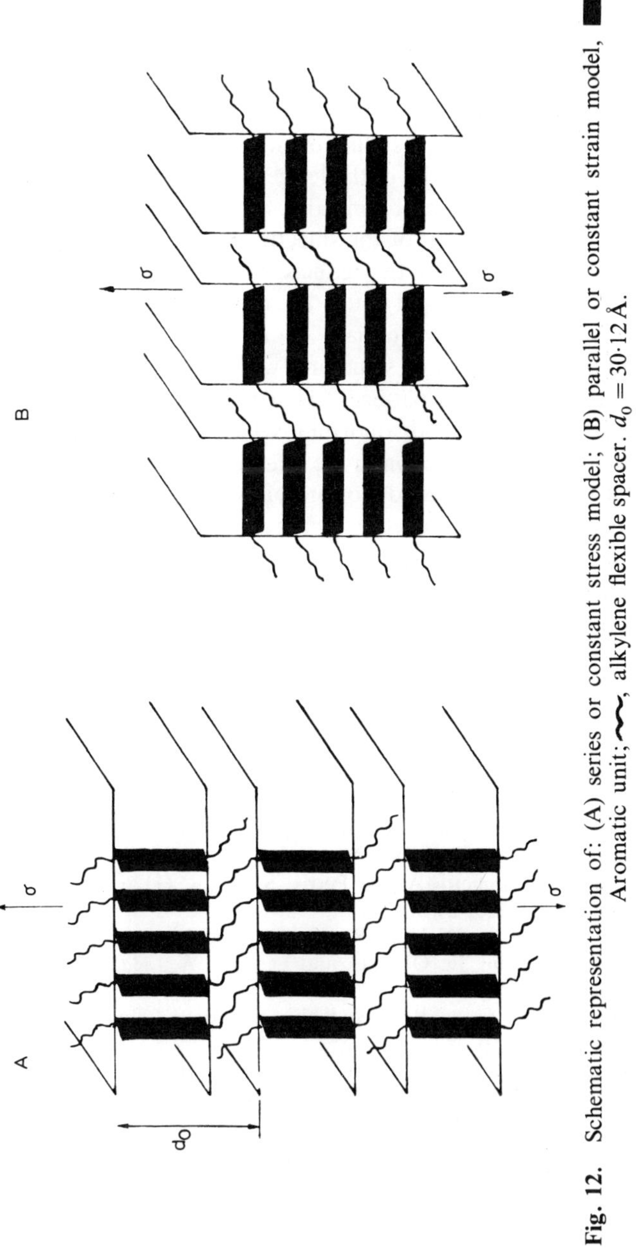

Fig. 12. Schematic representation of: (A) series or constant stress model; (B) parallel or constant strain model, ▬, Aromatic unit; ∼, alkylene flexible spacer. $d_0 = 30{\cdot}12$ Å.

samples is not so evident could be due to the fact that the supercooled mesophase of this copolymer would not possess the high degree of ordering which is required to obtain a clear-cut separation of two distinct smectic layers. In fact, in order to accommodate the rigid moieties connected to the aliphatic segments of different lengths in a layered structure, distortion of the strata, produced by the sliding of two adjacent chain molecules along the chain axis, or puckering of the longer sequences of the flexible spacers should occur. Under such circumstances, structures with a degree of order intermediate between that of the smectic and nematic phases may result. This can also account for the reduced values of the transition strength observed in these two copolyesters as compared to HTH–C_{10} polyester. Accordingly, the bulk mesophase containing long aliphatic sequences arranged in ordered layered structures will exhibit two separate processes in their mechanical loss patterns. That is to say that smectic ordering, in contrast to crystal and amorphous lattices, may induce a valuable anisotropy of internal friction in the γ-region owing to different modes by which the thermally activated motions of $-(CH_2)_n-$ segments can be biased by an external stress. Evidence in favour of this hypothesis may be obtained by dynamic-mechanical experiments performed on oriented samples, in which the direction of an applied stress is either parallel or perpendicular to the plane of the smectic layers.

REFERENCES

1. Frosini, V., Marchetti, A. and de Petris, S., *Makromol. Chem., Rapid Commun.*, 1982, **3**, 795.
2. Frosini, V., de Petris, S., Chiellini, E., Galli, G. and Lenz, R. W., *Mol. Cryst. Liq. Cryst.*, 1983, **98**, 223.
3. Frosini, V., Marchetti, A., de Petris, S., Galli, G. and Chiellini, E., *ACS Polym. Prep.*, 1983, **24**(2), 302.
4. Ober, C., Jin, J. I. and Lenz, R. W., *Polymer J.*, 1982, **14**, 9.
5. Galli, G., Chiellini, E., Ober, C. and Lenz, R. W., *Makromol. Chem.*, 1982, **183**, 2693.
6. Kelker, H. and Hatz, R., *Handbook of Liquid Crystals*, Verlag Chemie, Weinheim, 1980, p. 362.
7. Westrum, E. F. and McCullough, *Physics and Chemistry of the Organic Solid State*, Vol. I, Wiley, New York, 1965, p. 1.
8. Liang, C. Y. and Krimm, S., *J. Mol. Spectrosc.*, 1959, **3**, 554.
9. Yokouchi, M., Sakakibara, Y., Chatani, Y., Tadokoro, H., Tanaka, T. and Yoda, K., *Macromolecules*, 1976, **9**, 266.
10. Hutchinson, I. J., Ward, I. M., Willis, H. A. and Zichy, V., *Polymer*, 1980, **21**, 55.

11. Trappeniers, N., *Changement de Phase*, Herman et Cie., Paris, 1952, p. 241.
12. McCrum, N. G., Read, B. E. and Williams, G., *Anelastic and Dielectric Effects in Polymeric Solids*, Wiley, London, 1967.
13. Frosini, V., Magagnini, P. L. and Newman, B. A., *J. Polym. Sci., Polym. Phys. Ed.*, 1977, **15**, 2239.
14. Helfand, E., Wasserman, Z. R. and Weber, T. A., *ACS Polym. Prep.*, 1981, **22**(2). 279.

4

THERMOTROPIC LIQUID CRYSTAL AROMATIC COPOLYESTERS CONTAINING CYCLOALIPHATIC UNITS

C. BHASKAR,* J. KOPS,† B. MARCHER and H. SPANGGAARD

Instituttet for Kemiindustri, Technical University of Denmark, Lyngby, Denmark

INTRODUCTION

Thermotropic liquid crystal behaviour of polymeric systems has been intensively investigated in recent years. Aromatic polyesters having main chain mesogenic units are among the most promising materials; however, it is recognized that it is necessary to disrupt the perfect regularity of simple *para*-linked aromatic polymers in order to lower the melting ranges to manageable levels. Various polyester systems have been synthesized by Blumstein,[1] Jackson,[2] Lenz[3] as shown also in previous publications by these authors and by others. The various approaches, as summarized by Griffin and Cox,[4] to disrupt the chain packing have been investigated. The most common modification is to incorporate flexible links consisting of short aliphatic chains. With increasing length of these spacers the overall chain cohesion in the melt is lowered, resulting in a corresponding lower melting point often with distinct odd–even effects related to the number of methylene groups in the spacer.

The objective of our work is to prepare materials which may be processed as a mesogenic melt phase in order to obtain good mechanical properties. Thus the modification should not lower the melting point excessively and the structure should govern the properties of the resulting polymer, e.g. with respect to thermal and oxidative stability. We investigate here the effect of using a flexible unit containing a cycloaliphatic ring in aromatic polyesters. Previously, the effect of cycloaliphatic structures in

* Present address: Division of Polymer Chemistry, National Chemical Laboratory, Poona 411 008, India.

† To whom correspondence should be addressed.

low molecular weight liquid crystals has been investigated.[5] It was found that replacement of a benzene ring in various nematogens by a *trans*-1,4-disubstituted cyclohexane ring would enhance the nematic–isotropic transition unless this was already rather high (>220 °C). We wish to report here on copolyesters prepared from various proportions of 1,4-cyclohexanedimethanol (CDM) and 2-methyl- or 2,5-dimethylhydroquinone (2-MHQ or 2,5-DMHQ) with terephthaloyl dichloride:

X HOCH$_2$–C$_6$H$_{10}$–CH$_2$OH + Cl–C(=O)–C$_6$H$_4$–C(=O)–Cl + (1−X) HO–C$_6$H$_2$(CH$_3$)(CH$_3$)–OH →

[–(–CH$_2$–C$_6$H$_{10}$–CH$_2$–O–C(=O)–C$_6$H$_4$–C(=O)–O–)$_X$–(–C$_6$H$_2$(CH$_3$)(CH$_3$)–O–C(=O)–C$_6$H$_4$–C(=O)–O–)$_{1-X}$]

EXPERIMENTAL

Materials

Cyclohexanedimethanol, *cis–trans*, approx. 30/70 mixture (Fluka), was used after drying over conc. H_2SO_4. Methylhydroquinone, commercial 99% material was recrystallized from xylene, m.p. 125–126 °C. 2,5-Dimethylhydroquinone was synthesized from the corresponding substituted phenol by a published procedure;[6] it had a m.p. 215–218 °C after recrystallization from ethanol–toluene. Terephthaloyl dichloride (98%) was recrystallized from dry hexane. The solvents tetrachloroethane, pyridine and acetone were all distilled prior to use.

Polymerization Procedure

Solution polymerization was carried out essentially according to the procedure published by Lenz.[7] Terephthaloyl dichloride (0·01 mol) dissolved in 12·5 ml of tetrachloroethane was added dropwise to a solution of a mixture of cyclohexanedimethanol and methyl-substituted hydroquinone (total amount 0·01 mol of the two components) in 12·5 ml tetrachloroethane containing 3 ml pyridine. The reaction was carried out in a three-neck flask with a vigorous stirring at room temperature. After completion of the reaction, when a viscosity increase was observed and partial precipitation of polymer had occurred, 130 ml of acetone was added after two hours. The polymer was isolated by filtration and after washing with acetone was dried in vacuum at 125–130 °C.

Characterization

Inherent viscosities were measured in *p*-chlorophenol (0·5 g per 100 ml solvent) at 45 °C using an Ubbelohde viscometer.

DSC thermograms were recorded on a DuPont Thermal Analysis Unit Model 900. The heating rate was 20 °C per min in a nitrogen atmosphere. TGA thermograms for thermal stability were obtained on the same unit with attachment of the Model 950 TGA module also with a heating rate of 20 °C per min.

Melt behaviour was observed on a Mettler FP-82 hot stage used in connection with a Reichert–Jung Micro-Star polarizing microscope equipped with an ExpoStar photographic unit.

RESULTS AND DISCUSSION

Copolyesters were prepared with cyclohexanedimethanol and the two methyl-substituted hydroquinones in the following mole ratios: 15/85, 30/70, 40/60, 50/50, 60/40, 70/30 and 85/15. The polymerization data for the two series of polymers are summarized in Table 1. The inherent viscosities and yields are highest in the case of the copolyesters prepared with 2,5-DMHQ. Generally, more material is lost in connection with the work-up of the polymers with 2-MHQ, hazy filtrates, etc., which leads to

TABLE 1
Preparation of Copolyesters with CDM and 2-MHQ or CDM and 2,5-DMHQ by Solution Polymerization[a]

Mole ratio (CDM/subst. HQ)	*2-MHQ*		*2,5-DMHQ*	
	Yield (%)[b]	η_{inh}[c]	*Yield (%)*[b]	η_{inh}[c]
15/85	59	0·21	89	0·24
30/70	76	0·22	86	0·27
40/60	78	0·24	81	0·25
50/50	73	0·18	87	0·24
60/40	77	0·21	84	0·13
70/30	80	0·18	83	0·19
85/15	63	0·14	90	0·17

[a] Conditions given in Experimental section.
[b] After precipitation in acetone.
[c] In *p*-chlorophenol at 45 °C.

TABLE 2
Melt Behaviour of CDM–2-MHQ and CDM–2,5-DMHQ Copolyesters

Composition (mole ratio[a])	*DSC*		*Microscopy*	
	Annealing (60 min) temp. (°C)	*Endotherm temp. (°C)*	*Nematic clearing (°C)[b]*	*Isotropic clearing (°C)[c]*
0/100	290	348	—	None
15/85	275	325	276–304	None
30/70	250	302	258–265	None
40/60	250	275	255–268	None
50/50	225	255	246–259	268
60/40	250	267	223–254	268
70/30	275	285	266–273	280
85/15	275	292	247–281	287
0/100	260	309	—	None
15/85	280	300	336–347	None
30/70	250	280	242–300	None
40/60	225	250	256–314	342
50/50	215	235	228–287	332
60/40	225	245	227–234	243
70/30	240	260	225–243	250
85/15	275	290	264–274	280

[a] The upper series were prepared with 2-MHQ and the lower series with 2,5-DMHQ.
[b] The initial temperature for flow and the temperature at which a melt pool is formed is given in each case.
[c] Only a part of the sample showed clearing due to compositional inhomogeneity; however, the observed isotropic behaviour is reversible.

generally lower yields. Although the values of η_{inh} are low, they are within ranges previously reported for some modified aromatic polyesters.[8,9]

The thermal properties for the two series of polymers are summarized in Table 2. The copolyesters with 2,5-DMHQ generally melt at somewhat lower temperatures than the corresponding copolyesters with 2-MHQ. The melting point of the pure CDM polyester of 292 °C had previously been reported,[10] while we determined the melting points of the pure aromatic polyesters.[11] The lowest melting composition is found where there is an equal molar ratio between the cycloaliphatic and the aromatic components. This effect is illustrated in the plot of the DSC endotherms in Fig. 1. All the DSC measurements were performed on annealed samples since it was found that the crude polymers showed broad or ill-defined melting endo-

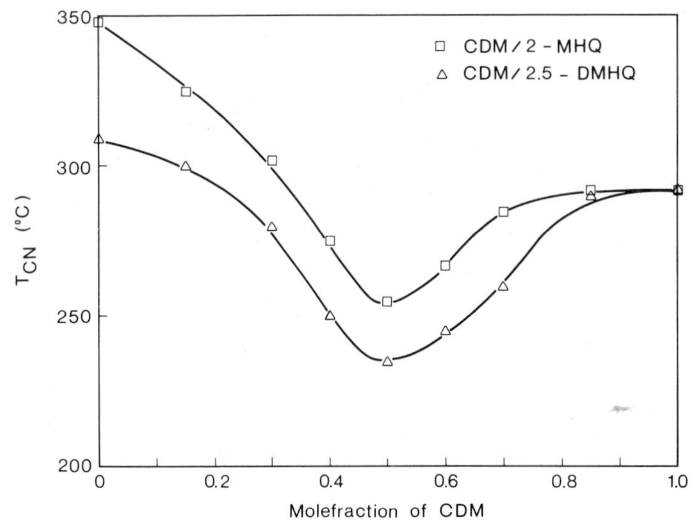

Fig. 1. Melting temperatures as determined by DSC of annealed samples of copolyesters of CDM and 2-MHQ or CDM and 2,5-DMHQ of varying composition.

therms, particularly those with the lowest content of the cycloaliphatic component. In all cases the melting point was raised after annealing. Reasonably good agreement is seen between the DSC observations and the results obtained by hot stage polarized light microscopy, although the microscopy was generally performed on the crude polymers.

Higher contents of CDM (above 50%) led to sharper melting endotherms and clearing points could be observed. Actually only a portion of the mesophase melted isotropically while the remaining portion remained unchanged. We ascribe this to the inhomogeneity of the copolymer samples. The diol components will deviate in reactivity in the copolyester preparation resulting in compositional variations in the products prepared by the batch polymerizations. The sequence distributions in the polymers are not known. In addition to the variation between the aliphatic and the aromatic component the *cis/trans* distribution of the former may also vary.

It has been found previously that isomerization reactions may occur in *cis/trans* CDM terephthalate polyesters just below the melting point.[12] These crystallization-induced reactions may lead to replacement of *cis* by *trans* units and an increase in the degree of crystallinity and melting point. Thus caution should be exercised in interpreting the result of annealing since chemical changes may actually be introduced. Also, these effects have

Fig. 2. Photomicrograph of the copolyester CDM–2-MHQ (60/40) at 259 °C after initial melting where the streaks are due to mobility in the melt during exposure (× 240 magnification).

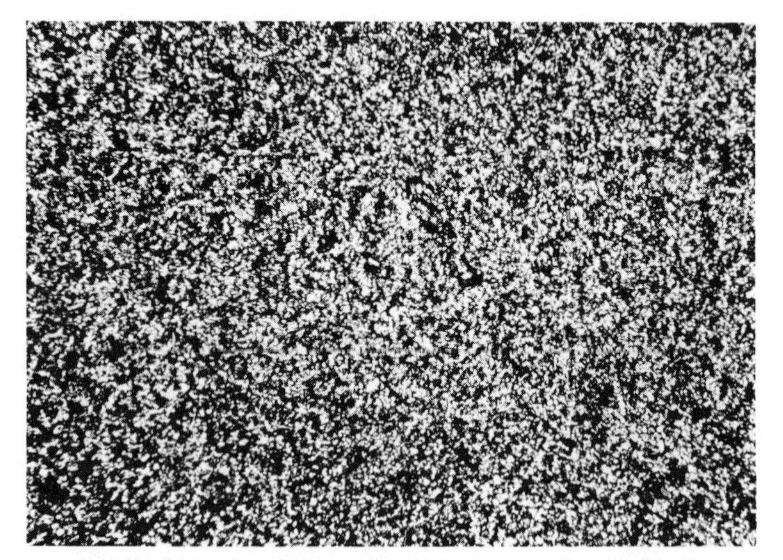

Fig. 3. Photomicrograph of the copolyester CDM–2-MHQ (60/40) after cooling and reheating to 250 °C just under the melting point. A nematic phase with a schlieren structure is observed (× 240 magnification).

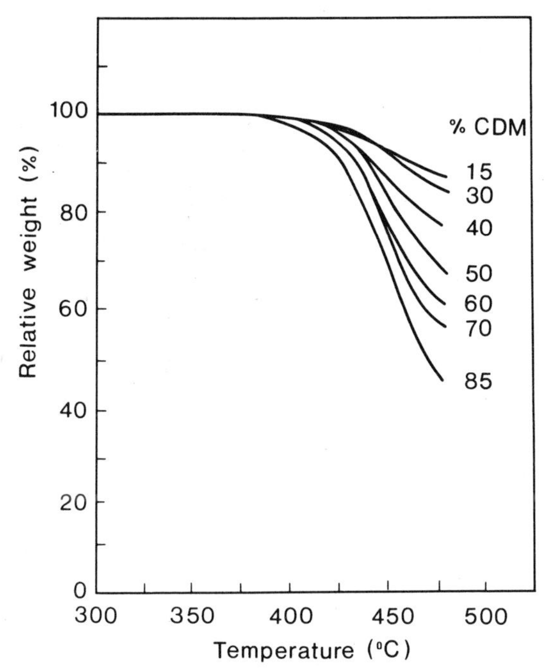

Fig. 4. Thermogram showing weight-loss curves by TGA of copolyesters of CMD and 2-MHQ of varying composition. Heating rate 20 °C per min, in a nitrogen atmosphere.

recently been considered for blends of polyesters,[13] where interchange reactions between the components must be considered.

In our future work attempts will be made to obtain these copolyesters with a uniform structure by suitable coupling reactions.

In Fig. 2 a photomicrograph is shown of a sample of polyester CDM–2-MHQ (60/40) which is held at 259 °C on the polarizing microscope. Streaks are seen due to mobility in the melt phase under the exposure. In Fig. 3 the same sample is seen after cooling and reheating to 250 °C where it is held to form a nematic phase with a schlieren structure. Further heating leads to an isotropic transition for a major part of the sample and this part reforms the mesophase upon cooling in a reversible manner.

An important question is whether the materials are thermally stable under the conditions under which they show mesophase behaviour. Therefore, the thermal stability was investigated by TGA and the thermograms are shown for the two copolyester series in Figs 4 and 5. No significant

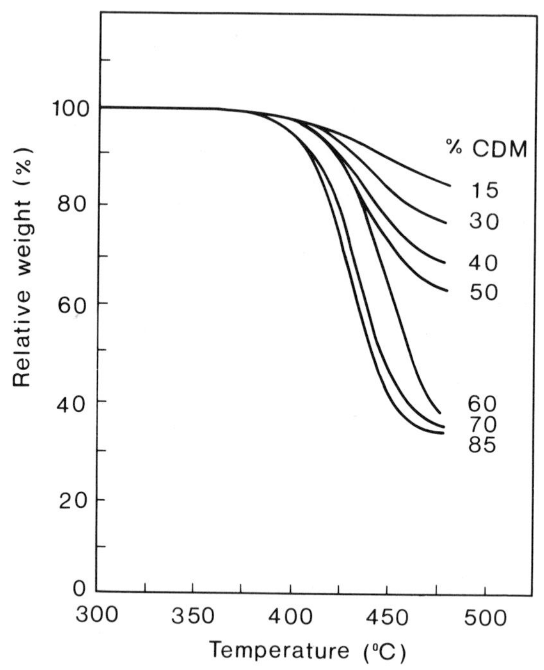

Fig. 5. Thermogram showing weight-loss curves by TGA of copolyesters of CMD and 2,5-DMHQ of varying composition. Heating rate 20 °C per min, in a nitrogen atmosphere.

weight loss is recorded below 375 °C which is well above the melting ranges for the polymers. The thermal stability is seen to decrease with increasing content of the cycloaliphatic comonomer.

CONCLUSIONS

The melting ranges of fully aromatic polyesters may be lowered by modification of the systems by copolymerization with a cycloaliphatic monomer (CDM) to obtain anisotropically melting polymers. With high contents of CDM clearing points of the melts could be observed demonstrating that such behaviour may be attained at moderate temperatures for polymers having structures dominated by cyclic units and devoid of straight chain aliphatic or siloxan spacers. The polymers are thermally stable as determined by thermogravimetric analysis to temperatures well above the melting intervals.

ACKNOWLEDGEMENT

The financial support to one of the authors (CB) by the Danish International Development Agency (DANIDA) is gratefully acknowledged.

REFERENCES

1. Blumstein, A., Thomas, O., Asrar, J., Gauthier, M. M., Makris, P., Clough, S. B. and Blumstein, R. B., *Polymer Preprints*, 1983, **24**(2), 258.
2. Jackson, W. J. Jr., *Macromolecules*, 1983, **16**, 1027.
3. Zhou, Qi-F. and Lenz, R. W., *J. Polym. Sci., Polym. Chem. Ed.*, 1983, **21**, 3313.
4. Griffin, B. P. and Cox, M. K., *Brit. Polym. J.*, 1980, **12**, 147.
5. Gray, G. W. in *Polymer Liquid Crystals*, Ciferri, A., Krigbaum, W. R. and Meyer, R. B. (Eds), Academic Press, New York, 1982, pp. 13–22.
6. Nilsson, J. L. G., Sievertsson, H. and Selander, H., *Acta Pharm. Suecica*, 1968, **5**, 215.
7. Jin, J.-I., Anton, S., Ober, C. and Lenz, R. W., *Brit. Polym. J.*, 1980, **12**, 132.
8. Antown, S., Lenz, R. W. and Jin, J.-I., *J. Polym. Sci., Polym. Chem. Ed.*, 1981, **19**, 1901.
9. Ober, C., Jin, J.-I. and Lenz, R. W., *Polymer J.*, 1982, **14**, 9.
10. Hasek, R. H. and Knowles, M. B., US Patent 2 917 549 (1959); CA 14154d (1960).
11. Kops, J., Patil, D. R., Shenoy, M. A. and Shinde, B. M., *Proceedings IUPAC Makromolecular Symposium*, Amherst, 1982, p. 219.
12. Lenz, R. W. and Go, S., *J. Polym. Sci., Polym. Chem. Ed.*, 1973, **11**, 2927.
13. Barnum, R. S., Barlow, J. W. and Paul, D. R., *J. Appl. Polym. Sci.*, 1982, **27**, 4065.

5

POLYMERIZATION OF ALLYLDIMETHYL-DODECYLAMMONIUM BROMIDE LIQUID CRYSTALLINE MONOMER TO ITS LIQUID CRYSTALLINE POLYMER

C. M. PALEOS

Nuclear Research Center 'Demokritos', Aghia Paraskevi, Attikis, Greece

and

PHOTIS DAIS

Department of Chemistry, McGill University, Montreal, PQ, Canada

INTRODUCTION

Recently we have introduced[1] a novel class of thermotropic liquid crystalline compounds comprised of quaternary ammonium salts, bearing the dimethyldodecylamine moiety as the basic unit. Specifically, it

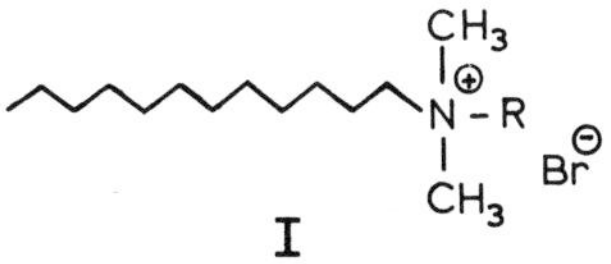

CH_3
$N^{\oplus}-R$ $\quad Br^{\ominus}$
CH_3

I

was found that compounds of formula **I** in which R was substituted by diversified groups such as $—CH_3$, $—CH_2CH_3$, $—CH_2CH_2CH_3$, $—CH_2CH_2OH$, $—CH_2CH_2CH_2OH$, $—CH_2COOH$, $—CH_2CH_2COOH$, $—CH_2CH_2CH_2CN$, $—CH_2COOCH_3$ or $—CH_2COOCH_2CH_2OH$, exhibited liquid crystalline properties. In this paper the liquid crystalline character of a novel monomer, obtained by the substitution of R with the allyl group, is investigated in conjunction with the liquid crystalline properties of the polymer prepared by its polymerization. It should be noted that the findings on the thermotropic liquid crystalline behaviour of these quaternary ammonium salts extend in the melt phase, their well-known organizational ability of forming micelles in solution.

EXPERIMENTAL

Monomer Synthesis

Allyldimethyldodecylammonium bromide was prepared instantaneously by the interaction of dimethyldodecylamine with a slight excess of allyl bromide in ethyl acetate. The precipitated salt was recrystallized from the same solvent and the quaternary salt, which was extremely hygroscopic, was dried over phosphorus pentoxide. Analysis: Found (%): C, 60·80; H, 10·99; N, 4·29; Calculated for $C_{17}H_{36}NBr$: C, 61·11; H, 10·77; N, 4·19.

Polymerization of Allyldimethyldodecylammonium Bromide

The critical micelle concentration (cmc) of the monomer was determined by the conductivity method and it was found to equal $1{\cdot}25 \times 10^{-2}$ M. Polymerization was conducted in a micellar solution, i.e. a solution whose concentration exceeded that of cmc. A typical polymerization experiment leading to 100% polymerization is the following:

An 0·1 M aqueous solution of the monomer was sealed in glass ampoules, under vacuum, after the application of the freeze–thaw technique. The ampoules were kept at room temperature for 24 h before being irradiated for 48 h in a cobalt-60 source (dose rate 2300 rad min^{-1}). Polymerization was completed as determined by proton NMR spectroscopy and the content of the ampoules after water removal was characterized for liquid crystalline properties.

Optical microscopy was performed with a Reichert 'Thermopan' polarizing microscope equipped with a Leitz camera. The thermal studies were conducted with a Dupont 910 Differential Scanning Calorimeter.

RESULTS AND DISCUSSION

Quaternization of dimethyldodecylamine with allyl bromide for the synthesis of the monomer was performed in ethyl acetate, in which the reaction proceeds relatively fast[2] and in addition the quaternary salt precipitates readily.

Polymerization of the monomer was conducted under micellar and isotropic conditions, radiolytically or catalytically. In the latter case azobis

Fig. 1. Photomicrograph of the monomer taken on the cooling cycle at 55 °C.

(isobutyronitrile) was employed as the catalyst. Radiation-induced micellar polymerization was, however, the preferred mode since it proceeded relatively fast as compared to catalytic polymerization. Furthermore, γ-rays had a pronounced destructive effect on the polymer when polymerization was conducted isotropically. In addition, due to the concentrated solution employed in micellar experiments, it was possible to follow the course of polymerization directly by NMR spectroscopy without concentrating the respective solutions.

In discussing the liquid crystalline properties of the polymer it is constructive to investigate, at first, the mesomorphic character exhibited by the monomer and to rationalize its origin. Thus, on heating a sample of the monomer, by slightly pressing the cover-slip, a primarily homeotropic texture appears at about 40 °C which becomes isotropic at ~64 °C. On cooling from the isotropic melt a smectic fan-type texture, Fig. 1, appears at the clearing point transition which it supercools down to 30 °C.

On second heating the clearing point is observed at a slightly lower temperature; this is apparently attributable to a slight decomposition and/or polymerization of the monomer. The liquid crystalline character of the monomer was also established by thermal analysis. As seen from DSC cooling traces, in Fig. 2, there is substantial supercooling of the mesomorphic phase. The polymer used for liquid crystalline characterization

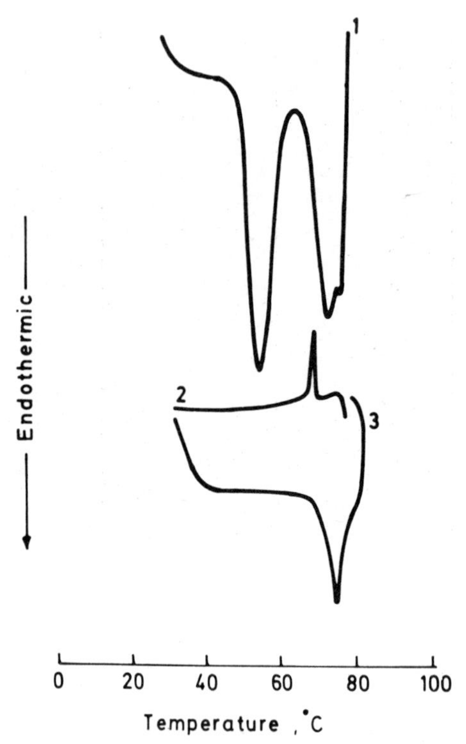

Fig. 2. DSC traces of the monomer: (1) first heating run; (2) first cooling run; (3) second heating run.

was obtained from the irradiated micellar solution by water removal through lyophilization. The remaining material was completely polymerized as determined by NMR spectroscopy.[3] It was also found[3] by carbon-13 NMR spectra that peaks due to the side chain carbons in the polymer spectrum appear in similar positions to those of the monomer. This indicates that the side-chain remained intact and it was not affected by irradiation under micellar conditions.

The polymer so-obtained was a waxy material exhibiting a smectic fan-type texture, shown in Fig. 3, from room temperature up to 155 °C, at which it completely disappears. The polymer, therefore, exhibits a clearing point about 90 °C higher than that of the monomer. On cooling from the isotropic melt the liquid crystalline phase only partially appears; this is indicative probably of some decomposition of the polymer. The complexity of the system is also demonstrated in the polymer DSC traces in Fig. 4.

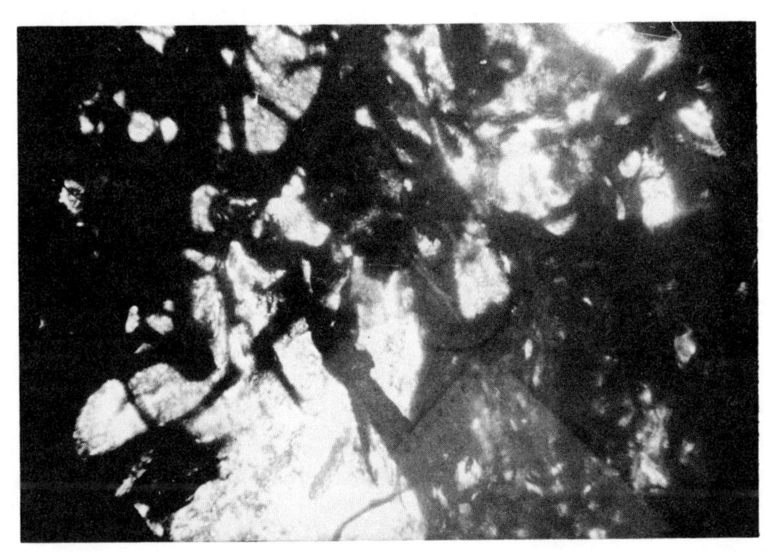

Fig. 3. Photomicrograph of the polymer taken on the heating cycle at 25 °C.

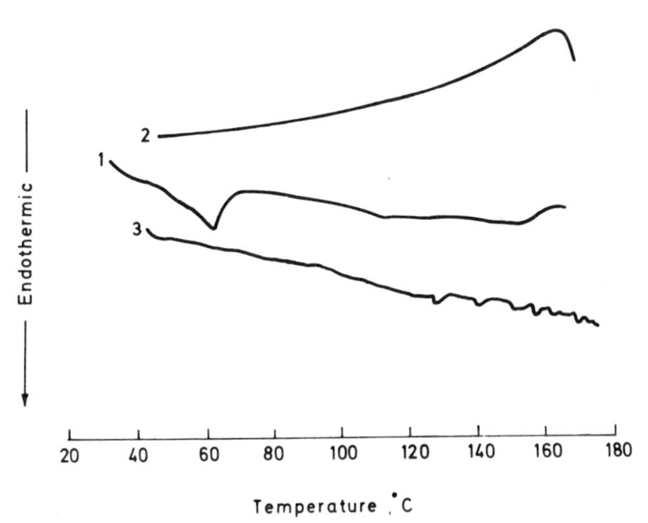

Fig. 4. DSC traces of the polymer: (1) first heating run; (2) first cooling run; (3) second heating run.

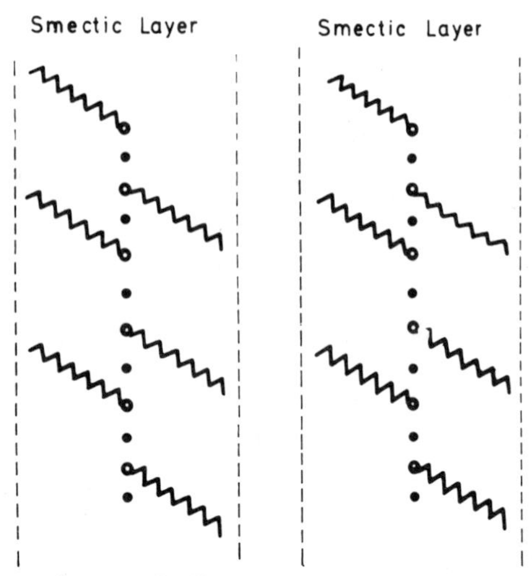

Fig. 5. Diagrammatic smectic liquid crystalline structure of the monomer. ⩘⩘, C_{12}-chain; ○, $CH_2{=}CH{-}CH_2{-}\overset{+}{N}{-}(CH_3)_2$; ●, Br.

The mesomorphic character of the monomer may be rationalized by considering the crystal and mesomorphic structure of similar molecules, i.e. of *n*-alkyltrimethylammonium halides[4] bearing long aliphatic chains. Thus, in analogy with the previously mentioned quaternary salts we may assume that the crystal structure of the allyl quaternary monomer consists of a polar layer within which ionic-type bonding between the quaternary centre and bromide anions occurs, and a non-polar layer formed by the C_{12}-alkyl chain and held together by van der Waals' forces. In this context the formation of the smectic phase can be seen as a process involving the breakage of the terminal van der Waals' bonding between neighbouring non-polar layers forming a structure shown in Fig. 5, which consists of ionic and non-polar layers that can now slide with respect to each other while still retaining their internal structure due to the rigidness of the polar layer. It has however to be noticed here that this is not a typical smectic phase, since although the layered structure has been postulated and the material as a whole is in the molten state due to the 'liquid-like' state of the alkyl-chains occupying the greatest part of the molar volume, the solid state structure of the polar layers remains practically unchanged.[4] At higher temperature, i.e. at the clearing point when the thermal energy is sufficient to break the ionic bonds, the liquid crystalline structure must collapse forming the isotropic melt. By considering the molecular structure of the polymer a similar liquid crystalline structure may be proposed for the polymer. The higher clearing point of the polymer liquid crystal is justified by the fact that in addition to the ionic bonding between quaternary centres and bromide ions, polymer bonding is involved due to the presence of the allyl group located at the polar head of the monomer.

In conclusion, polymerization of allyldimethyldodecylammonium bromide liquid crystalline monomer leads to a liquid crystalline polymer possessing higher stability and broader mesomorphic range than its monomeric counterpart.

ACKNOWLEDGEMENT

We thank Dr G. Margomenou-Leonidopoulou for obtaining the DSC thermograms.

REFERENCES

1. Malliaris, A., Christias, C., Margomenou-Leonidopoulou, G. and Paleos, C. M., *Mol. Cryst. Liq. Cryst.*, 1982, **82**, 161; and our own unpublished results.
2. Goerdeler, J., in *Methoden der Organischen Chemie* (Houben Weyl), Müller, E. (Ed.), Vol. XI/2, Georg Thieme Verlag, Stuttgart, 1958, p. 591.
3. Paleos, C. M., Dais, P. and Malliaris, A., *J. Polym. Sci., Polym. Chem. Ed.*, in press.
4. Iwamoto, K., Ohnuki, Y., Sawada, K. and Senó, M., *Mol. Cryst. Liq. Cryst.*, 1981, **73**, 95.

PART II

THEORY

6

THEORETICAL BASIS FOR LIQUID CRYSTALLINITY IN POLYMERS

P. J. FLORY

Department of Chemistry, Stanford University, California, USA

The molecular features responsible for liquid crystallinity are: (i) asymmetry of molecular shape and (ii) anisotropy of intermolecular forces. The former feature is dominant, especially among polymers.[1,2] Feature (ii) is most prominent in liquid crystalline compounds of low molecular weight; it is doubtless important also in many polymers containing groups, such as *p*-phenylene, whose polarizabilities are highly anisotropic. The latter feature, although of lesser importance than (i) in promoting liquid crystallinity, is usually responsible for thermotropic behaviour.

Lattice theories[3-5] have been most successful in treating the role of molecular shape, as manifested in the axial ratios of highly elongated molecules or of rigid and semi-rigid polymer chains.[6] The Maier–Saupe theory,[7] put forward in 1959–60, addresses feature (ii) exclusively. The lattice theory has been extended recently[8] to accommodate orientation-dependent interactions of the kind comprehended in the Maier–Saupe theory.

The lattice theory in its most rudimentary form treats so-called hard rods; i.e. particles of rod-like shape which are devoid of interactions with neighbouring molecules, with the exception of repulsions of infinite magnitude which arise when two such particles overlap in space. Principal predictions[3-5,9] of the theory are the following:

(a) Above a critical concentration that depends on the axial ratio x the system adopts a state of partial order relative to a preferred axis, or director. The transformation is discontinuous: below a critical concentration the directions of the particles are completely uncorrelated, whereas above that concentration the degree of order is large, although imperfect.

(b) The critical ordering described above is bridged by a phase transition, and hence obscured from experimental observation. The coexisting phases consist of one that is isotropic and the other highly anisotropic, or nematic. The difference between their concentrations is relatively small; they are distinguished more strikingly by the degree of order in one of them.

(c) For hard rods of axial ratio x, phase separation is predicted to occur at a volume fraction v_2 given by[3,9]

$$v_2 \approx (8/x)(1 - 2/x)$$

in close approximation for axial ratios $x >$ about 10. The predicted axial ratio for phase separation in the neat liquid is $x = 6{\cdot}4$.

Predictions (a), (b) and (c) find abundant verification in experiments on lyotropic solutions of liquid crystalline polymers.[9] α-Helical polypeptides dissolved in various solvents exhibit separation of a cholesteric phase at concentrations in close agreement with the equation above.[9–13] Degraded DNA dissolved in aqueous solution likewise induces the formation of a nematic phase above a well-defined concentration[14] that is in good agreement with the equation above. In both instances, the ratio of the volume fractions in the two phases is about 1·3–1·4, in satisfactory agreement with theory. Observations on the onset of phase separation in solutions of polyaramides[15–20] are also in approximate agreement with theoretical predictions.[9] Further predictions of the theory are as follows:

(d) Polydispersity should broaden the biphasic gap.[21,22] Larger species are predicted to occur preferentially in the nematic phase. These predictions are confirmed qualitatively by experiments,[23] but not quantitatively. Discrepancies between theory and experiments probably are due to the slow rate of transport of species between the coexisting ordered and disordered phases.[9,24]

(e) According to theory,[25] solutions containing both rigid rods and random coils should yield nematic phases from which the random coils are totally excluded within limits of experimental detection. The coexisting isotropic phase should contain a relatively small proportion of the rod-like species. These predictions have been fully confirmed by experiments.[26–28]

(f) Hypothetical semi-flexible chains consisting of rigid rods connected consecutively by completely flexible joints should exhibit nematic behaviour that virtually coincides with that of independent rods

> having the same axial ratio as the rigid segments in the hypothetical chains.[29] This result, based on the lattice theory, suggests that a broad class of semi-flexible polymers exhibiting nematic (or cholesteric) behaviour may be treated according to the classical Kuhn model chain, the length of a segment and the number of them in the chain being so chosen as to duplicate the full extension and characteristic ratio of the real chain.

Results on the lyotropic behaviour of various solutions of cellulose esters[30,31] and esters[32–34] are in good agreement with this scheme.[9]

Recent developments clearly indicate that, in the case of chain molecules in which some, at least, of the units admit of a degree of conformational disorder, the formation of a nematic phase is accompanied by significant changes in conformation.[35,36] These changes confer greater stability on the nematic phase, and may be reflected in higher transition temperatures. Investigations along this line also provide the key to understanding so-called 'odd–even' effects both in low-molecular nematogens and in polymers comprising rod-like units joined by flexible sequences,[2,37,38] e.g. by polymethylene chains.

Departing from the case of hard rods, one may first consider isotropic 'soft' interactions, these being of the kind prevalent in non-mesogenic solutions. In the case of a solvent which interacts favourably with the nematogenic molecule, such interaction has relatively little effect on the transition yielding a nematic phase. If, however, the interaction between solvent and solute is unfavourable, with the result that contacts between mesogenic molecules are favoured relative to those between the solute and the solvent, the biphasic gap may be greatly widened, even for a relatively small magnitude of the interaction.[3] For axial ratios $x >$ about 50, a second pair of nematic phases is predicted to occur at higher concentrations. These predictions have been confirmed by experiments.[11,12,20] They constitute a further confirmation of the validity of the lattice theory.

The anisotropic interactions cited above are of broader importance. According to theory,[7,8,39] they depend directly on the square of the anisotropy $\Delta\alpha$ of the polarizability of the particle relative to its mean polarizability $\bar{\alpha}$. The anisotropy $\Delta\alpha$ can be evaluated from depolarized Rayleigh scattering combined with other measurements of optical anisotropy; the mean polarizability $\bar{\alpha}$ can be deduced from the refractive index.[39] The magnitude of the orientation-dependent interaction can be expressed by a characteristic temperature T^* that denotes the magnitude of the interaction for perfect orientation.[8,9]

Orientation-dependent interactions may affect the phase behaviour of a lyotropic system in much the same way as the isotropic interactions described above.[40] More importantly, they appear to account for the fact that low-molecular nematogens having axial ratios substantially less than 6·4 exhibit nematic behaviour.

Detailed studies on homologous *p*-phenylenes[39] and on homologous *p*-oxybenzoates[41] confirm most of the predictions of the theory, including the correlation between the orientation-dependent interactions measured by T^* and the polarizability anisotropy.

REFERENCES

1. Blumstein, A. (Ed.), *Liquid Crystalline Order in Polymers*, Academic Press, New York, 1978.
2. Ciferri, A., Krigbaum, W. R. and Meyers, R. B. (Eds), *Polymer Liquid Crystals*, Academic Press, New York, 1982.
3. Flory, P. J., *Proc. Roy. Soc., London*, 1956, **A234**, 73.
4. Flory, P. J. and Ronca, G., *Mol. Cryst. Liq. Cryst.*, 1979, **54**, 289.
5. Flory, P. J. and Abe, A., *Macromolecules*, 1978, **11**, 1119.
6. Flory, P. J., *Proc. Roy. Soc., London*, 1956, **A234**, 60.
7. Maier, W. and Saupe, A., *Z. Naturforsch.*, 1959, **14a**, 882; *ibid.*, 1960, **15a**, 287.
8. Flory, P. J. and Ronca, G., *Mol. Cryst. Liq. Cryst.*, 1979, **54**, 311.
9. Flory, P. J., *Advan. Polym. Sci.*, 1984, **59**, 1.
10. Hermans, J. Jr., *J. Coll. Sci.*, 1962, **17**, 638.
11. Nakajima, A., Hayashi, T. and Ohmori, M., *Biopolymers*, 1968, **6**, 973.
12. Wee, E. L. and Miller, W. G., *J. Phys. Chem.*, 1971, **75**, 1446; Miller, W. G., Wu, C. C., Wee, E. L., Santee, G. L., Rai, J. H. and Goebel, K. D., *Pure Appl. Chem.*, 1974, **38**, 37.
13. Kiss, G. and Porter, R., *J. Polym. Sci., Polym. Symp.*, 1978, **65**, 193.
14. Brian, A., Frisch, H. L. and Lerman, L. S., *Biopolymers*, 1981, **20**, 1305.
15. Papkov, S. P., Kulichikhin, V. G., Kalmykhova, V. D. and Malkin, A. Ya., *J. Polym. Sci., Polym. Phys. Ed.*, 1974, **12**, 1753; Kulichikhin, V. G., Papkov, S. L., *et al.*, *Polym. Sci. USSR*, 1976, **18**, 672; Papkov, S. P., *Polym. Sci. USSR*, 1977, **19**, 1.
16. Schaefgen, J. R., *et al.*, *Polym. Prepr. Am. Chem. Soc. Div. Polym. Chem.*, 1976, **17**(1), 69.
17. Kwolek, S. L., Morgan, P. W., Schaefgen, J. R. and Gulrich, L. W., *Macromolecules*, 1977, **10**, 1390.
18. Balbi, C., Bianchi, E., Ciferri, A. and Tealdi, A., *J. Polym. Sci., Polym. Phys. Ed.*, 1980, **18**, 2037.
19. Bair, T. I., Morgan, P. W. and Kilian, F. L., *Macromolecules*, 1977, **10**, 1396.
20. Nakajima, A., Hirai, T. and Hayashi, T., *Polym. Bull.*, 1978, **1**, 143.
21. Abe, A. and Flory, P. J., *Macromolecules*, 1978, **11**, 1122.

22. Flory, P. J. and Frost, R. S., *Macromolecules*, 1978, **11**, 1126.
23. Conio, G., Bianchi, E., Ciferri, A. and Tealdi, A., *Macromolecules*, 1981, **14**, 1084.
24. Ballauff, M. and Flory, P. J., *Ber. Bunsenges. Phys. Chem.*, 1984, in press.
25. Flory, P. J., *Macromolecules*, 1978, **11**, 1138.
26. Aharoni, S. M., *Polymer*, 1980, **21**, 21.
27. Bianchi, E., Ciferri, A. and Tealdi, A., *Macromolecules*, 1982, **15**, 1268.
28. Hwang, W. F., Wiff, D. R. and Benner, C. L., *J. Macromol. Sci. Phys.*, 1983, **B22**, 231.
29. Flory, P. J., *Macromolecules*, 1978, **11**, 1141.
30. Aharoni, S. M., *Mol. Cryst. Liq. Cryst. Lett.*, 1980, **56**, 237.
31. Dayan, S., Maissa, P., Vellutini, M. J. and Sixou, P., *J. Polym. Sci., Polym. Lett. Ed.*, 1982, **20**, 33.
32. Bheda, J., Fellers, J. F. and White, J. L., *Coll. Polym. Sci.*, 1980, **258**, 1335.
33. Werbowj, R. S. and Gray, D. G., *Macromolecules*, 1980, **13**, 69.
34. Conio, G., Bianchi, E., Ciferri, A., Tealdi, A. and Aden, M. A., *Macromolecules*, 1983, **16**, 1264.
35. Abe, A., *Macromolecules*, 1984, **17**, in press.
36. Matheson, R. S. Jr, and Flory, P. J., *J. Phys. Chem.*, 1984, **88**, in press.
37. Roviello, A. and Sirigu, A., *Makromol. Chem.*, 1982, **183**, 895; see also references cited therein.
38. Jin, J. I., Antoun, S., Ober, C. and Lenz, R. W., *Brit. Polym. J.*, 1980, 132; Ober, C., Jin, J. I. and Lenz, R. W., *Polym. J.*, 1982, **14**, 9.
39. Irvine, P. A. and Flory, P. J., *J. Chem. Soc. Faraday Trans.*, 1984, *I*, **80**, in press.
40. Warner, M. and Flory, P. J., *J. Chem. Phys.*, 1980, **73**, 6327.
41. Ballauff, M., Wu, D. and Flory, P. J., *Ber. Bunsenges. Phys. Chem.*, 1984, in press.

7

NEMATIC LIQUID CRYSTALS FORMED FROM FLEXIBLE MOLECULES: A MOLECULAR FIELD THEORY

G. R. LUCKHURST

Department of Chemistry, University of Southampton, UK

1. INTRODUCTION

Thermotropic liquid crystal polymers may be formed by linking rigid mesogenic units, such as an azoxybenzene, with flexible spacers which are normally alkyl chains. The liquid crystalline properties of these materials are found to depend critically on the length of the alkyl chain. For example, both the nematic–isotropic transition temperature and the entropy of transition exhibit a marked alternation with the number of methylene groups in the flexible spacer, as the results for the poly α,ω-[4,4′-(2,2′-dimethylazoxyphenyl)]alkandioates given in Figs 1 and 2 clearly demonstrate.[1] In consequence any molecular theory of such liquid crystal polymers must allow specifically for the effects of the flexible spacer. This is a formidable task for a polymeric system but one which may not be necessary for certain mesogenic properties of low molar mass materials are found to be analogous to those of the liquid crystal polymers. Thus the variation of the transition temperatures and entropies for the α,ω-bis(4,4′-cyanobiphenyloxy)alkanes with the length of the flexible spacer is strikingly similar to those found for polymeric materials as the results in Figs 3 and 4 show.[2] It is clearly easier to develop a molecular theory for these simpler systems and this should enable us also to understand the liquid crystal behaviour of the thermotropic liquid crystal polymers. Indeed such a theory has been produced for nematics such as the 4-n-alkyl-4′-cyanobiphenyls for which an alkyl chain is attached to a single rigid core.

The first attempt to develop a theory for non-rigid mesogens was made by Marcelja[3] who extended the Maier–Saupe theory for nematics composed of cylindrically symmetric particles[4] to include molecular flexibility. The advent of studies of the variation of the orientational order

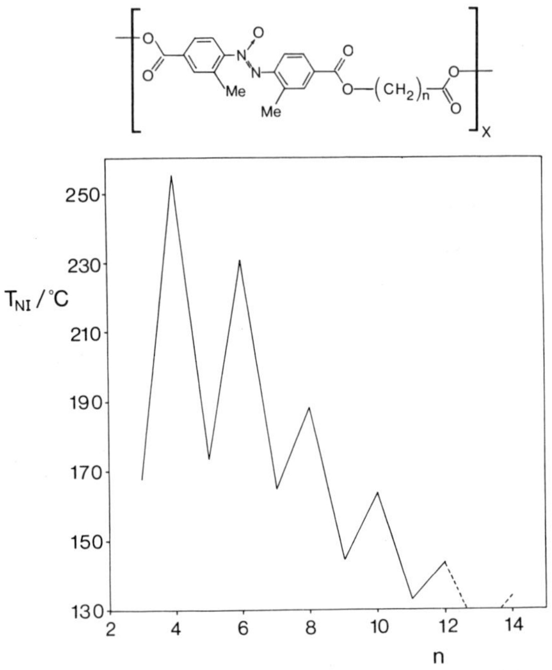

Fig. 1. The molecular structure of the thermotropic liquid crystal polymers, poly α,ω-[4,4′-(2,2′-dimethylazoxyphenyl)]alkandioates and their nematic–isotropic transition temperatures T_{NI} as a function of the number of methylene groups n in the flexible spacer.

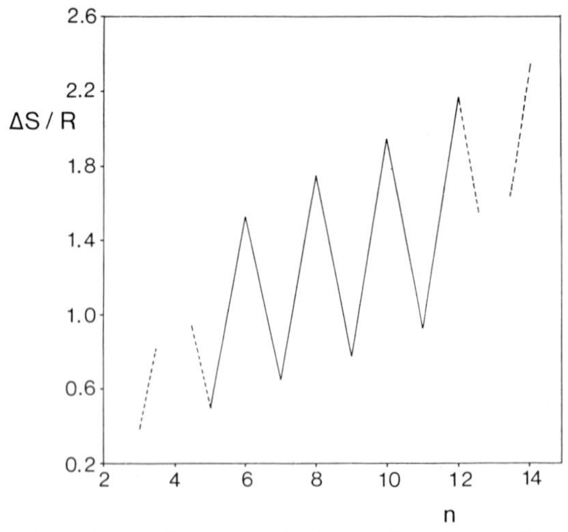

Fig. 2. The dependence of the nematic–isotropic entropy of transition $\Delta S/R$ on the number of methylene groups in the flexible spacer for the poly α,ω-[4,4′-(2,2′-dimethylazoxyphenyl)]alkandioates.

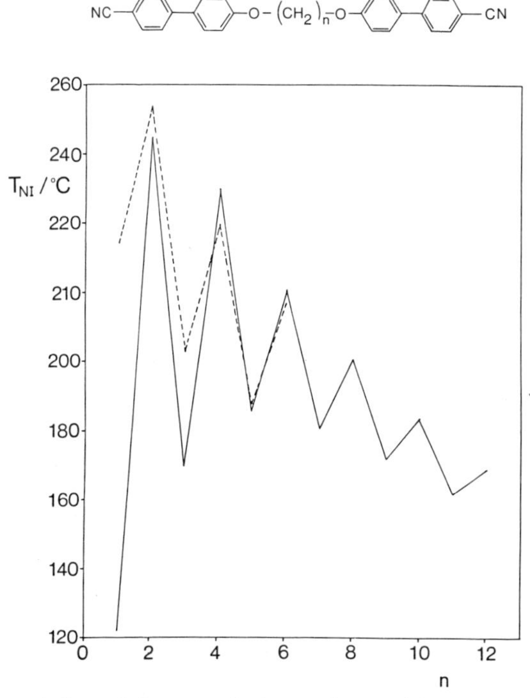

Fig. 3. The variation of the nematic–isotropic transition temperature with the number of methylene groups in the flexible spacer for the α,ω-bis(4,4′-cyanobiphenyloxy)alkanes together with their molecular structure. The theoretical predictions of T_{NI}, for the early members of the homologous series, are joined by the dashed lines. Results are not yet available for higher members because the computational time required for the calculations increases dramatically with the number of methylene groups in the flexible spacer.

along the alkyl chain, by deuterium NMR spectroscopy,[5] has allowed searching tests of the Marcelja theory to be made and has stimulated its development.[6] Here we shall describe this new theory for nematics composed of flexible molecules. We begin, in the following sections, with a purely qualitative account of the theory. This starts (Section 2) with the model used to describe the conformational states of the alkyl chain. It progresses in Section 3 to a calculation of the orientational order for each conformer and hence to the variation of the observed order parameter along the chain. The success of this aspect of theory is demonstrated by comparing its predictions with the results for 4-*n*-hexyloxy-4′-cyanobiphenyl. Use of the molecular field approximation in Section 4 allows the

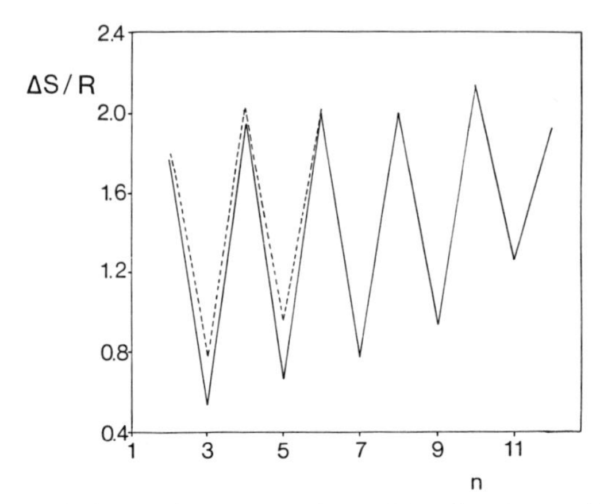

Fig. 4. The nematic–isotropic entropy of transition for the α,ω-bis(4,4′-cyanobiphenyloxy)alkanes as a function of the number of methylene groups in the flexible spacer. The theoretical predictions of $\Delta S/R$ are joined by the dashed lines.

free energy and hence the nematic–isotropic transition temperature as well as the entropy of transition to be determined. The resulting theory is able to predict the variation in these transitional properties with the length of the flexible spacer in quite good agreement with the behaviour of the α,ω-bis(4,4′-cyanobiphenyloxy)alkanes. The mathematical details of the theory are confined to Section 6 for those interested in its quantitative development.

2. CONFORMATIONAL STATES

The first stage in the development of any molecular theory for flexible molecules is to describe the conformational states available and to determine their relative weights. The non-rigidity of a mesogenic molecule such as 4-*n*-hexyloxy-4′-cyanobiphenyl (6OCB) stems from many internal modes but we shall ignore all of these except for rotations about O–C and C–C bonds in the alkoxy chain (cf. Fig. 5). The rotameric state model proposed by Flory for alkanes can be used to describe the conformations resulting from these rotations.[7] Consider a C–C bond in the chain where the remainder of the chain is represented by groups R and R′ attached to these two carbon atoms. The non-bonded interactions result in three conformations for which the internal or intramolecular energy is a minimum; these are shown as Newman projections along the C–C bond

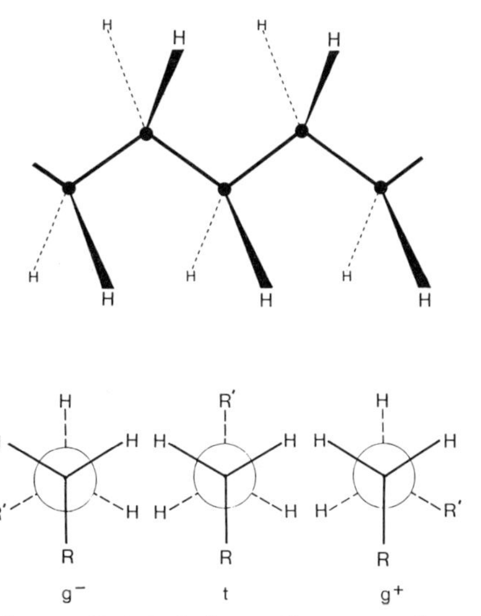

Fig. 5. A section of an *n*-alkyl chain in an all-*trans* conformation together with *trans* and *gauche* linkages shown as Newman projections.

in Fig. 5. The lowest energy is for the *trans* conformer (t) while the two *gauche* conformers ($g^{\pm}$) are equal in energy but less stable than the *trans* arrangement. The rotameric state model assumes that only these three conformations are allowed and so the conformation of an alkyl chain can be described by the number and location of the *gauche* linkages in the chain. Such conformations are more readily shown in two rather than three dimensions and some typical conformations of 6OCB in two dimensions are sketched in Fig. 6.

In the simplest version of the model the internal energy of an alkyl chain depends only on the number and not the location of the *gauche* linkages in the chain. As a consequence the conformations of 6OCB, which we denote by *tgtgt* and *gtttg* with the link nearest the rigid core written first, would occur in equal amounts in the isotropic phase. This is not expected to be so in the nematic phase where the more elongated conformations should be more compatible with the long-range orientational order of the phase. Thus, there should be more of the conformer *tgtgt* (cf. Fig. 6) than *gtttg*. The difference in the statistical weights for the two conformers is clearly influenced by the extent of the orientational order for the nematic phase and in the limit of the isotropic phase this difference must vanish. Some theories have allowed for this discrimination by the liquid crystal phase in a rather crude manner. For example,

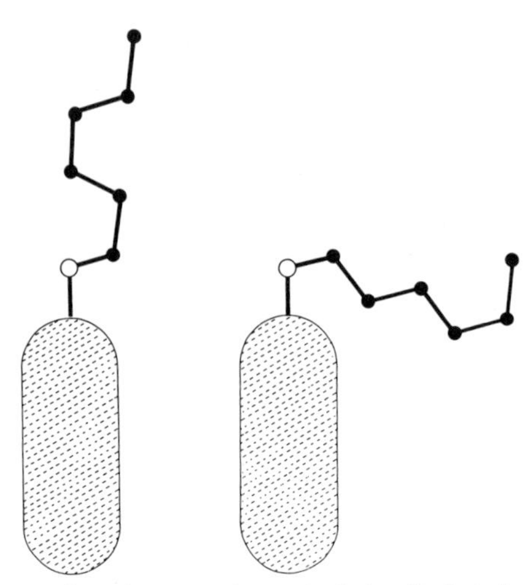

Fig. 6. Two conformations (*tgtgt* and *gtttg*) of the alkyl chain of 4-*n*-hexyloxy-4′-cyanobiphenyl restricted to a plane.

according to the kink model for membranes, it is argued that, because only the kink sequence $g^{\pm}tg^{\mp}$ tends to preserve the linearity of the chain, every other conformation is not allowed except, of course, for the all-*trans* form.[8] Such dramatic discrimination between conformers is unrealistic and in Section 6 we derive expressions for the statistical weights in an orientationally ordered phase which are a natural consequence of this order.

3. ORIENTATIONAL ORDER

The molecules in a nematic liquid crystal tend to be parallel to a unique direction known as the director which is identical with the optic axis of the phase. The molecular orientation fluctuates with respect to the director and the extent of these fluctuations is reflected by the orientational order parameters;[9] they are defined in Section 6. As we shall see thc order parameters for cylindrically symmetric particles are chosen to be unity in a perfectly ordered phase or crystal while for the disordered phase or liquid the order parameters vanish. In a nematic phase the order parameters are intermediate between these two extremes. The magnitude of the orientational fluctuations are controlled by the energy of the molecule as it changes its orientation with respect to the director. For a cylindrically symmetric molecule this energy is determined entirely by the angle between the director and the molecular symmetry axis.

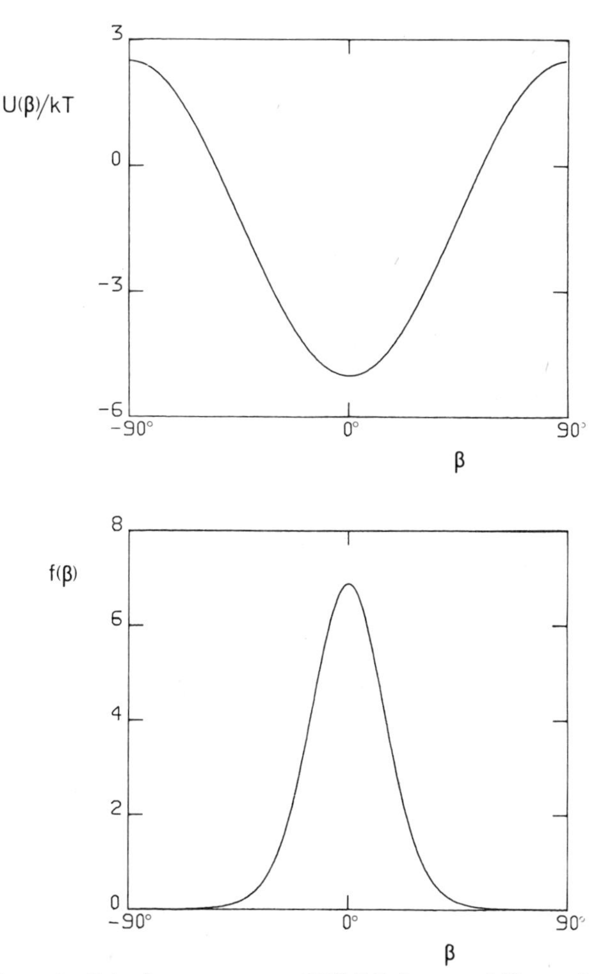

Fig. 7. The potential of mean torque $U(\beta)/kT$ for a rod-like molecule as a function of the angle between the molecular symmetry axis and the director; the corresponding singlet orientational distribution function $f(\beta)$ is also shown.

The orientational dependence of the energy is depicted in Fig. 7; it resembles that of a molecule, with an anisotropic diamagnetic susceptibility, interacting with a magnetic field. However, in a nematic liquid crystal this orientational energy results from the interactions of a molecule with its neighbours which may be thought of as providing a molecular field. The strength of this molecular field depends on many factors, including the molecular packing and the orientational order in the system.[4] The form of

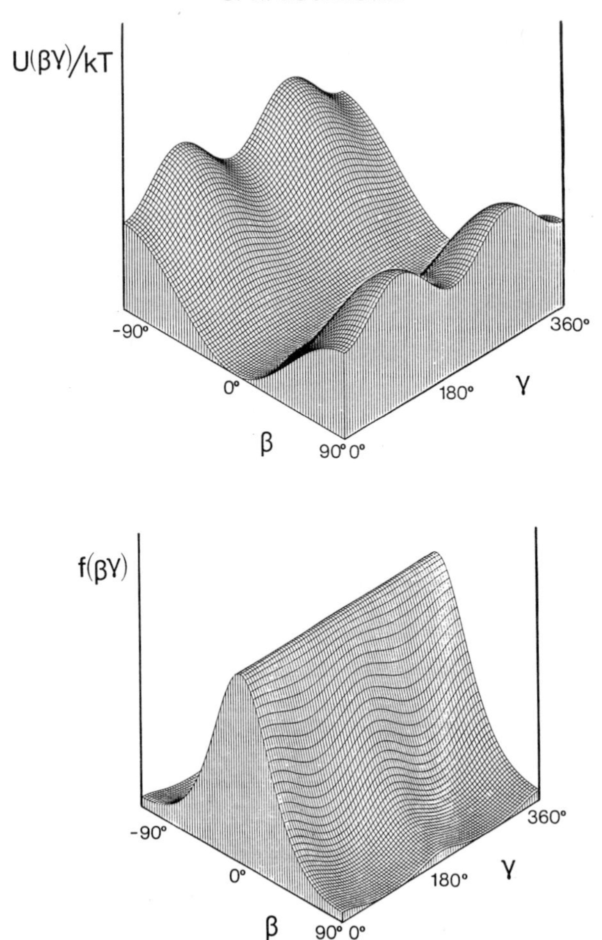

Fig. 8. The potential of mean torque $U(\beta\gamma)/kT$ for a biaxial molecule where $\beta\gamma$ are the spherical polar coordinates of the director in a molecular frame (cf. Fig. 9). The corresponding singlet orientational distribution function $f(\beta\gamma)$ is also shown.

the orientational energy or, as it is known, the potential of mean torque can be used to determine the probability of finding the molecule at a particular angle to the director from the appropriate Boltzmann factor; this distribution function is also shown in Fig. 7. This probability distribution can then be employed to determine the average of any function of the molecular orientation which has been chosen to define the orientational order parameter.

The situation is more complicated for any conformation of a mesogenic

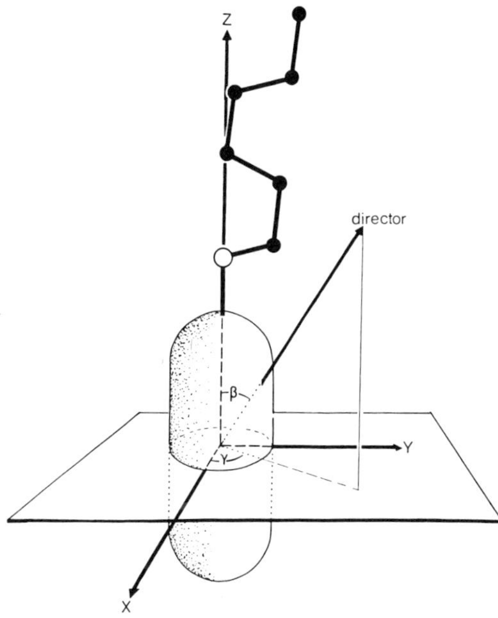

Fig. 9. A conformation of 4-*n*-hexyloxy-4′-cyanobiphenyl with its molecular frame used to define the orientation of the director.

molecule because this is not cylindrically symmetric. In consequence the potential of mean torque resulting from the molecular field depends on two angles. These are the spherical polar angles made by the director in a frame set in the molecule. The dependence of the potential of mean torque and the singlet orientational distribution function calculated from it for a typical conformation of 6OCB are shown in Fig. 8, while the conformation, the molecular axis system and the spherical polar angles for the director are given in Fig. 9. The orientational distribution function is clearly a maximum when the director is parallel to the molecular z-axis; however, the difference in alignment of the x- and y-axes is discernible as ripples along the side of the distribution function. This difference in alignment may not appear to be large but it can have a dramatic influence on certain of the nematic properties; it would be quite wrong to ignore this molecular biaxiality.[4] The orientational order parameters for the conformer can be evaluated from this distribution function. However because the rate of exchange between conformers is fast on the NMR time-scale the order parameters determined from the experiment are a weighted average of those for all conformers.[10]

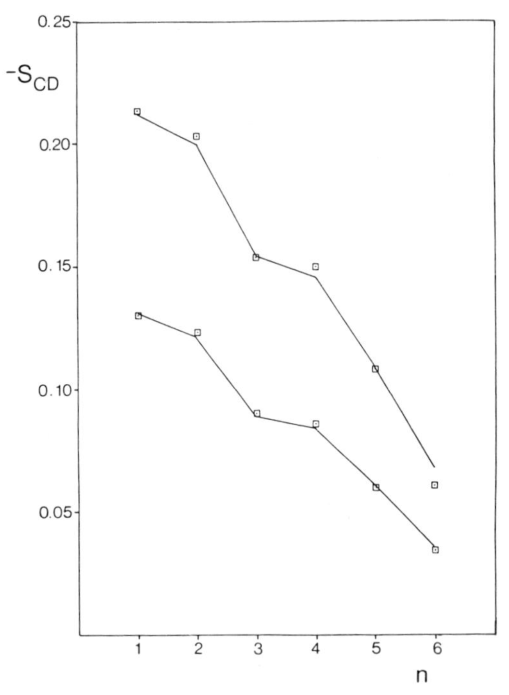

Fig. 10. The variation of the order parameter S_{CD} for each methylene group along the alkyl chain for 4-*n*-hexyloxy-d_{13}-4′-cyanobiphenyl at two temperatures in the nematic phase. The solid lines show the order parameter profile predicted by the theory.

We need therefore to calculate the order parameters for each conformer but this would introduce too many arbitrary variables into the theory if the orientational distribution functions were unrelated; for example there are 243 conformers of 6OCB. To overcome this problem it is assumed that the strength of the molecular field is made up of a limited number of segmental interactions.[6] For 6OCB the simplest choice of segments would be the aromatic core with a strength parameter of X_a and the C–C segments of the chain with their parameter X_c. This assumes that the C–H bond segments do not contribute to the molecular field and that the O–C bond in the chain is identical to the C–C bonds. From a knowledge of X_a and X_c the order parameters of each conformation can be determined. To obtain the conformational average we need to know the difference in energy E_{tg} between a *trans* and a *gauche* linkage but estimates of E_{tg} are available from independent experiments.[7] Using typical values of E_{tg} and varying the

TABLE 1

The Eleven Most Common Conformers Adopted by 4-*n*-Hexyloxy-4′-cyanobiphenyl in the Nematic and Isotropic Phases Together with Their Statistical Weights, Expressed as Percentages, Calculated with $X_a/kT = 2{\cdot}78$, $X_c/kT = 0{\cdot}38$ and $E_{tg}/kT = 1{\cdot}36$

Conformation[a]	*Statistical weight*, p_n	
	Isotropic	*Nematic*
ttttt	12·72	20·02
$g^{\pm}tttt$	3·25	2·10
$tg^{\pm}ttt$	3·25	4·17
$ttg^{\pm}tt$	3·25	2·68
$tttg^{\pm}t$	3·25	4·51
$ttttg^{\pm}$	3·25	3·63

[a] The first letter denotes the conformation of the chain segment nearest to the core.

molecular field strength parameters X_a and X_c it has been possible to fit the order parameter profile for the chain in 6OCB.[11]

The very good agreement found for 6OCB is shown for two temperatures in Fig. 10 where the experimental order parameters, denoted by S_{CD} and defined in Section 6, are indicated by open squares; the solid lines link the predicted values of S_{CD}. This accord lends support to the assumptions made in developing the theory and allows us to use it to determine, with some confidence, the extent of the discrimination between conformers by the anisotropic environment of a nematic. Some typical results are given in Table 1 for the statistical weights of the eleven most popular conformers of 6OCB just above and just below the nematic–isotropic transition. These results show that there are significant changes in the statistical weights on going from the isotropic to the nematic phase with the more elongated conformations being favoured. According to this theory conformations with kinks ($g^{\pm}tg^{\mp}$) have only a small probability of occurrence in both the isotropic and nematic phases, in marked contrast to the kink model which would take them to be dominant in the nematic.[8]

4. THERMODYNAMIC PROPERTIES

One of the most important properties of a nematic liquid crystal is the nematic–isotropic transition temperature and one of our main objectives is to predict how this varies with the length of the flexible spacer, at least for the relatively simple α,ω-bis(4,4′-cyanobiphenyloxy)alkanes. To determine the T_{NI} we require the free energy for the nematic and isotropic phases because at the transition these free energies must be equal. The statistical mechanical calculation of the free energy via the configurational partition function is straightforward for a set of non-interacting particles. In consequence, use of the molecular field approximation, which replaces the complex network of molecular interactions with a field acting on single particles, simplifies the calculation considerably.[4] This approximation is equivalent to ignoring direct orientational correlations between pairs of molecules. Of course molecular orientations in the nematic phase are highly correlated but according to the molecular field approximation this arises indirectly because both molecules adopt orientations which are directly correlated with the director and hence indirectly with each other.

The situation is analogous to the application of a magnetic field to a set of non-interacting particles. Here the magnetic field induces indirect orientational correlations between particles because they all tend to be parallel to the applied field but clearly there can be no direct correlations since the particles are non-interacting. However, unlike the interaction with a magnetic field, the molecular field strength parameters vary with temperature because of their dependence on the orientational order; indeed they must vanish in the isotropic phase. Accordingly it is assumed that the strength parameters are linear in the second-rank order parameters for the chain and core segments, some of which are available from experiment.

The contribution of each order parameter to a particular molecular field strength parameter introduces a variable into the theory. Thus for X_a we shall need a core–core interaction parameter as well as a core–chain parameter to relate the strength parameter to the core order parameter and the chain order parameter respectively; similarly for X_c we require a chain–chain interaction parameter together with the same chain–core parameter (cf. Section 6). We expect the core–core interaction parameter to be larger than that for the chain–chain interactions with the chain–core parameter intermediate between these two. Indeed to reduce the number of variables it is convenient to assume that the core–chain parameter is the geometric mean of the other two.[12] In addition it is possible to obtain a good estimate for the ratio of these by fitting the theory to the order parameter profile observed for the alkyl chain.[6] The remaining unknown can then be used to scale the calculated transition temperatures and hence optimize the agreement with the variation of T_{NI} observed for a homologous series. Typical results are shown for the α,ω-bis(4,4′-cyanobiphenyloxy)alkanes in Fig. 3. It is clear that the agreement between theory and experiment is good especially as the theory is so simple. The observed strong alternation in the nematic–isotropic transition temperature is correctly predicted by the theory, as is the damping of this alternation with increasing length of the flexible spacer. In addition the theory accounts for the higher nematic–isotropic transition temperatures for those compounds with an even number of methylene groups in the alkyl chain.

Calculation of the free energy also allows the entropy of transition to be determined and the results of these calculations are compared in Fig. 4 with the observed values. The agreement between theory and experiment is seen to be extremely good. Indeed the quantitative nature of the agreement is somewhat surprising since for nematics composed of rigid particles use of the molecular field approximation is known to overestimate the entropy of transition because the orientational correlations are underestimated in the isotropic phase.[4] Nonetheless the theory does predict the dramatic alternation in the entropy of transition with the number of methylene groups in the flexible spacer. In contrast the variation in the entropy of transition for the homologous series of 4-*n*-alkoxy-4′-cyanobiphenyls is far less pronounced and this feature is also predicted by the theory.[3,13]

The orientational order of the α,ω-bis(4,4′-cyanobiphenyloxy)alkanes has yet to be studied experimentally. However it is of some considerable interest to see what variations are predicted by the theory, especially as it has been suggested that the large alternations in the entropy of transition found for main-chain liquid crystal polymers imply a large alternation in their orientational order.[1] The core order parameter is necessarily obtained as part of the calculation for the free energy and its value at the transition is

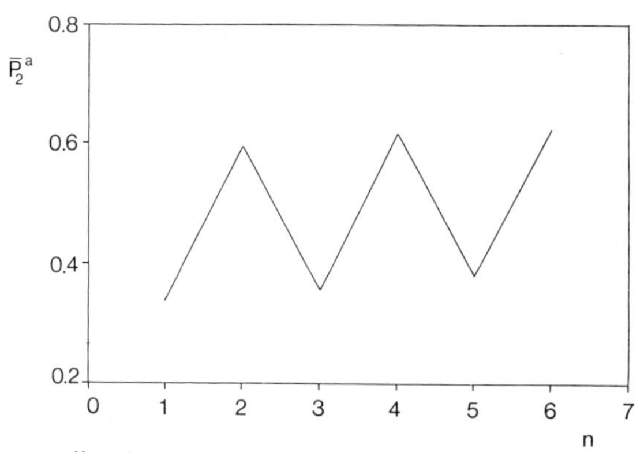

Fig. 11. The predicted dependence of the core order parameter $\bar{P}_2^a$ on the number of methylene groups n in the flexible spacer for the α,ω-bis(4,4′-cyanobiphenyloxy)alkanes.

given, as a function of the length of the flexible spacer, in Fig. 11. We see that this order parameter is indeed predicted to vary dramatically, with members of the series having an odd number of methylene groups in the flexible spacing being significantly less ordered than those with an even number. This behaviour contrasts with that for mesogenic molecules containing a single rigid core where the order parameter exhibits a weak dependence on the length of the alkyl chain, again in accord with theoretical predictions.[3,13]

5. SUMMARY

This relatively simple molecular theory for liquid crystals composed of flexible molecules has three essential elements. The first of these is the rotational isomeric state model which is used to describe the conformational states of an alkyl chain and their statistical weights. The next element is the form assumed for the potential of mean torque, responsible for the alignment of a conformer and the use of segmental interactions to relate these potentials for different conformers. In combination the two elements allow the variation of the order parameters for the methylene groups along an alkyl chain to be predicted successfully. They also permit a unique investigation of the stabilization or de-stabilization of conformers by the orientational order of a nematic liquid crystal. The third major element of the theory is the molecular field approximation which is employed to calculate the free energy and hence the other thermodynamic properties of the system. The dependence of the nematic–isotropic transition temperature and the entropy of transition on the length of the flexible spacer is in good qualitative and even semi-quantitative agreement with experiment. The theory would therefore appear to provide a valuable tool with which to understand the behaviour of low molecular mass nematic liquid crystals associated with the molecular flexibility and possibly even the analogous characteristics of thermotropic liquid crystal polymers.

6. MATHEMATICAL DETAILS OF THE THEORY

(i) Conformational States

In its simplest form the rotational isomeric state model gives the internal energy of an *n*-alkyl chain with N_g *gauche* linkages as[7]

$$U_{\text{int}}(n) = N_g E_{tg} \tag{1}$$

where E_{tg} is the difference in energy between a *gauche* and a *trans* state; the label n is used to indicate the conformational variables needed to characterize the state of the chain. The probability of finding the chain in its nth conformation in the isotropic phase is

$$p_n = \exp\{-U_{\text{int}}(n)/kT\}/Z_{\text{int}} \tag{2}$$

where the conformational partition function is

$$Z_{\text{int}} = \sum_n \exp\{-U_{\text{int}}(n)/kT\} \tag{3}$$

This result is not valid in an orientationally ordered phase, such as a nematic, because the molecular conformation and orientation are coupled; in other words the potential of mean torque U_{ext} responsible for the alignment of a molecule depends on its conformational state n. The total energy can be approximated by[10]

$$U(n, \omega) = U_{\text{int}}(n) + U_{\text{ext}}(n, \omega) \tag{4}$$

where ω denotes the spherical polar angles made by the director in a reference frame set in some rigid unit of the molecule (cf. Fig. 9). Equation (4) is approximate because $U_{\text{int}}(n)$ is identified as the energy of the conformer resulting from intramolecular interactions while $U_{\text{ext}}(n, \omega)$ is the potential of mean torque for the nth conformer which comes from the averaged anisotropic intermolecular interactions. The normalized probability of finding a molecule in conformation n making an orientation ω with the director is

$$f(n, \omega) = \exp\{-U_{\text{int}}(n)/kT\}\exp\{-U_{\text{ext}}(n, \omega)/kT\}/Z \tag{5}$$

where the conformational–orientational partition function is

$$Z = \sum_n \exp\{-U_{\text{int}}(n)/kT\}Q_n \tag{6}$$

and the orientational partition function for the nth conformer is

$$Q_n = \int \exp\{-U_{\text{ext}}(n, \omega)/kT\}\,d\omega \tag{7}$$

The probability of finding the molecule in conformation n irrespective of its orientation is obtained by integrating the joint probability $f(n, \omega)$ over all orientations, as[10]

$$p_n = \exp\{-U_{\text{int}}(n)/kT\}Q_n/Z \tag{8}$$

The orientational partition functions Q_n cause the statistical weights p_n to differ from their values in the isotropic phase. Thus for well-ordered conformations $U_{ext}(n, \omega)$ will be large and negative which means that Q_n will be large and hence p_n will be enhanced as we saw in Section 3. In the limit that the potential of mean torque vanishes, as for the isotropic phase, the orientational partition functions will all tend to 4π and we recover eqn (2) for the statistical weights.

(ii) Orientational Order

The orientational order in a uniaxial phase, such as a nematic, composed of rigid particles of arbitrary shape is characterized by an infinite set of order parameters[9] defined as the averages of modified spherical harmonics $C_{L,m}(\omega)$

$$\bar{C}_{L,m} = \int C_{L,m}(\omega) f(\omega)\, d\omega \tag{9}$$

Here the bar indicates an ensemble average and $f(\omega)$ is the normalized singlet orientational distribution function for finding the director at an orientation ω in a molecular frame. Of this infinite set only the second rank quantities $\bar{C}_{2,m}$ can readily be measured and these five components, corresponding to $m = 0, \pm 1, \pm 2$, form the ordering tensor which is the irreducible analogue of the more familiar Saupe ordering matrix. This matrix is defined by

$$S_{\alpha\beta} = (3\overline{l_\alpha l_\beta} - \delta_{\alpha\beta})/2 \tag{10}$$

where l_α is the direction cosine between the α molecular axis and the director; $\delta_{\alpha\beta}$ is the Kronecker delta function.[9] The diagonal elements $S_{\alpha\alpha}$ measure the tendency of the director to align parallel to the α-axis; it is this quantity which can be determined from deuterium NMR experiments for the C–D bond direction and provides the order parameter (S_{CD}) profile along an alkyl chain in a mesogenic molecule.

When the molecule is non-rigid we need to define an ordering tensor for each rigid sub-unit if we are to begin to characterize the orientational order of the mesophase.[10] The orientation of the director in an axis system set in a sub-unit changes because of fluctuations in both the conformational state of the molecule and its orientation. The ordering tensor for the sub-unit is therefore calculated from the joint singlet distribution function as

$$\bar{C}_{2,m} = \sum_n \int C_{2,m}(\omega) f(n, \omega)\, d\omega \tag{11}$$

Our previous assumption, contained in eqn (4), concerning the separation of the total energy into internal and external contributions allows us to write the joint distribution as the simple product

$$f(n, \omega) = p_n f_n(\omega) \tag{12}$$

Here p_n is the probability of finding a molecule in the nth conformation irrespective of its orientation and $f_n(\omega)$ is the singlet orientational distribution function for this conformer. Such a factorization is equivalent to assuming that the intramolecular energy is independent of the molecular orientation—an assumption which is eminently reasonable for the interactions governing the conformations of mesogenic molecules. Equation (11) can be rewritten using the factorized joint distribution function as

$$\begin{aligned}\bar{C}_{2,m} &= \sum_n p_n \int C_{2,m}(\omega) f_n(\omega)\,\mathrm{d}\omega \\ &= \sum_n p_n \bar{C}^n_{2,m}\end{aligned} \tag{13}$$

where $\bar{C}^n_{2,m}$ is the ordering tensor for the nth conformer. The observed order parameters are therefore the conformationally weighted averages of those for each conformer.

Calculation of $\bar{C}^n_{2,m}$ requires the potential of mean torque with which to determine the orientational distribution function

$$f_n(\omega) = \exp\{-U_{\mathrm{ext}}(n, \omega)/kT\}/Q_n \tag{14}$$

for conformer n. Formally $U_{\mathrm{ext}}(n, \omega)$ may be expanded in a complete basis set of modified spherical harmonics as

$$U_{\mathrm{ext}}(n, \omega) = -\sum_{\substack{L(\mathrm{even}) \\ q}} (-)^q X^n_{L,q} C_{L,-q}(\omega) \tag{15}$$

where the summation is restricted to even values of L by the symmetry of the nematic phase. The expansion coefficients $X^n_{L,q}$ are components of irreducible spherical tensors whose magnitudes depend on factors such as the orientational order, the molecular distribution and the anisotropic molecular interactions.[4,12] Since our knowledge of the anisotropic interactions is sparse for mesogenic molecules the expansion coefficients cannot be evaluated and so to reduce the number of adjustable parameters in the theory the expansion is truncated at the second-rank term

$$U_{\text{ext}}(n, \omega) = -\sum_{q} (-)^q X^n_{2,q} C_{2,-q}(\omega) \tag{16}$$

Such an early truncation is consistent with both theory[4,14] and experiment[15] at least for rod-like particles. The number of unknowns may still be large if the molecule can exist in many conformations and to overcome this problem we assume that a molecule may be divided into a number of segments each of which makes its own contribution to the interaction tensors. The form of this contribution is further assumed to be independent of the conformation but to depend on the nature of the segment.[6] In consequence the $X^n_{2,q}$ change with the conformation for purely geometric reasons. Thus the total interaction tensor is related to the segmental tensors $X^j_{2,r}$ in their local frames by

$$X^n_{2,q} = \sum_{j,r} D^2_{r,q}(\Omega_{n,j}) X^j_{2,r} \tag{17}$$

where $D^2_{r,q}$ is a Wigner rotation matrix[16] used to transform from the segmental frame to some reference frame for the conformer. The Euler angles for the necessary rotations are $\Omega_{n,j}$.[16] The conformational interaction tensor is then diagonalized and its principal axis system used as a reference frame in which to describe the orientation of the director. This has the considerable advantage that the potential of mean torque reduces to

$$U_{\text{ext}}(n, \omega) = -\{X^n_{2,0} P_2(\cos\beta) + X^n_{2,2}(3/2)^{1/2} \sin^2\beta \cos 2\gamma\} \tag{18}$$

where $\beta\gamma$ are the spherical polar angles for the director in the reference frame and $P_2(\cos\beta)$ is the second Legendre polynomial. Perhaps more importantly the principal axes for $X^n_{2,q}$ are also principal axes for the ordering tensor $\bar{C}_{2,q}$ of the conformation and these principal components are given by[17]

$$\bar{C}_{2,0} = 2\pi Q_n^{-1} \int P_2(\cos\beta) I_0\{(3/8)^{1/2} b_n \sin^2\beta\} \exp\{a_n P_2(\cos\beta)\} \sin\beta \, d\beta \tag{19}$$

and

$$\bar{C}_{2,2} = \bar{C}_{2,-2} = 2\pi(3/8)^{1/2} Q_n^{-1} \int \sin^2\beta I_1\{(3/8)^{1/2} b_n \sin^2\beta\} \times \exp\{a_n P_2(\cos\beta)\} \sin\beta \, d\beta \tag{20}$$

where $I_n(x)$ is an nth order modified Bessel function

$$I_n(x) = \pi^{-1} \int_0^{\pi} \cos n\gamma \exp(x \cos \gamma)\, d\gamma \tag{21}$$

The conformational parameters a_n and b_n are simply $X^n_{2,0}/kT$ and $2X^n_{2,2}/kT$ $(\equiv 2X^n_{2,-2}/kT)$ respectively.

The ordering tensor for a particular rigid segment in the nth conformer is then obtained by transforming from the principal axis system to the frame set in this segment. Thus

$$\bar{C}'_{2,n} = \sum_m D^2_{m,n}(\Omega)\bar{C}_{2,m} \tag{22}$$

where the prime indicates the ordering tensor in the segmental frame. These segmental ordering tensors are evaluated for each conformer and then averaged over all conformers using the statistical weights given in eqn (8). Some typical results are shown in Fig. 10 for 6OCB. In these calculations the segmental interaction tensors for the core $(X^{\mathrm{a}}_{2,m})$ and the carbon–carbon bonds were assumed to be cylindrically symmetric, $X^{\mathrm{a}}_{2,m}$ about the para-axis and $X^{\mathrm{c}}_{2,m}$ about the C—C bond.[6] The non-zero components in these local frames are denoted by X_{a} and X_{c}; the values giving the best fit to experiment vary with temperature but the ratio $X_{\mathrm{c}}/X_{\mathrm{a}}$ is found to be approximately constant at about 0·13.

(iii) Thermodynamic Properties

The Helmholtz free energy for N independent particles is obtained from the single particle partition function Z via

$$A = -NkT \ln Z \tag{23}$$

This result is not appropriate when the particles interact with a molecular field because the thermodynamic internal energy (not to be confused with the intramolecular energy) caused by molecular interactions is counted twice.[4] This situation obtains because particles both generate and experience the molecular field. To correct this expression for the free energy we need to evaluate the contribution to the internal energy caused by the anisotropic interactions. The starting point is eqn (16) for the potential of mean torque for the nth conformer which can be rewritten in terms of the segmental interactions as

$$U_{\mathrm{ext}}(n, \omega) = -\sum_{j,r} (-)^r X^j_{2,r} C_{2,-r}(\omega_j) \tag{24}$$

by using eqn (17) for the $X^n_{2,q}$ together with the closure relation for Wigner rotation matrices.[16] Here the ω_j denote the spherical polar coordinates for the director in the local frame set in the jth rigid fragment. Our previous assumption of cylindrically symmetric segmental interaction tensors reduces eqn (24) to

$$U_{\text{ext}}(n, \omega) = -\{X_{\text{a}}P_2(\cos\beta_{\text{a}}) + X_{\text{c}}\sum_j P_2(\cos\beta_j)\} \tag{25}$$

where β is the angle between the director and the symmetry axis for a local interaction tensor. The average energy of a particle is now obtained by averaging $U_{\text{ext}}(n, \omega)$ over both orientations and conformations to give the simple result

$$\langle \bar{U}_{\text{ext}} \rangle = -\{X_{\text{a}}\langle \bar{P}_2^{\text{a}} \rangle + X_{\text{c}}\langle \bar{P}_2^{\text{c}} \rangle\} \tag{26}$$

where the angular brackets are used here to denote a conformational average and, as before, the bars indicate an orientational average. The order parameter for the chain $\langle \bar{P}_2^{\text{c}} \rangle$ is defined as the sum of the order parameters for each carbon–carbon bond. The contribution to the internal energy of the system with N particles from the molecular interactions is then just $N\langle \bar{U}_{\text{ext}} \rangle/2$ where the factor of 1/2 corrects for the over-counting of these interactions. The correct expression for the Helmholtz free energy of the nematic is then

$$A_{\text{N}} = N\{X_{\text{a}}\langle \bar{P}_2^{\text{a}} \rangle + X_{\text{c}}\langle \bar{P}_2^{\text{c}} \rangle\}/2 - NkT\ln Z \tag{27}$$

where the partition function is given by eqn (6). The phase transition occurs when the free energy of the nematic, A_{N}, is equal to that of the isotropic; the free energy of the isotropic phase, A_{I}, is obtained from eqn (27) by setting the order parameters equal to zero. Thus

$$A_{\text{I}} = -NkT\ln Z_{\text{I}} \tag{28}$$

where

$$Z_{\text{I}} = 4\pi\sum_n \exp\{-U_{\text{int}}(n)/kT\} \tag{29}$$

and 4π is the orientational partition function for a biaxial particle.

The entropy of transition is just the change in the internal energy at the transition divided by the nematic–isotropic transition temperature. We have already evaluated the internal energy resulting from the anisotropic molecular interactions and the contribution from the intramolecular energy is obtained by averaging $U_{\text{int}}(n)$ over all conformations. This gives

$$U_{\text{N}} = -N\{X_{\text{a}}\langle\bar{P}_2^{\text{a}}\rangle + X_{\text{c}}\langle\bar{P}_2^{\text{c}}\rangle\}/2 + N\langle U_{\text{int}}\rangle_{\text{N}} \tag{30}$$

for the nematic phase and

$$U_{\text{I}} = N\langle U_{\text{int}}\rangle_{\text{I}} \tag{31}$$

for the isotropic phase. The entropy of transition is obtained, by evaluating both quantities at the nematic–isotropic transition, as

$$\Delta S/R = -\{X_{\text{a}}^{\text{NI}}\langle\bar{P}_2^{\text{a}}\rangle_{\text{NI}} + X_{\text{c}}^{\text{NI}}\langle\bar{P}_2^{\text{c}}\rangle_{\text{NI}}\}/2kT_{\text{NI}} + \{\langle U_{\text{int}}\rangle_{\text{N}}^{\text{NI}} - \langle U_{\text{int}}\rangle_{\text{I}}^{\text{NI}}\}/kT_{\text{NI}} \tag{32}$$

We see therefore that the entropy of transition is predicted to result not only from the change in the orientational order but also in the variation of the statistical weights for the conformers produced by the anisotropic environment.

To locate the nematic–isotropic transition it is necessary to determine the temperature dependence of the free energy but this is not possible without making further approximations concerning the temperature variation of the segmental interaction parameters X_{a} and X_{c}. This variation with temperature results primarily through their dependence on the orientational order of the system. A rigorous derivation of this dependence is extremely difficult and so we adopt a semi-intuitive approach. As we have seen, the orientational order in a mesophase is characterized by an infinite set of order parameters but the most important of these are the second-rank order parameters, at least close to the nematic–isotropic transition. Indeed for cylindrically symmetric particles both theory and experiment agree that the potential of mean torque is proportional to $\bar{P}_2$.[4,18] When the mesophase contains a mixture of different rod-like particles then the molecular field experienced by one particle will depend both on the order parameters for the other species as well as the strength of the anisotropic intermolecular interactions and the composition.[19] According to a molecular field theory for such mixtures the strength parameter for species i is given by[20]

$$X_i = \sum_j \varepsilon_{ij}\phi_j v_j^{-1}\bar{P}_2^j \tag{33}$$

where v_j is the molecular volume of component j and ϕ_j is its volume fraction. The parameters ε_{ij} are determined by the average anisotropic interaction between species i and j. It seems reasonable to treat a single component nematic composed of flexible molecules as a mixture of core and chain segments and so by analogy with the rigorous result for multicomponent mixtures we write[3,12]

$$X_a = \varepsilon_{aa}\phi_a v_a^{-1}\langle \bar{P}_2^a \rangle + \varepsilon_{ac}\phi_c v_c^{-1}\langle \bar{P}_2^c \rangle \tag{34}$$

and

$$X_c = \varepsilon_{ac}\phi_a v_a^{-1}\langle \bar{P}_2^a \rangle + \varepsilon_{cc}\phi_c v_c^{-1}\langle \bar{P}_2^c \rangle \tag{35}$$

Here ϕ_a is the volume fraction of the core and ϕ_c is the volume fraction of a single chain segment; the molecular volumes v_a and v_c of the two basic segments, necessary to evaluate these are available from several tabulations.

Nonetheless, the theory still contains three unknowns, namely the strength parameters, ε_{aa}, ε_{ac} and ε_{cc} which result from core–core, core–chain and chain–chain interactions, respectively. It is possible to remove one of these variables by assuming that they obey Berthelot's combining rule

$$\varepsilon_{ac} = (\varepsilon_{aa}\varepsilon_{cc})^{1/2} \tag{36}$$

which has proved to be particularly successful for binary mixtures of nematogenic molecules.[19] Given this approximation the ratio of the strength parameters X_c/X_a reduces to

$$X_c/X_a = (\varepsilon_{cc}/\varepsilon_{aa})^{1/2} \tag{37}$$

and so is temperature-independent; this independence produces considerable simplifications in the numerical evaluation of the free energy. In addition, X_c and X_a are available from fitting the observed order parameter profile and this provides good estimates of the ratio $(\varepsilon_{cc}/\varepsilon_{aa})^{1/2}$ needed in the prediction of the thermodynamic properties.

This completes the mathematical description of the theory which, as we have already seen, provides a good account of the orientational order and thermodynamic properties of nematogens composed of flexible molecules of low molecular weight.

REFERENCES

1. Blumstein, A. and Thomas, O., *Macromolecules*, 1982, **15**, 1264.
2. Emsley, J. W., Luckhurst, G. R., Shilstone, G. N. and Sage, I., *Mol. Cryst. Liq. Cryst. Lett.*, in press.

3. Marcelja, S., *J. Chem. Phys.*, 1974, **60**, 3599.
4. See, for example, Luckhurst, G. R., in *The Molecular Physics of Liquid Crystals*, Luckhurst, G. R. and Gray, G. W. (Eds), Academic Press, London, 1979, Chapter 4.
5. See, for example, Charvolin, J. and Deloche, B., in *The Molecular Physics of Liquid Crystals*, Luckhurst, G. R. and Gray, G. W. (Eds), Academic Press, London, 1979, Chapter 15.
6. Emsley, J. W., Luckhurst, G. R. and Stockley, C. P., *Proc. Roy. Soc. Lond.*, 1982, **A381**, 117.
7. Flory, P. J., *Statistical Mechanics of Chain Molecules*, Interscience, New York, 1969.
8. Seelig, J. and Niederberger, W., *J. Amer. Chem. Soc.*, 1974, **96**, 2069.
9. See, for example, Zannoni, C., in *The Molecular Physics of Liquid Crystals*, Luckhurst, G. R. and Gray, G. W. (Eds), Academic Press, London, 1979, Chapter 3.
10. Emsley, J. W. and Luckhurst, G. R., *Molec. Phys.*, 1980, **41**, 19.
11. Counsell, C. R., Emsley, J. W. and Luckhurst, G. R., to be published.
12. Luckhurst, G. R., in *Cristalli Liquidi*, Cooperativa Libraria Universitaria Torinese, 1982, Chapter II, 3.
13. Counsell, C. R., Emsley, J. W. and Luckhurst, G. R., to be published.
14. Gelbart, W. M. and Gelbart, A., *Molec. Phys.*, 1977, **73**, 1387.
15. Leadbetter, A. and Norris, E. K., *Molec. Phys.*, 1979, **38**, 669; Luckhurst, G. R., Simpson, P. and Zannoni, C., *Chem. Phys. Lett.*, 1981, **78**, 429.
16. See, for example, Rose, M. E., *Elementary Theory of Angular Momentum*, John Wiley, New York, 1957.
17. Luckhurst, G. R., Zannoni, C., Nordio, P. L. and Segre, U., *Molec. Phys.*, 1975, **30**, 1345.
18. Luckhurst, G. R. and Simpson, P., *Molec. Phys.*, 1982, **47**, 251.
19. Humphries, R. L., James, P. G. and Luckhurst, G. R., *Symp. Faraday Soc.*, 1971, **5**, 107.
20. Martire, D. E., in *The Molecular Physics of Liquid Crystals*, Luckhurst, G. R. and Gray, G. W. (Eds), Academic Press, London, 1979, Chapter 11.

8

ORDER AND ODD–EVEN EFFECTS IN THERMOTROPIC NEMATIC POLYESTERS

A. Blumstein and R. B. Blumstein

University of Lowell, Polymer Program, Lowell, Massachusetts, USA

Odd–even effects were investigated for the series based on mesogen 2,2′-dimethyl-4,4′-dioxyazoxybenzene (mesogen-9) and alkanedioyl spacers[1]

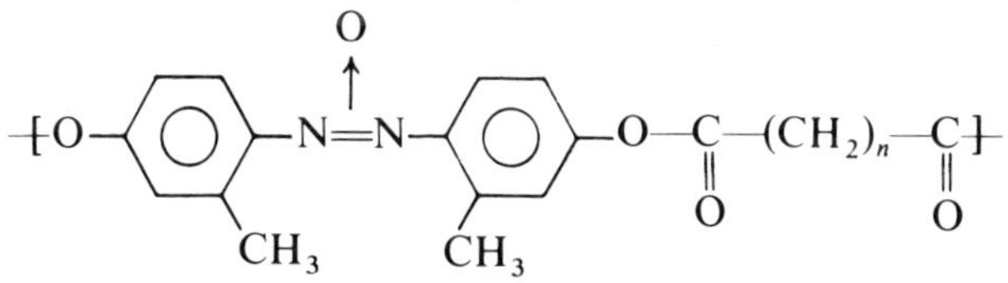

$$\left[O - C_6H_3(CH_3) - N(\rightarrow O){=}N - C_6H_3(CH_3) - O - \underset{\|}{\overset{}{C}}(=O) - (CH_2)_n - C(=O) \right] \quad (n \text{ up to } 14)$$

In addition to homopolymers, copolymers of mesogen-9 with mixed spacers ($n = 10$ and $n = 7$) were obtained. The polymers were characterized by DSC, polarizing microscopy, NMR (order parameters) and x-ray diffraction (mesophases aligned and quenched). Alignment was achieved by extrusion and application of high magnetic fields (10–16 tesla). The study has revealed the existence of two distinct levels of nematic order in these PLC. Homopolymers containing an even number of methylene units in the spacer (n = even) form nematic mesophases with a high degree of order (cybotactic nematics); homopolymers in which n = odd, and some copolymers, form a nematic phase with a lower degree of order ('ordinary' nematics). Both types of PLC are totally miscible with each other and with standard low-molecular mass liquid crystals (LMLC).[2,3]

The alignment and, hence, the overall order parameter S of these polymers, while levelling off for $\bar{M}_n \sim 4000$–5000,[4,5] oscillate following the odd–even alternation of the nematic/isotropic phase transition entropy ΔS_{NI}. The alignment of the flexible part of the polyester DDA-9 ($n = 10$) was analysed by PMR and found to match the alignment of the rigid mesogenic core of the repeating unit,[6] indicating spacer extension in the nematic phase.

In order to obtain a direct insight into the alkyl chain order and mobility, a model compound (9DDA-9-d_{20}) and a polymer (DDA-9-d_{20}) were selectively deuterated in the spacer moiety. Model 9DDA-9-d_{20}:

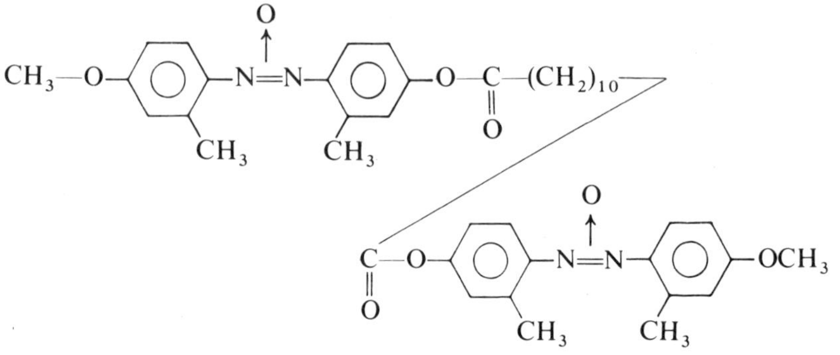

O
↑
CH_3—O—⟨◯⟩—N=N—⟨◯⟩—O—C—$(CH_2)_{10}$—
‖
O
CH_3 CH_3

O
↑
C—O—⟨◯⟩—N=N—⟨◯⟩—OCH_3
‖
O
CH_3 CH_3

Quadrupolar splittings of the spacer methylene groups were measured as a function of temperature and DMR was found to confirm spacer extension[7] previously inferred from PMR spectra.[6] Spacer flexibility was found to decrease considerably as the temperature was lowered in the nematic phase.

Mixtures of polymer DDA-9 with a standard LMLC (perdeuterated *p*-azoxyanisole, PAA-d_{14}) were investigated by PMR and DMR.[8] Preliminary results indicate that the guest and host molecules have the same order parameter, within experimental error: $S_{\mathrm{DDA\text{-}9}} \approx S_{\mathrm{PAA\text{-}d14}}$ (varying continuously with composition).

The temperature variation of the nematic order parameter $S(T)$ was investigated in a number of systems. This is not represented by the standard monotonic curve observed in pure LMLC: pronounced undulations are observed in the $S(T)$ curve. This effect is tentatively explained by the interplay between the natural increase of S with decreasing temperature and the sequential incorporation into the nematic phase of molecules with decreasing chain length and decreasing order parameter.[9,10] However, detailed NMR investigation of the nematic-isotropic biphase and of sharp fractions will be required before this effect can be fully explained.

Investigations of electric instabilities show that Williams' domains and dynamic scattering modes can be easily observed for both cybotactic and 'ordinary' nematic PLC.[11] As expected, the influence of molecular mass is pronounced and the threshold voltage increases with the increase in molecular mass.[12]

The systems of mesogenic polyesters described briefly appear particularly interesting from several points of view: theoretically, they contrast the behaviour and properties of PLC and LMLC; practically, they suggest which structural characteristics are most likely to bring about optimum sample alignment and orientation and provide an additional incentive to study mixed systems.

REFERENCES

1. Blumstein, A. and Thomas, O., *Macromolecules*, 1982, **15**, 1264.
2. Blumstein, A., Blumstein, R. B., Gauthier, M. M., Thomas, O. and Asrar, J., *Mol. Cryst. Liq. Cryst. Lett.*, 1983, **92**, 87.
3. Blumstein, A., Thomas, O., Asrar, J., Makris, P., Clough, S. B. and Blumstein, R. B., *J. Polym. Sci., Polym. Lett. Ed.*, 1984, **22**, 13.
4. Blumstein, A., Vilasagar, S., Ponrathnam, S., Clough, S. B. and Blumstein, R. B., *J. Polym. Sci., Polym. Phys. Ed.*, 1982, **20**, 877.
5. Blumstein, R. B., Stickles, E. M. and Blumstein, A., *Mol. Cryst. Liq. Cryst. Lett*, 1982, **82**(6), 205.
6. Martins, A. F., Ferreira, J. B., Volino, F., Blumstein, A. and Blumstein, R. B., *Macromolecules*, 1983, **16**(2), 279.
7. Samulski, E. T., Gauthier, M. M., Blumstein, R. B. and Blumstein, A., *Macromolecules*, 1984, **17**, 479.
8. Blumstein, R. B., Blumstein, A., Stickles, E. M., Poliks, M. D., Giroud, A. M. and Volino, F., *Polymer Preprints*, 1983, **24**(2), 275.
9. Volino, F., Allonnear, J. M., Giroud-Godquin, A. M., Blumstein, R. B., Stickles, E. M. and Blumstein, A., *Mol. Cryst. Liq. Cryst. Lett.*, 1984, **102**, 21.
10. Volino, F., Giroud, A. M., Dianoux, A. J., Poliks, M., Blumstein, R. B. and Blumstein, A., European Science Foundation Workshop on Liquid Crystal Polymer Systems, Lyngby, Denmark, 1983.
11. Blumstein, A., Schmidt, H. W., Thomas, O., Kharas, G. B., Blumstein, R. B. and Ringsdorf, H., *Mol. Cryst. Liq. Cryst. Lett.*, 1984, **92**, 271.
12. Gilli, J. M., Schmidt, H. W., Pinton, J. F., Sixou, P., Thomas, O., Kharas, G. and Blumstein, A., *Mol. Cryst., Liq. Cryst.*, 1984, **102**, 49.

PART III

CHARACTERIZATION

9

STRUCTURE AND CHARACTERIZATION OF THERMOTROPIC LIQUID CRYSTALLINE POLYMERS

CLAUDINE NOËL

Laboratoire de Physicochimie Structurale et Macromoléculaire, ESPCI, Paris, France

INTRODUCTION

The unique physical and chemical properties of liquid crystalline polymers make them attractive to chemists, physicists, electrical engineers, mechanical engineers and chemical engineers. We have limited ourselves here to a description of thermotropic liquid crystals—those prepared by heating certain polymers.

In the first part of this paper we discuss the ease and reliability with which mesophase transitions can be detected by DSC. In the second part we describe the optical phenomena (textures) observable during microscopic investigation of liquid crystals. The textures are summarized according to common optical features and the different groups of textures are compared with the system established for low molecular weight mesogens. The third part of the paper deals with comprehensive studies of the miscibility relations of liquid crystalline modifications in binary mixtures (isobaric temperature–concentration diagrams). Then we consider the possibility of inducing molecular orientation by surface treatment or external field. The tendency to order is compared for various polymeric structures, molecular weights, flexible spacer lengths. The paper ends with a short description of the structures of the different types of liquid crystals evidenced by x-ray studies.

In this chapter an attempt is made to integrate the information available from the thermal data, textural phenomena, miscibility tests and x-ray diffraction patterns in a discussion of the specific features of the nematic, cholesteric and smectic phases exhibited by liquid crystalline 'main-chain' and 'side-chain' polymers. Results obtained by these methods on liquid crystalline polymers are compared with those obtained for low molecular weight mesogens.

DIFFERENTIAL SCANNING CALORIMETRY

As pointed out by Finkelmann[1] in his recent review, the essential features of the phase behaviour of liquid crystalline side-chain polymers are relatively well established. Usually, the DSC trace exhibits a glass transition, characteristic of the polymer backbone, and a first-order transformation from the mesophase to the isotropic phase due to the mesogenic side chains. Complementary studies with a polarizing microscope reveal that the texture observed in the liquid crystalline state can be frozen without change in the glassy state. A given polymer may have several mesomorphous phases and, so far as can be gathered from observations, these are then separated from one another by first-order transition points.

Systematic studies of a number of types of liquid crystalline side-chain polymers have established that certain systematic trends in liquid crystal behaviour accompany particular changes in molecular structure, and this suggests that it would be profitable to discuss the way in which changes in chemical structure and molecular weight affect the thermal stabilities of liquid crystals. It has been demonstrated that, when mesogenic side groups of the same chemical structure are attached via flexible spacers of approximately the same length to different polymer backbones, the thermal stability of the liquid crystalline phase diminishes with increasing flexibility of the polymer backbone.[1] Thus, comparison of polymethacrylates, polyacrylates and polysiloxanes shows that both the glass transition and the mesophase–isotropic liquid transition temperatures are lowered as the flexibility of the backbone is increased.[1] As for many conventional liquid crystals, the mesophase–isotropic phase transition temperatures alternate in a regular manner as the flexible spacer is extended.[2] However, the senses of alternation are neither totally consistent nor everywhere in agreement with expectations based on elementary liquid crystals. Comparison of mesogenic monomers and corresponding polymers shows that polymerization stabilizes the mesophase: the phase transition temperatures of polymers are shifted to higher temperatures.[1]

The thermal behaviour of liquid crystalline main-chain polymers is more complicated. In most cases, samples show a glass transition (positive ΔC_p), melting (endotherm) and mesophase–mesophase and/or mesophase–isotropic liquid transitions (endotherms). Quickly cooled samples may

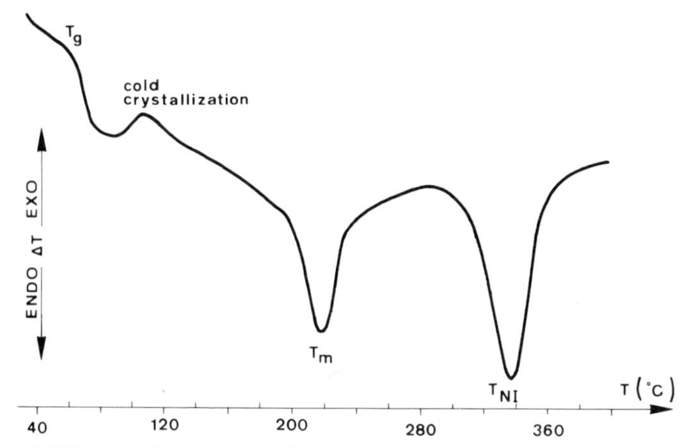

Fig. 1. DSC curve of copolyester formed by transesterification of poly(ethylene-1,2 diphenoxyethane-*p*-*p*′-dicarboxylate) with *p*-acetoxybenzoic acid. From Ref. 3.

demonstrate so-called cold crystallization (exotherm) when heated above the glass transition[3,4] (Fig. 1). However, as observed by Grebowicz and Wunderlich[5] for low molecular weight liquid crystals, the low temperature behaviour depends on the cooling rate: the slower the cooling rate, the smaller are the glass transition and the cold crystallization exotherm. Additional first-order transitions may be observed at low temperature.[6–11] They may result from:

(i) A recrystallization of the polymer chains during the melting of the crystalline structure originally formed. This implies the existence of two interconvertible forms of polymer, which differ from each other only in degree of crystal size and perfection. The relative peak area is strongly dependent on the heating rate. Indeed, such a recrystallization will only occur if its rate is larger or equal to the scanning rate. Higher heating rates allow for less recrystallization.

(ii) Fundamental differences in crystal morphologies (for example folded-chain crystals and partially extended chain crystals). The two morphologies are assumed to be so different that no structural changes will occur during thermal scanning in the DSC. As a consequence, the melting endotherms are assumed to be dependent only on the structure of the material prior to the scan.

(iii) 'True' polymorphism.

This makes it difficult to interpret the DSC curves of liquid crystalline main-chain polymers and the nature of the transition can be established

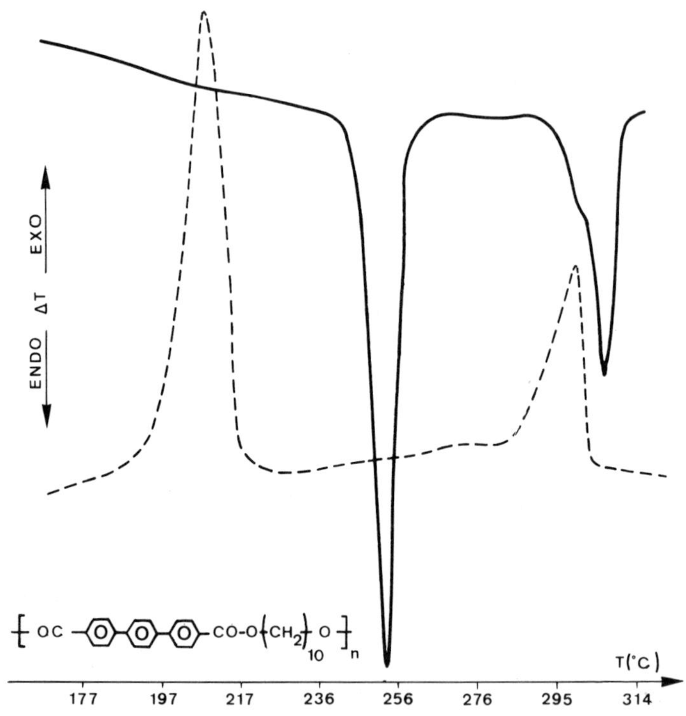

Fig. 2. DSC curves for poly(terphenyl-4,4″ dicarboxylate). From Ref. 12. ——, First heating run; ----, first cooling run.

only through a combination of optical observations and x-ray investigations.

On cooling of the isotropic liquid, the mesophase–isotropic phase and mesophase–mesophase transitions are almost reversible while a marked supercooling is observed for the solid–mesophase and solid–solid transitions[4,6,7,12] (Fig. 2). Only the lowest temperature mesophase can be frozen in the glassy state. Such behaviour is common to many types of small molecule liquid crystals whose crystal–crystal and crystal–mesophase transitions may be identified by DSC since they exhibit supercooling in cooling cycles while mesophase–mesophase and mesophase–isotropic liquid transitions occur at only slightly different temperatures on cooling and heating.[13]

Usually, lower clearing points occur for the polymers incorporating longer flexible spacers.[6,12,14–19] This is consistent with results for elementary liquid crystals and reflects the decreasing thermal stability of the mesophases with decreasing polarity and molecular rigidity. Plots of mesophase–isotropic transition temperature versus number of carbon atoms in the flexible spacer sometimes reveal an even–odd alternation reminiscent of trends in homologous series of conventional liquid crystals.[6,16,19–24] Many polymers show clearing entropies and enthalpies larger than those determined for the low molecular weight model compounds.[4,6,19,25,26] This is an indication that at least a portion of the soft segments participate in the ordered regions of the macromolecules. In some instances a regular ascending odd–even alternation of clearing enthalpies and entropies was reported.[19–21] This confirms the previous statement as to the extended nature of the flexible spacer in the mesophase.

The melting point vs. flexible spacer length curves are much less regular than those for the mesophase–isotropic liquid transition. It is to be noted that with small molecule liquid crystals, regularity of the melting points is encountered only when the members of the homologous series possess the same or very similar crystal structures.[27] The enthalpy changes associated with the melting of polymers are usually lower than those determined for small molecules of similar chemical structure.[6,12] This difference reflects the lack of ability of polymers to form crystals having regularity and perfection.

MICROSCOPIC OBSERVATIONS

Nematic Modifications

Several polymers when examined under the polarizing microscope exhibit threaded and/or schlieren textures indicative of nematic phases[3,8,12,14,22,24,25,28–34] (Figs 3 and 4).

Upon cooling an isotropic melt, the nematic phase begins to separate at the clearing point in the form of typical droplets which, after further cooling, grow and coalesce to form large domains.[35] Nematic droplets characterize a type-texture of the nematic phase since they occur nowhere else.[36] When observed between crossed polarizers, the schlieren texture shows dark brushes which join at certain points. In some cases, four extinction bands are seen to radiate from the centre. Such points indicate disclinations of strength ± 1 in the structure.[3,24,33,34] However, there are also disclinations of strength $\pm 1/2$ which appear as points at which only two dark brushes meet.[3,24,25,33] From the observation of the latter defects, the mesophase can be unambiguously identified as a nematic phase since these singularities occur nowhere else.[37,38]

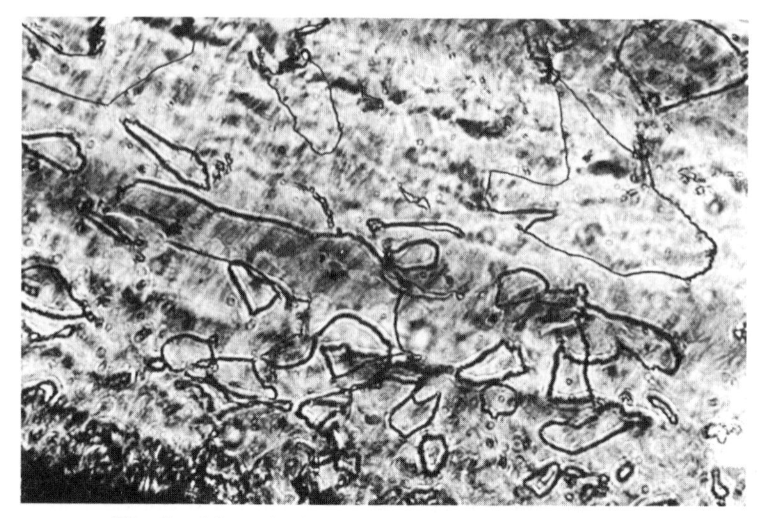

Fig. 3. Nematic threaded texture. Crossed polarizers.

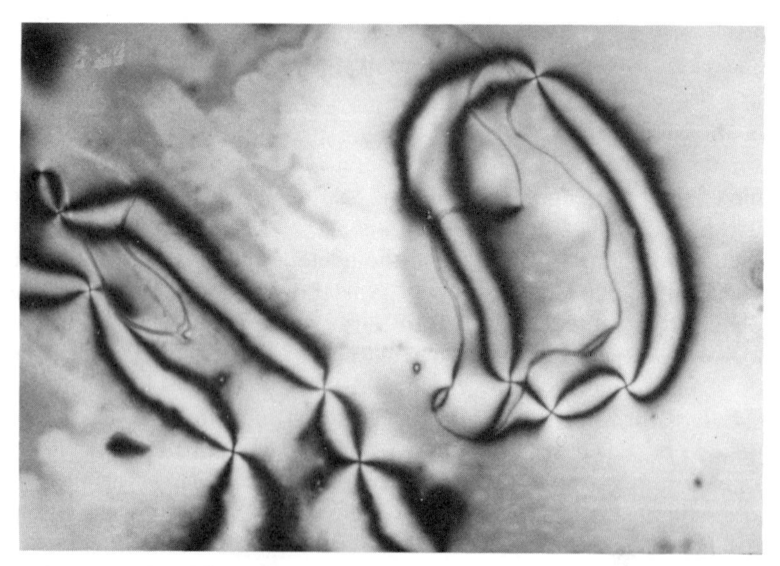

Fig. 4. Nematic schlieren texture. Both integral nuclei (four brushes) and half-nuclei (two brushes) can be seen.

By a suitable surface treatment, it is also possible to obtain uniformly aligned samples with the optic axis normal to the glass surfaces.[31,34,35] Such samples with a homeotropic texture show no birefringence in orthoscopic observation when viewed vertically. However, if the cover glass is touched, the originally dark field of view brightens instantly, thus distinguishing between a homeotropic and an isotropic texture. At high temperatures, perfect layers of this kind, observed between completely crossed polarizers, show a grainy appearance which restlessly changes. These scintillation effects are due to a directly observable Brownian motion.

As mentioned in the preceding section, quenching the polymers from their nematic phases to room temperature succeeds in 'freezing in' these nematic-like textures.[1,3,12,22,31,33,34,39,40] In spite of the thermal shock, both the homeotropic alignment and the planar one can be retained in the glassy state.

Cholesteric Modifications

Conventional nematic liquids can be changed into cholesteric ones by doping them with small amounts of optically active materials.[36–38] Chiral compounds on addition to polymers in the nematic state also yield cholesteric liquids[3,29,31,32] (Fig. 5). They exhibit typical planar textures with oily streaks, moiré fringes and/or Grandjean lines. In the planar texture, these cholesterics can show bright reflection colours. The wavelength of the light at the centre of the reflection band is, for perpendicular incidence, equal to the length of the pitch multiplied by mean refractive index.

Another possibility is offered by introducing chirality into the molecular structure. Copolymerization or copolycondensation of a monomer, capable of forming a thermotropic nematic homopolymer, with a chiral compound yields cholesteric copolymers.[15,25,32,41–47] The optical properties of these cholesteric copolymers resemble those of conventional cholesteric compounds.[1,44,45]

More recently it has been proved that it is possible to prepare cholesteric homopolymers.[32,48,49,50] Krigbaum *et al.*[32] observed fan-like texture for polyester synthesized from 4,4′-dihydroxy-α-methylstilbene and (+)-3-methyladipic acid. These authors suggested the formation of a strongly twisted cholesteric. Indeed, in low molecular weight cholesterics with high twist, the defects and textures resemble those of smectics, especially smectic A.[37] Such cholesteric substances may yield non-planar textures. They may appear in fan-shaped, focal conic or polygonal textures. In the case of

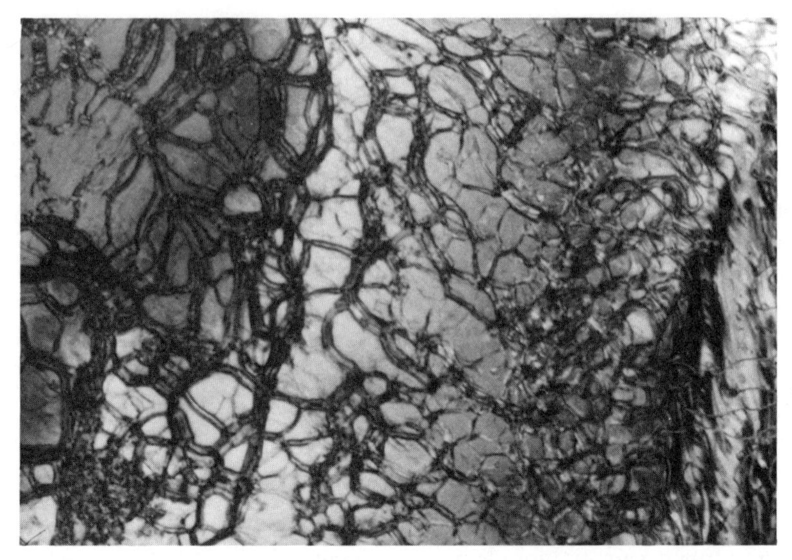

Fig. 5. Cholesteric texture obtained by dissolving a chiral compound in copolyester prepared from terephthalic acid, methylhydroquinone and pyrocatechol. From Ref. 31.

Fig. 6. Smectic fan-shaped texture with focal conics. Crossed polarizers.

homopolymers prepared from chiral monomers, this makes it difficult to interpret the observed textures in terms of cholesterics or smectics. However, by adding either a low molecular weight nematogen or a nematic polymer, the cholesteric pitch can be increased and planar textures can be observed.

Smectic Modifications

In the case of smectic polymers, observation of specific textures may be difficult. Often textures occur whose characteristics are somewhat obscure and observable only with difficulty even at large magnification. This might be due to the high viscosities of the smectic melts.[17] However, several variants occur which closely resemble the focal conic and fan-shaped textures of A and C modifications in conventional liquid crystals[12,17,23,29,51–55] (Figs 6 and 7). Thus, polyesters prepared from di-*n*-propyl-*p*-terphenyl-4,4″ carboxylate and linear or branched aliphatic diols formed smectic A phases which were easily identifiable from their simple focal conic and fan textures.[12,51,56] Occasionally it was possible to get uniaxially ordered layers characterized by homeotropic textures and to observe oily streaks starting from air bubbles. On the other hand, poly (terphenyl-4,4″ dicarboxylates) incorporating 'ether' flexible spacers form smectic C phases which show 'broken' focal conic textures.[12,29,56,57]

Fig. 7. S_A, fan-shaped texture. Crossed polarizers.

Compared to the corresponding textures in smectic A phases, they are less regular and disturbed by additional disclinations. These C structures can be twisted by the addition of optically active compounds.[29]

Transition Phenomena

Transitions with the participation of liquid crystals sometimes show characteristic phenomena. If a nematic modification turns to a smectic A or smectic C phase, transient stripes in the form of a myelinic texture (also called chevron texture or striated texture) are often visible. Typically for the polyester prepared from di-*n*-propyl-*p*-terphenyl-4,4″ carboxylate and tetramethylene glycol, the nematic phase separates from the isotropic liquid on cooling in droplets which coalesce and form large domains. Cooling of the threaded-schlieren texture produces a transition to the smectic A phase; this change is characterized by transition phenomena, mostly stripes, which broaden into larger areas ('transition bars').[12]

For conventional liquid crystals, E modifications can appear from smectic A in the form of a fan-shaped texture with concentric arcs.[38] Optical observations suggest that some poly(terphenyl-4,4″ dicarboxylates) also form an S_E phase upon cooling from the S_A state.[51,56] The A phase exhibits a fan-shaped texture; at the transition to the second smectic

Fig. 8. S_E, paramorphic fan-shaped texture with concentric arcs. Crossed polarizers.

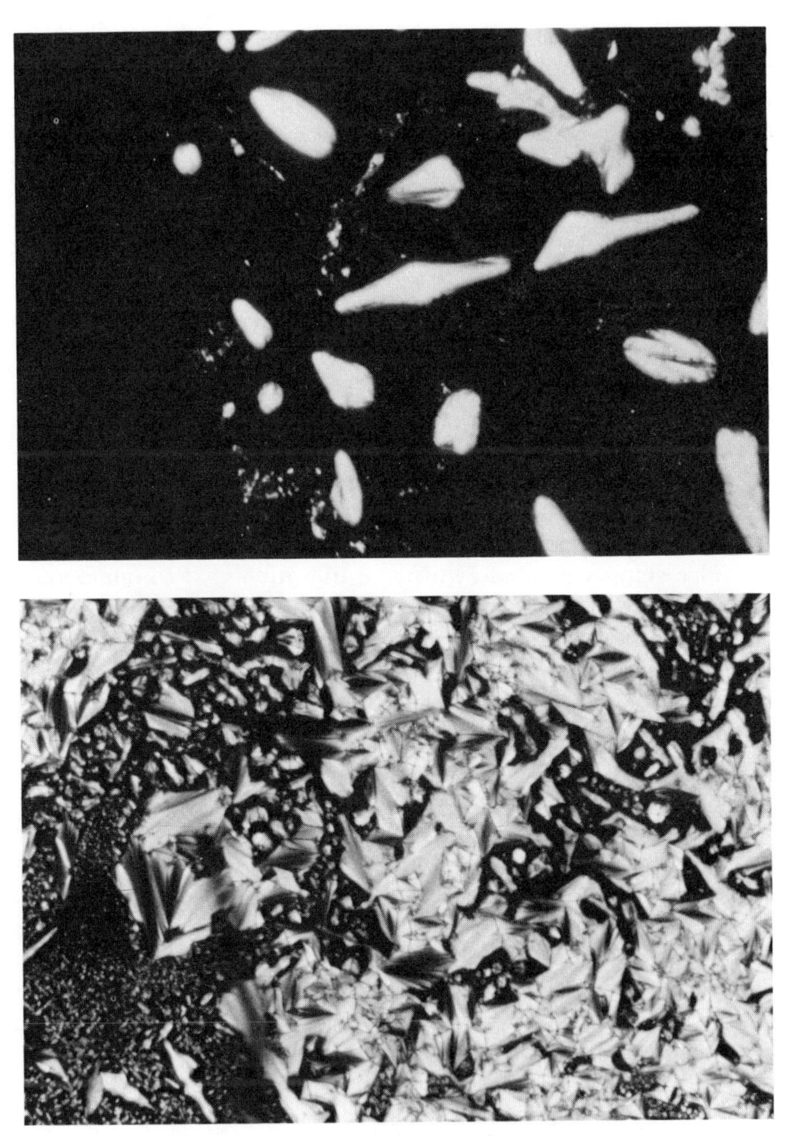

Fig. 9. S_A bâtonnets. Crossed polarizers.

phase the fans are lightly chequered and some arcs run laterally across them as in a transient fan-shaped texture of smectic E (Fig. 8).

The transition phenomena may point to the symmetry of the new modification. For instance, the formation of bâtonnets out of an isotropic melt indicates the existence of a smectic phase.[36] Bâtonnets correspond to nematic droplets with regard to formation conditions. They are closely related to the focal conic texture. Bâtonnets were observed for polyesters of *p,p'*-bibenzoic acid.[12,17,23] Figure 9 illustrates this for the smectic A phase of polyester based on *p,p'*-bibenzoic acid and hexamethylene glycol. The bâtonnets join together to form larger domains from which the compact focal–conic texture finally forms. Elongated particles, reminiscent of the bâtonnets seen in conventional smectics, were also observed for a polyester based on 4,4'-dihydroxybiphenyl.[18]

MISCIBILITY STUDIES

A complete classification of smectic phases by texture is not always possible. It can happen that similar textures are observed with two liquid crystalline states separated by a phase transition.[37] If so, an extremely useful and powerful tool for assessing the type of mesophase is the determination of the isobaric temperature–concentration diagrams of binary mixtures. According to Sackmann and Demus[38] isomorphous liquid crystals are considered as equivalent and characterized by the same symbol. While uninterrupted miscibility establishes isomorphism, the converse is not necessarily true. Temperature–composition phase diagrams for liquid crystalline mixtures can be generated from thermal data or, because of the various optical features characteristic of each mesophase structure, from observations of microscopic textures of the mesophases between crossed polarizers. The latter method (also called the contact method[58]) allows great rapidity in the assessment of the phase diagram.

The applicability of Sackmann and Demus' miscibility rules for high molecular weight compounds was confirmed for several nematogenic main-chain polymers[8,29,59–61] and copolymers.[3,31] It was possible to get diagrams of state with an uninterrupted series of mixed crystals between modifications of the same type. An example is given in Fig. 10. The eutectic composition and temperature can be calculated for a binary system using the Schroeder–van Laar equation.[62,63] However, the calculated values do not agree with those obtained experimentally, the discrepancy being explained as arising from difficulties in packing long macromolecular

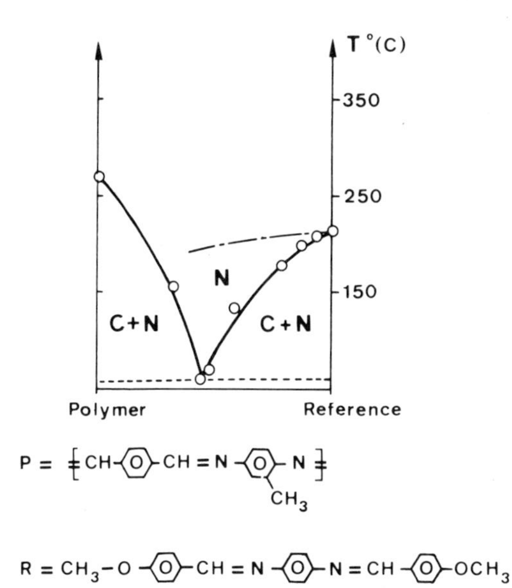

Fig. 10. Isobaric phase diagram of an LC main-chain polymer and a reference compound. From Ref. 59.

chains with small molecules.[60] More recently, cholesteric polyesters based on 4,4′-azoxybenzene and 4,4′-azoxy-2,2′-methylbenzene were found miscible with nematic polyesters of similar structure and an example of perfect solid, cholesteric and isotropic liquid solutions formed by polymeric enantiomers was reported.[64] Mutual miscibility tests have also been used successfully to identify S_A[17,56] and S_C[29] phases in liquid crystalline main-chain polymers. As shown in Fig. 11, confirmation of the classification of the nematic and smectic A phases of poly(terphenyl-4,4″ dicarboxylate) was obtained by co-miscibility studies with standard materials. On the other hand, smectic A and smectic E regions are connected by a heterogeneous region in which the mixed liquid crystals are in equilibrium. This heterogeneous region is an argument for the different nature of these two smectic modifications. It is to be noted that Krigbaum *et al.*[65] and Bosio *et al.*[56] were unsuccessful in their attempt to identify from miscibility tests the S_H phase of poly(*p*-biphenylsebacate) and the S_E phase of poly(terphenyl-4,4″ dicarboxylate), respectively. Thus, it appears that, for the identification of polymeric mesophases, mutual miscibility is the method of choice for nematics and smectic phases of low order while x-ray diffraction may be required for the more ordered smectic phases.

The applicability of the rule of selective miscibility has also been

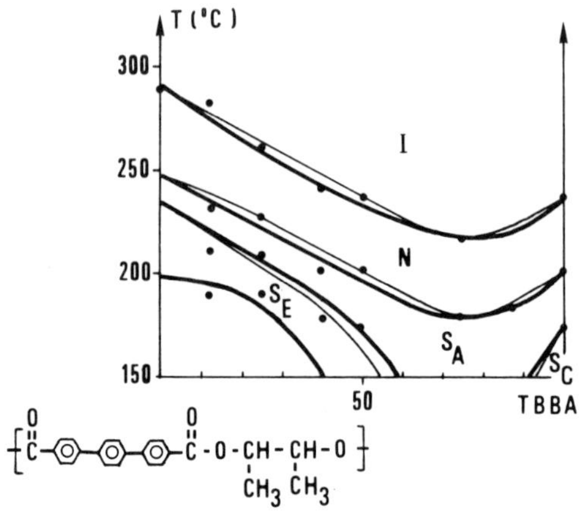

Fig. 11. Isobaric phase diagram of poly(terphenyl 4,4″-dicarboxylate) and TBBA. From Ref. 56.

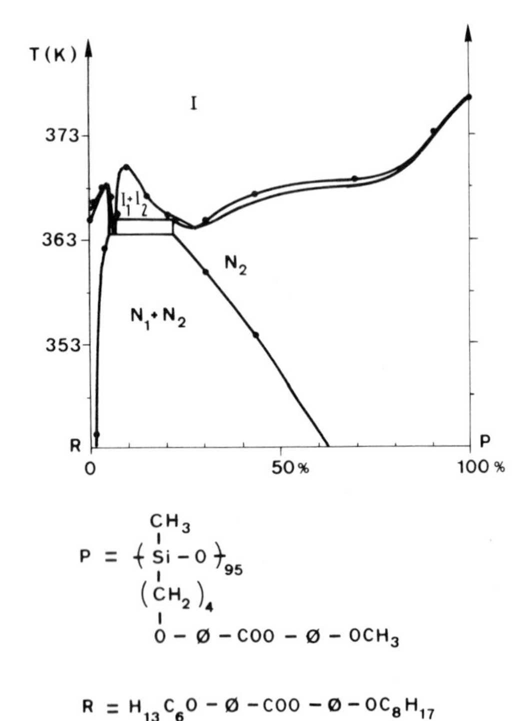

$$P = \left(\underset{\substack{| \\ (CH_2)_4 \\ | \\ O-\emptyset-COO-\emptyset-OCH_3}}{\overset{\substack{CH_3 \\ |}}{Si}} - O \right)_{95}$$

$$R = H_{13}C_6O - \emptyset - COO - \emptyset - OC_8H_{17}$$

Fig. 12. Isobaric phase diagram of an LC side-chain polymer and a reference compound. From Ref. 70.

examined for mixtures of nematic side-chain polymers with conventional nematogens.[66–70] The diagrams of state can be rather complex. An example is shown in Fig. 12. It has been chosen because of a remarkable feature not commonly observed in the case of small molecules and liquid crystalline main-chain polymers; a heterogeneous region is observed between two nematics. Contrary to small molecules and liquid crystalline main-chain polymers, a complete series of mixed liquid crystals seems to exist only if the chemical structure of the polymer-bound side groups is similar to that of the low molecular weight reference compound.

ALIGNMENT OF NEMATIC PHASES

Macroscopic alignment of conventional mesophases by surface treatments, magnetic or electric fields and/or viscous flow is well known.[71] Alignment mechanisms, field-induced structural changes and instabilities produced by flow are examples of phenomena so rich in variety that they are still being explored actively after years of study. There is every reason to believe that polymeric nematics and cholesterics will prove to be just as interesting as the low molecular weight materials.[72] The rôle of flow and other external fields on polymeric nematic and cholesteric phases was reviewed by Krigbaum.[73] Thus, we shall concern ourselves with the alternative approach for obtaining aligned samples: that in which the molecular alignment is controlled by surface treatment of a substrate such as a glass slide. We shall also consider the effects of magnetic field since a few reports appeared after the review was published.

Alignment of Polymeric Nematics on Treated Surfaces

Two terms which will be mentioned often in this section are homeotropic and homogeneous. In homeotropic alignment, the molecules are oriented with their long axes perpendicular to the surface. Homogeneous alignment results when liquid crystal molecules lie parallel to a surface; it becomes uniform when all the molecules at the surface point in the same direction. A polarizing microscope can be used to observe the uniformity of alignment and the absence of defects. Homeotropic preparations are characterized by appearing dark between crossed polarizers and, for this reason, are also termed 'pseudo-isotropic'. On the other hand, viewing planar samples from the top between crossed polarizers results in the observation of four positions of extinction. Such tests are insufficient to check the uniformity of

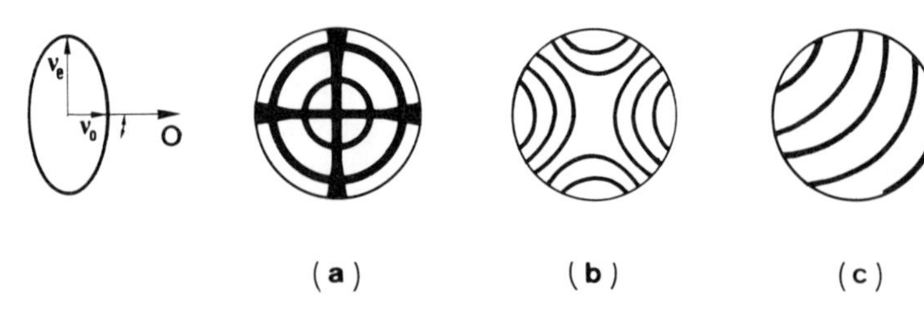

Fig. 13. Ellipsoid of indices for a positive uniaxial nematic. The typical conoscopic images with average beam direction perpendicular to the plates are shown for (a) homeotropic, (b) planar and (c) oblique alignment. From Ref. 71.

alignment and observations of the conoscopic images formed in a monochromatic beam converging in the sample are needed[71] (Fig. 13).

Boiling a glass plate in chromic–sulphuric acid is supposed to unpolish it which may help homeotropic orientation. For a copolyester synthesized from terephthalic acid and an equimolecular mixture of the bis acetates of methylhydroquinone and pyrocatechol, a homeotropic alignment was achieved by simple treatment of glass slides with boiling chromic–sulphuric acid, acetone and methanol (sequentially interspersed with water rinses) and rinsing with hot distilled water.[31] The conoscopic pattern corresponded to that of a uniaxial system showing the optically positive character of conventional nematic phases with a half-wave plate being used. With the aid of conoscopic observations, Finkelmann[74] has also proved the positive uniaxial character of liquid crystalline side-chain polymers. It is interesting to note that sometimes homeotropic texture forms spontaneously when the substance is heated at high temperature between glass slides cleaned without the use of any agent which might etch the glass.[31,35,50] As has been proposed by Meyer,[72] short chains, i.e. abundant chain ends, may favour a homeotropic texture. Indeed, while entropy tends to distribute chain ends randomly, energy tends to place them so as to relieve strains.

A number of techniques have been used to create reproducible homogeneous alignment. It is convenient to separate them into two categories: (a) organic alignment layer materials and (b) inorganic films and substrates.

(a) Noël *et al.*[31] found that planar alignment of copolyester prepared from terephthalic acid and an equimolecular mixture of the bis acetates of methylhydroquinone and pyrocatechol was produced in experiments using surface-deposited hexadecyltrimethylammonium bromide (HMAB). In this case, HMAB films were obtained by slowly pulling glass plates from a

10^{-6} mole litre^{-1} chloroform solution. Such conditions are consistent with HMAB molecules lying flat on the substrate.[71] The weak effect due to the flow action on the HMAB molecules can provide a preferential alignment of these molecules. Such treated surfaces lead to planar samples in which the director lies along the axis of pulling of the plates.

Polymer coatings on glass substrates can be used to align liquid crystals homogeneously. However, since a non-degenerate planar alignment is generally required, rubbing of the substrate after deposition of the polymer is used. Polyimide layers were reported to align along the direction of rubbing—a polyester having the repeating unit:[75]

$$-\!\!\left[OC-(CH_2)_5-CO-O-\langle\bigcirc\rangle-CO-O-\langle\bigcirc\rangle-O\right]\!\!-$$

Our investigation proved that the film uniformity and the choice of substrate influence the observed results.

(b) The first known report of surface alignment of liquid crystals was provided by Mauguin[76] in 1911 when he rubbed a raw glass plate with a piece of paper and obtained uniform homogeneous alignment of *p*-azoxyanisole. This technique was successfully used by Casagrande *et al.*[77] who made cells suitable to determine the viscoelastic coefficients K_{11} and γ_1 of liquid crystalline side-chain polymer. In this particular case, the rubbing was done with diamond paste. However, attempts to orient liquid crystalline main-chain polymers by the rubbing technique were unsuccessful.[41,78,79] On the other hand, SiO_x layers evaporated obliquely to a substrate promoted uniform homogeneous alignment of copolyester based on terephthalic acid, methylhydroquinone and pyrocatechol.[31] Well-aligned samples were also obtained by using freshly cleaved mica surfaces.[3]

Coupling with Magnetic Fields

The effect of a magnetic field on a nematic main-chain copolyester was first investigated by Noël *et al.*[80] They inferred that the molecules have a preferred alignment parallel to the field from the change of the diamagnetic susceptibility as a function of the magnetic field strength (Fig. 14). The threshold field for the onset of magnetic alignment was found to be 5 kG. Practically complete orientation was approximated at 6 kG in approximately 2 min for a sample having a weight-average molecular weight of 220 000 and an apparent viscosity of 300 Pa s at a shear rate of 32 s^{-1}. The anisotropy of the magnetic susceptibility $\Delta\chi$ was $1{\cdot}4 \times 10^{-7}$ emu g^{-1}, compared with a typical value for conventional liquid crystals of

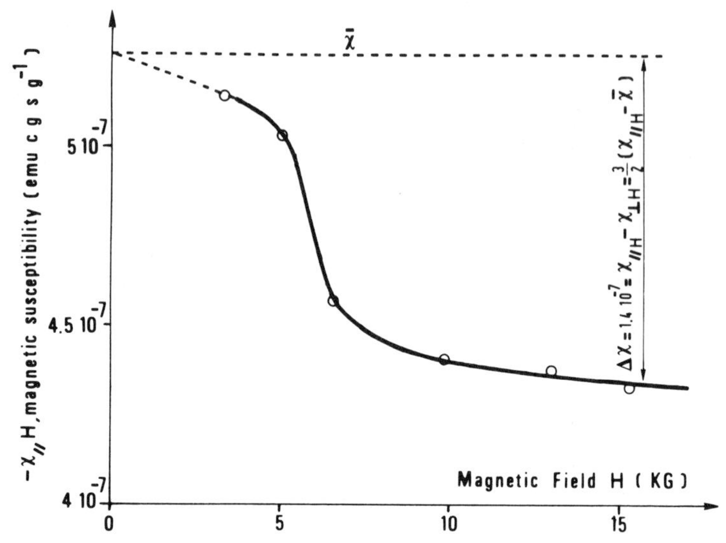

Fig. 14. Magnetic susceptibility χ of an LC main-chain copolyester vs. the magnetic field strength H. From Ref. 80.

10^{-7} emu g^{-1}.[36] This study was extended by Hardouin *et al.*[81] to polyesters:

$$-\!\!\left[\overset{\overset{\displaystyle O}{\|}}{C}-(CH_2)_n-\overset{\overset{\displaystyle O}{\|}}{C}-O-\langle\bigcirc\rangle-\overset{\overset{\displaystyle O}{\|}}{C}-O-\langle\bigcirc\rangle-O\right]\!\!- \qquad \mathbf{I}$$

For this series a rather strong influence of the number of carbon atoms in the flexible spacer on the nematic orientation was reported. Increasing n from 5 to 9 produced an increase in the threshold field strength from 4 to 15 kG for a 1-min orientation time. No evidence of alignment was observed after 200 min for the polyester with $n = 12$. Hardouin *et al.*[81] suggested that the higher viscosity of the latter polyester might hinder the molecular alignment. If the difficulty of obtaining aligned samples was due to the high viscosities of the nematic melts, this problem could be overcome by studying polymer samples of low molecular weight. Investigation of magnetic field effects for the polyester with $n = 5$ revealed that the orientation time increases noticeably as the sample inherent viscosity is increased from 0·29 dl g^{-1} to 0·54 dl g^{-1} and that the effects saturate only for the sample of lowest inherent viscosity. Generally, low molecular weight polymers orient easily even in low intensity magnetic fields, while high molecular weight polymers are difficult to orient.[82,83] It is interesting to note that cholesteric copolyesters prepared from 4,4′-dihydroxyazoxybenzene and mixtures of (+)-3-methyladipic acid and dodecanedioic acid could not be oriented with fields exceeding 12 T, presumably due to the very high critical fields necessary to produce an untwisted nematic structure.[25,84,85]

Hardouin *et al.*[81] examined the effect of a rotating magnetic field on polyesters of type **I**. Conventional nematics in a rotating magnetic field exhibit viscous behaviour in the bulk characterized by the twist viscosity coefficient γ_1. In contrast, polyesters of series **I** behave as solids even at low rotational frequencies of the magnetic field.

The first experimental studies of the magnetic field effects on liquid crystalline side-chain polymers were reported by Casagrande and co-workers.[77,86] They determined the diamagnetic anisotropy by using the Faraday method and they evaluated the splay elastic constant K_{11}, the bend elastic constant K_{33} and the twist viscosity coefficient γ_1 from the dynamics of a Fredericks' transition. The diamagnetic anisotropy, the threshold field and the elastic constants approach those of conventional nematics. Contrary to the results obtained for main-chain polymers,[87] they are in no way systematically higher in side-chain polymers than in small molecules of similar chemical structure. As might be anticipated, only the mesogenic side groups are responsible for the liquid crystalline order. However, some differences also exist between side-chain polymers and conventional nematogens: the response time and the twist viscosity coefficient are several orders of magnitude larger than those obtained for the model compounds. The occurrence of strong kinetic effects of orientation was also reported by Achard *et al.*[88] for side-chain polymers of the same type. The shorter the flexible spacer is, the stronger the effects are. Casagrande and co-workers concluded that the rotation of the director involves a cooperative motion of the backbone accompanying the rotation of the mesogenic side groups.

Order Parameters

The order parameter $S = 1/2\langle 3\cos^2\theta - 1\rangle$ is the single quantity necessary to specify completely the degree of order in nematics; θ is the angle between the individual molecular long axis and the director. The brackets indicate the average value. This relation is also valid for liquid crystalline main-chain and side-chain polymers. With comb-like polymers, however, one refers to the mesogenic side groups. The techniques used to determine the order parameters of polymers in the nematic state include: analysis of refractive index data,[1,48,74,90] magnetic susceptibility measurements,[88,89] PMR and DMR lineshape analysis,[83,91–96] ESR spectra simulation,[97,98] x-ray scattering,[3,82] IR dichroism[3] and UV dichroism.[99,100]

From the comparison of the order parameters for liquid crystalline side-chain polymers with those for small molecules of similar chemical structure, the following points were established:

1. The temperature dependence of *S* is similar for the side-chain polymers and their low molecular weight analogues.[88,90,96,97,99,100] For comb-like polymers, however, *S* is significantly smaller than the corresponding parameter for monomers. From a detailed analysis of deuteron-NMR spectra, Boeffel *et al.*[95] established that the order parameter of the spacer (methylene group adjacent to the polymer chain) is reduced to 50% of the value of the mesogenic group whereas the corresponding methylene group in the low molecular weight analogue exhibits a reduction by only 25%. This indicates a substantially higher fraction of gauche conformers in the spacer of the polymeric liquid crystal than in the alkyl chains of conventional liquid crystals.
2. Extending the flexible spacer does not change *S*. An increase in the spacer length is only associated with less hindered rotation of the mesogenic groups around their long molecular axes.[1,90]
3. At the nematic–smectic transition, a jump in the order parameter is observed[90,95,97] which is consistent with results for conventional liquid crystals.
4. *S* can be frozen-in at the glass transition.[90,95,97] No change in the order parameter is observed over a period of one year, keeping the sample at room temperature.[97] This result opens new applications as electro-optical memory and display devices.

Liebert *et al.*[82] performed the earliest measurements of the order parameter in the nematic phase of the polyester **I** with $n = 5$. After exposure to a magnetic field of 3 kG for 24 h the order parameter *S* was found to be 0·64 for a sample having an inherent viscosity of 0·29 dl g^{-1} and 0·54 for a sample having a higher inherent viscosity (0·54 dl g^{-1}). Subsequent work, however, revealed unusually large values for the order parameter of main-chain polymers macroscopically aligned by imposing high magnetic fields (>1 T) or by solid-state extrusion. Mueller *et al.*[98] obtained from ESR study values of *S* in the range 0·8–0·9, significantly larger than those for side-chain polymers and ordinary liquid crystals. Volino and co-workers[83,91–93] and Mueller *et al.*[94] also deduced from NMR investigations order parameters ranging from 0·64 to 0·87. This very high degree of order was required for the mesogenic group as well as for the spacer for a satisfactory agreement between the experimental and the simulated spectra. These authors concluded that the flexible spacer aligns with a degree of order comparable to that of the mesogenic moiety and adopts a highly extended conformation, evidenced by a *trans* population of $n_t \sim 0{\cdot}8$. It is interesting to note that the nematic order parameter of main-chain polymers was found to increase with molecular weight.[89,93] However, at the isotropic–nematic transition, values falling in the range 0·4–0·7 were reported.[3,83,89,93] Such results confirm the recent theory of Ronca and Yoon:[101] the isotropic–nematic phase transition of semiflexible polymers does not entail a very high degree of order in the resultant nematic phase.

X-RAY DIFFRACTION PATTERNS

X-ray diffraction provides information concerning the arrangement and mode of packing of molecules and the types of order present in a mesophase. Detailed reviews[102–109] have appeared laying stress upon the liquid crystalline order in side-chain polymers (see also Refs 46, 52–54, 110–117). Thus, we shall confine our attention mainly to structural studies performed on main-chain polymers.

X-Ray Diffraction Patterns for Powder Samples

It is often possible to distinguish between the nematic and smectic phases of conventional liquid crystals from the x-ray diffraction patterns of unoriented samples but this is not always possible (e.g. S_A and S_C may be confused). The diffraction pattern of a powder sample can be divided into inner rings at small diffraction angles and outer rings at large angles. The inner rings are indicative of longer layer spacings. The outer rings correspond to shorter preferred spacings occurring in the lateral packing arrangement of the molecules. The appearance of a broad halo or a sharp ring furnishes a qualitative indication of the degree of order. The most common mesophases can be divided into three groups according to the characteristics of their x-ray diffraction patterns at large angles.[118–120] In the first group we find the nematic, smectic A and smectic C phases. They give only one diffuse outer halo which indicates that the lateral arrangement of the molecules is disordered: the distribution of the molecular centres of mass is random. The second group is composed of the smectic phases which exhibit three-dimensional order: the smectic E and smectic B and their tilted modifications—the smectic H and smectic G. Their diffraction patterns show a single or several sharp outer rings which are related to the high degree of order within the layers. The S_F and S_I phases are intermediate between these two groups.

Owing to the existence of polycrystalline and amorphous material, x-ray patterns of main-chain polymers are sometimes too diffuse to be of help for the identification of mesophases. However, x-ray patterns were reported which are consistent with a nematic structure.[24,31,39,42,43,121–124] At large diffraction angles they all present a diffuse, broad ring which corresponds to average intermolecular spacings of approximately 4–6 Å. This diffuse halo, which is quite similar to that given by the isotropic liquid phases, indicates a lack of periodic lateral order. At small angles, nematic patterns may present a diffuse ring, corresponding to distances close or equal to the repeat unit length which indicates that there is no order in the direction of the molecular long axes.

Other x-ray patterns obtained from main-chain polymers in the smectic state are characteristic of a disordered lamellar structure.[55,56,122c,124,125] They present a sharp inner ring corresponding to the lamellar thickness and a diffuse outer ring reflecting the absence of ordering within the layer planes. For S_C phases, diffraction patterns of powder samples are essentially the same as those for S_A except that the layer spacings are usually relatively smaller. In an S_A phase, the director is normal to the layers which are about one repeat unit thick. The accepted structure for the S_C is that the director is tilted with respect to the layers: the lamellar thickness is less than the repeat unit length.

Finally, it is to be noted that liquid crystalline main-chain polymers may also form smectic phases that have essentially long range three-dimensional (3-D) order. The x-ray patterns obtained by Blumstein *et al.*[121] for poly(4,4′-diphenylsebacate) in the liquid crystalline state consisted of a sharp inner ring and a sharp outer ring related to Bragg spacings of 16·4 Å and 4·46 Å. Such results suggested a layered organization of chains with a fair amount of order within layers. Krigbaum *et al.*[65] also used x-ray diffraction to identify the type of smectic phase of poly(4,4′-diphenyl-sebacate). Although they were not able to prepare monodomain samples, the x-ray data obtained with powder samples nevertheless argued for the existence of a smectic H phase: (i) the diffraction diagram consisted of one sharp inner ring (16·5 Å, s) and two sharp outer rings (4·99 Å, w; 4·50 Å, vs); (ii) the Bragg spacing corresponding to the inner ring was significantly shorter than the length of the repeat unit in its most extended conformation indicating a tilted phase; (iii) the lateral packing in the plane perpendicular to the molecules was represented by an oblique lattice. It is to be noted that such an assignment is in conflict with that made by Strzelecki and Van Luyen[14] from texture observations. These authors reported the poly(4,4′-diphenylsebacate) to be nematic.

Various types of evidence lend support to the conclusion that poly(terphenyl-4,4″ dicarboxylates) with short flexible spacers form a smectic E phase in addition to smectic A and nematic phases.[56] Cooling of the smectic A phase produced a phase which showed a simple fan texture. The fans were lightly chequered and some arcs ran laterally across them as in a transient fan-shaped texture of smectic E.[37,38] At low temperature, the x-ray diffraction patterns of powder samples were consistent with the highly ordered smectic E phase.[120] Besides a small interference at small angles (and its second order) which indicates the existence of the smectic layers, four sharp interferences were found at large angles. The lamellar spacing corresponding to the inner ring was close to the length of the repeat unit which eliminated tilted phases.

X-Ray Diffraction Patterns for Oriented Samples

If a sample can be obtained in the form of an oriented monodomain, it is possible to extract more detailed structural information from its diffraction diagram.[119,120] Oriented specimens can be prepared by cooling in a strong magnetic field from the isotropic liquid phase into the nematic phase. Aligned S_A and S_C may then be prepared by careful cooling from the aligned nematic phase. This may in fact be very difficult and the result depends on the specimen and the conditions (e.g. surface treatment). An alternative procedure, useful for preparing monodomains of the more ordered (and more viscous) phases is by careful melting of either a single crystal or an oriented fibre.

For well-aligned polymer samples, diffraction patterns characteristic of oriented nematics were reported.[3,11,20,25,31,82,85] The anisotropy is clearly shown (Fig. 15) and there are correlations of two distinct periods perpendicular and parallel to the director which correspond approximately to an average intermolecular spacing and repeat unit length, respectively. For $Q \perp n$ the diffraction pattern is very liquid-like. The outer diffuse halo is crescent-shaped. An average intermolecular spacing of approximately 4–6 Å is found for a variety of polymers. This closely corresponds to the lateral dimensions of a typical repeat unit which indicates that there must be strong local correlations in orientation about the long axes and that molecular rotation about the long axes must be

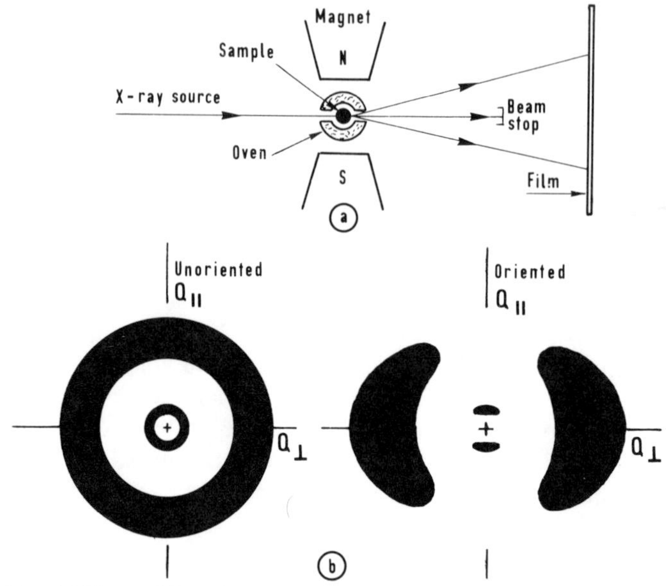

Fig. 15. (a) X-ray diffraction apparatus and (b) typical diffraction patterns for nematics. From Ref. 119.

strongly cooperative in nature. The dominant features of the scattering with $Q \| n$ are arcs or short bars of intensity comparable to or smaller than that of the strong equatorial crescents. Their position corresponds to a repeat distance of the order of the repeat unit length.

Secondary nematic structure was found by Blumstein and co-workers[25,85] during the x-ray investigation of polyester prepared from 4,4′-hydroxy(2,2′-methyl) azoxybenzene and dodecanedioic acid. The diffraction pattern of the nematic shows the development of enhanced order characteristic of smectic phase (Fig. 16): the first of the meridional arcs splits up into four sharp spots. This phenomenon is incompatible with the classical definition of the nematic phase and suggests an additional order of macromolecules within cybotactic groups for which a structural model is proposed.

Noël *et al.*[31] used x-ray diffraction to characterize the nematic phase of copolyester based on terephthalic acid, methylhydroquinone and pyrocatechol. Along the equator, in addition to the two pronounced crescents, there appear two diffuse spots at smaller diffraction angles. Similar diffraction features have been reported previously for certain ordinary nematic systems but there is no information at present with regard to their

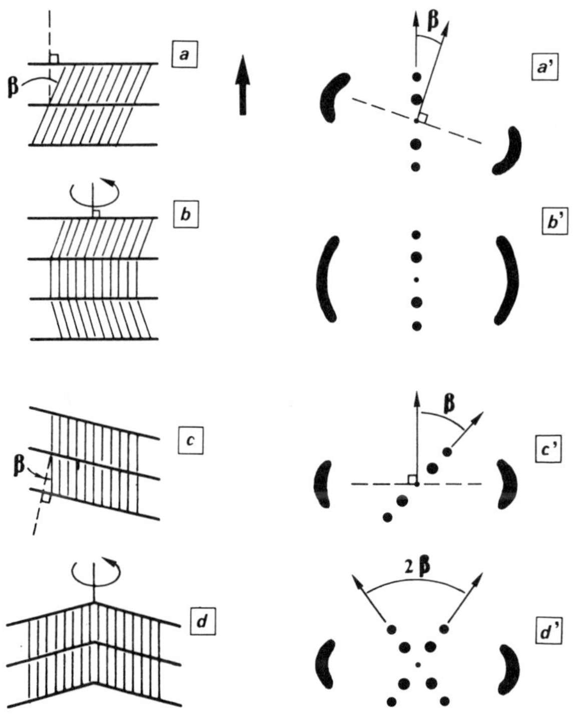

Fig. 16. Different S_C configurations. From Ref. 129.

origin.[126] Taking into account that such diffuse spots have been observed for helical structures,[127,128] Noël *et al.* considered the possibility of such arrangements of the chains but without long-range order. They proposed a structural model which bears a resemblance to the molecular arrays accepted for cybotactic nematic phases. The two diffuse spots would be expected for roughly parallel chain bundles in the form of two/four-stranded ropes.

Finally, well-aligned samples of polyester prepared from di-*n*-propyl-*p*-terphenyl-4,4″ carboxylate and $HO{-}(CH_2{-}CH_2{-}O)_4{-}H$ were obtained by careful melting of oriented fibres.[56] The x-ray pattern was essentially the same as that for a smectic A except that the layer spacing was significantly smaller than the repeat unit length suggesting a tilt angle relative to the layer normal. In fact, the smectic phase of polyester was found miscible with the smectic C phases of standard materials.[29] X-ray data and miscibility tests argue for the existence of a smectic C configuration which is azimuthally disordered with the layers orientationally ordered (Fig. 16(b)).

REFERENCES

1. Finkelmann, H., in *Polymer Liquid Crystals*, Ciferri, A., Krigbaum, W. R. and Meyer, R. B. (Eds), Academic Press, New York, 1982.
2. Gemmell, P. A., Gray, G. W. and Lacey, D., *Polym. Prepr., Am. Chem. Soc., Div. Polym. Chem.*, 1983, **24**(2), 253.
3. Noël, C., Lauprêtre, F., Friedrich, C., Fayolle, B. and Bosio, L., *Polymer*, 1984, **25**, 808.
4. Grebowicz, J. and Wunderlich, B., *J. Polym. Sci., Polym. Phys. Ed.*, 1983, **21**, 141.
5. Grebowicz, J. and Wunderlich, B., *Mol. Cryst. Liq. Cryst.*, 1981, **76**, 287.
6. Griffin, A. C. and Havens, S. J., *J. Polym. Sci., Polym. Phys. Ed.*, 1981, **19**, 951.
7. Roviello, A. and Sirigu, A., *J. Polym. Sci., Polym. Lett. Ed.*, 1975, **13**, 455.
8. Noël, C. and Billard, J., *Mol. Cryst. Liq. Cryst. Lett.*, 1978, **41**, 269.
9. Krigbaum, W. R. and Salaris, F., *J. Polym. Sci., Polym. Phys. Ed.*, 1978, **16**, 883.
10. Frosini, V., De Petris, S., Chiellini, E., Galli, G. and Lenz, R. W., *Mol. Cryst. Liq. Cryst.*, 1983, **98**, 223.
11. Roviello, A. and Sirigu, A., *Makromol. Chem.*, 1982, **183**, 409.
12. Meurisse, P., Noël, C., Monnerie, L. and Fayolle, B., *Brit. Polym. J.*, 1981, **13**, 55.
13. Petrie, S. E. B., in *Liquid Crystals, the Fourth State of Matter*, Saeva, F. D. (Ed.), Marcel Dekker, New York and Basle, 1979.
14. Strzelecki, L. and Van Luyen, D., *Eur. Polym. J.*, 1980, **16**, 299, 303.
15. Van Luyen, D., Liebert, L. and Strzelecki, L., *Eur. Polym. J.*, 1980, **16**, 307.
16. Ober, C., Jin, J. I. and Lenz, R. W., *Polymer J.*, 1982, **14**, 9.
17. Krigbaum, W. R. and Watanabe, J., *Polymer*, 1983, **24**, 1299.
18. Asrar, J., Toriumi, H., Watanabe, J., Krigbaum, W. R. and Ciferri, A., *J. Polym. Sci., Polym. Phys. Ed.*, 1983, **21**, 1119.
19. Blumstein, A. and Thomas, O., *Macromolecules*, 1982, **15**, 1264.
20. Roviello, A. and Sirigu, A., *Makromol. Chem.*, 1982, **183**, 895.
21. Griffin, A. C. and Britt, T. R., *Mol. Cryst. Liq. Cryst. Lett.*, 1983, **92**, 149.
22. Strzelecki, L. and Liebert, L., *Eur. Polym. J.*, 1981, **17**, 1271.
23. Krigbaum, W. R., Asrar, J., Toriumi, H., Ciferri, A. and Preston, J., *J. Polym. Sci., Polym. Lett. Ed.*, 1982, **20**, 109.
24. Antoun, S., Lenz, R. W. and Jin, J. I., *J. Polym. Sci., Polym. Chem. Ed.*, 1981, **19**, 1901.
25. Blumstein, A., Vilasagar, S., Ponrathnam, S., Clough, S. B. and Blumstein, R. B., *J. Polym. Sci., Polym. Phys. Ed.*, 1982, **20**, 877.
26. Jin, J. I., Antoun, S., Ober, C. and Lenz, R. W., *Brit. Polym. J.*, 1980, **12**, 132.
27. Gray, G. W. and Winsor, P. A., *Liquid Crystals and Plastic Crystals*, Vol. 1, Halsted, Chichester, 1974.
28. Millaud, B., Thierry, A., Strazielle, C. and Skoulios, A., *Mol. Cryst. Liq. Cryst. Lett.*, 1979, **49**, 299.
29. Fayolle, B., Noël, C. and Billard, J., *J. de Physique*, 1979, **40**, C3-485.
30. Jo, B. W., Jin, J. I. and Lenz, R. W., *Eur. Polym. J.*, 1982, **18**, 233.
31. Noël, C., Billard, J., Bosio, L., Friedrich, C., Lauprêtre, F. and Strazielle, C., *Polymer*, 1984, **25**, 263.

32. Krigbaum, W. R., Ciferri, A., Asrar, J., Toriumi, H. and Preston, J., *Mol. Cryst. Liq. Cryst.*, 1981, **76**, 79.
33. Viney, C. and Windle, A. H., *J. Mater. Sci.*, 1982, **17**, 2661.
34. Mackley, M. R., Pinaud, F. and Siekmann, G., *Polymer*, 1981, **22**, 437.
35. Galli, G., Chiellini, E., Ober, C. K. and Lenz, R. W., *Makromol. Chem.*, 1982, **183**, 2693.
36. Kelker, H. and Hatz, R., *Handbook of Liquid Crystals*, Verlag Chemie, Weinheim, 1980.
37. Demus, D. and Richter, L., *Textures of Liquid Crystals*, Verlag Chemie, Weinheim, 1978.
38. Sackmann, H. and Demus, D., *Mol. Cryst. Liq. Cryst.*, 1973, **21**, 239.
39. Lenz, R. W. and Jin, J. I., *Macromolecules*, 1981, **14**, 1405.
40. Talroze, R. V., Kostromin, S. G., Shibaev, V. P. and Platé, N. A., *Makromol. Chem., Rapid. Commun.*, 1981, **2**, 305.
41. Corazza, P., Sartirana, M. L. and Valenti, B., *Makromol. Chem.*, 1982, **183**, 2847.
42. Vilasagar, S. and Blumstein, A., *Mol. Cryst. Liq. Cryst. Lett.*, 1980, **56**, 263.
43. Blumstein, A. and Vilasagar, S., *Mol. Cryst. Liq. Cryst. Lett.*, 1981, **72**, 1.
44. Finkelmann, H., Koldehoff, J. and Ringsdorf, H., (a) *Angew. Chem.*, 1978, **90**, 992; (b) *Angew. Chem. Int. Ed. Engl.*, 1978, **17**, 935.
45. Finkelmann, H. and Rehage, G., *Makromol. Chem., Rapid Commun.*, 1980, **1**, 733.
46. Platé, N. A. and Shibaev, V. P., *J. Polym. Sci., Polym. Symp.*, 1980, **67**, 1.
47. Shibaev, V. P., Finkelmann, H., Kharitonov, A. V., Portugall, M., Platé, N. A. and Ringsdorf, H., *Polym. Sci. USSR*, 1981, **23**, 1029.
48. Finkelmann, H. and Rehage, G., *Makromol. Chem., Rapid Commun.*, 1982, **3**, 859.
49. Chiellini, E. and Galli, G., *Makromol. Chem., Rapid Commun.*, 1983, **4**, 285.
50. Krigbaum, W. R., Ishikawa, T., Watanabe, J., Toriumi, H., Kubota, K. and Preston, J., *J. Polym. Sci., Polym. Phys. Ed.*, 1983, **21**, 1851.
51. Noël, C., Meurisse, P., Friedrich, C. and Bosio, L., to be published.
52. Hahn, B., Wendorff, J. H., Portugall, M. and Ringsdorf, H., *Colloid Polym. Sci.*, 1981, **259**, 875.
53. Shibaev, V. P., Kostromin, S. G. and Platé, N. A., *Eur. Polym. J.*, 1982, **18**, 651.
54. Kostromin, S. G., Sinitzyn, V. V., Talroze, R. V., Shibaev, V. P. and Platé, N. A., *Makromol. Chem., Rapid. Commun.*, 1982, **3**, 809.
55. Ober, C. K., Jin, J. I. and Lenz, R. W., *Makromol. Chem., Rapid. Commun.*, 1983, **4**, 49.
56. Bosio, L., Fayolle, B., Friedrich, C., Lauprêtre, F., Meurisse, P., Noël, C. and Virlet, J., in *Liquid Crystals and Ordered Fluids*, Vol. 4, Griffin, A. and Johnson, J. F. (Eds), Plenum Press, New York, 1984, p. 401.
57. Noël, C., Friedrich, C., Bosio, L. and Strazielle, C., *Polymer*, in press.
58. Kofler, L. and Kofler, A., *Thermomikromethoden*, Verlag Chemie, Weinheim, 1954.
59. Millaud, B., Thierry, A. and Skoulios, A., *Mol. Cryst. Liq. Cryst. Lett.*, 1978, **41**, 263.
60. Griffin, A. C. and Havens, S. J., *J. Polym. Sci., Polym. Lett.*, 1980, **18**, 259.

61. Griffin, A. C. and Havens, S. J., *Mol. Cryst. Liq. Cryst. Lett.*, 1979, **49**, 239.
62. Schroeder, I., *Z. Phys. Chem.*, 1893, **11**, 449.
63. Van Laar, J. J., *Z. Phys. Chem.*, 1908, **63**, 216.
64. Billard, J., Blumstein, A. and Vilasagar, S., *Mol. Cryst. Liq. Cryst. Lett.*, 1982, **72**, 163.
65. Krigbaum, W. R., Watanabe, J. and Ishikawa, T., *Macromolecules*, 1983, **16**, 1271.
66. Nyitrai, K., Cser, F., Lengyel, M., Seyfried, E. and Hardy, Gy., *Eur. Polym. J.*, 1977, **13**, 673.
67. Cser, F., Nyitrai, K., Hardy, G., Menczel, J. and Varga, J., *J. Polym. Sci., Polym. Symp.*, 1981, **69**, 91.
68. Ringsdorf, H., Schmidt, H. W. and Schneller, A., *Makromol. Chem., Rapid Commun.*, 1982, **3**, 745.
69. Finkelmann, H., Kock, H. J. and Rehage, G., *Mol. Cryst. Liq. Cryst.*, 1982, **89**, 23.
70. Casagrande, C., Veyssié, M. and Finkelmann, H., *J. Phys. Lett.*, 1982, **43**, L-675.
71. Guyon, E. and Urbach, W., *4BBC Symp. Nomen. El. Disp.*, Kmetz and Von Willisen (Eds), Plenum Press, New York, 1976.
72. Meyer, R. B., in *Polymer Liquid Crystals*, Ciferri, A., Krigbaum, W. R. and Meyer, R. B. (Eds), Academic Press, New York, 1982.
73. Krigbaum, W. R., in *Polymer Liquid Crystals*, Ciferri, A., Krigbaum, W. R. and Meyer, R. B. (Eds), Academic Press, New York, 1982.
74. Finkelmann, H., in *Liquid Crystals of One- and Two-dimensional Order*, Helfrich, W. and Heppke, G. (Eds), Springer Verlag, Berlin, 1980.
75. Kléman, M., Liebert, L. and Strzelecki, L., *Polymer*, 1983, **24**, 295.
76. Mauguin, C., *Bull. Soc. Fr. Min.*, 1911, **34**, 71.
77. Casagrande, C., Veyssié, M., Weill, C. and Finkelmann, H., *Mol. Cryst. Liq. Cryst. Lett.*, 1983, **92**, 49.
78. Krigbaum, W. R., Lader, H. J. and Ciferri, A., *Macromolecules*, 1980, **13**, 554.
79. Krigbaum, W. R., Grantham, C. E. and Toriumi, H., *Macromolecules*, 1982, **15**, 592.
80. Noël, C., Monnerie, L., Achard, M. F., Hardouin, F., Sigaud, G. and Gasparoux, H., *Polymer*, 1981, **22**, 578.
81. Hardouin, F., Achard, M. F., Gasparoux, H., Liebert, L. and Strzelecki, L., *J. Polym. Sci., Polym. Phys. Ed.*, 1982, **20**, 975.
82. Liebert, L., Strzelecki, L., Van Luyen, D. and Levelut, A. M., *Eur. Polym. J.*, 1981, **17**, 71.
83. Martins, A. F., Ferreira, J. B., Volino, F., Blumstein, A. and Blumstein, R. B., *Macromolecules*, 1983, **16**, 279.
84. Maret, G., Blumstein, A. and Vilasagar, S., *Polym. Prepr. Am. Chem. Soc., Div. Polym. Chem.*, 1981, **22**, 246.
85. Maret, G. and Blumstein, A., *Mol. Cryst. Liq. Cryst.*, 1982, **88**, 295.
86. Fabre, P., Casagrande, C., Veyessié, M. and Finkelmann, H., in preparation.
87. Zheng-Min Sun and Kleman, M., *Mol. Cryst. Liq. Cryst.*, to be published.
88. Achard, M. F., Sigaud, G., Hardouin, F., Weill, C. and Finkelmann, H., *Mol. Cryst. Liq. Cryst. Lett.*, 1983, **92**, 111.

89. Sigaud, G., Yoon Do, Y. and Griffin, A. C., *Macromolecules*, 1983, **16**, 875.
90. Finkelmann, H., Benthack, H. and Rehage, G., *J. de Chimie Phys.*, 1983, **80**, 163.
91. Volino, F., Martins, A. F., Blumstein, R. B. and Blumstein, A., *C.R. Acad. Sci.*, 1981, **292**, II-829; *J. Phys.* (*Lett.*) *Paris*, 1981, **42**, L-305.
92. Volino, F. and Blumstein, R. B., *4th Winter Conference on Liquid Crystals of Low-Dimensional Order and Their Applications*, Bovec, 26–30 March 1984, *Mol. Cryst. Liq. Cryst.*, to be published.
93. Blumstein, R. B., Stickles, M. E., Gauthier, M. M., Blumstein, A. and Volino, F., *Macromolecules*, 1984, **17**, 177.
94. Mueller, K., Hisgen, B., Ringsdorf, H., Lenz, R. W. and Kothe, G., this volume, pp. 223–32.
95. Boeffel, Ch., Hisgen, B., Pschorn, U., Ringsdorf, H. and Spiess, H. W., *Israel J. Chem.*, 1983, **23**, 388.
96. Piskunov, M. V., Kostromin, S. G., Stroganov, L. B., Shibaev, V. P. and Platé, N. A., *Makromol. Chem. Rapid. Commun.*, 1982, **3**, 443.
97. Wassmer, K. H., Ohmes, E., Kothe, G., Portugall, M. and Ringsdorf, H., *Makromol. Chem., Rapid. Commun.*, 1982, **3**, 281.
98. Mueller, K., Wassmer, K. H., Lenz, R. W. and Kothe, G., *J. Polym. Sci., Polym. Lett. Ed.*, 1983, **21**, 785.
99. Ringsdorf, H., Schmidt, H. W., Baur, G. and Kiefer, R., *Polym. Prepr. Am. Chem. Soc., Div. Polym. Chem.*, 1983, **24**, 306.
100. Talroze, R. V., Shibaev, V. P., Sinitzyn, V. V., Platé, N. A. and Lomonosov, M. V., *Polym. Prepr. Am. Chem. Soc., Div. Polym. Chem.*, 1983, **24**, 309.
101. Ronca, G. and Yoon, D. Y., *J. Chem. Phys.*, 1982, **76**, 3295.
102. Wendorff, J. H., in *Liquid Crystalline Order in Polymers*, Blumstein, A. (Ed.), Academic Press, New York, 1978.
103. Wendorff, J. H., Finkelmann, H. and Ringsdorf, H., *Am. Chem. Soc. Polym. Prepr.*, 1978, **74**, 12.
104. Clough, S. B., Blumstein, A. and de Vries, A., *Am. Chem. Soc. Polym. Prepr.*, 1978, **74**, 1.
105. Blumstein, A. and Hsu, E. C., in *Liquid Crystalline Order in Polymers*, Blumstein, A. (Ed.), Academic Press, New York, 1978.
106. Shibaev, V. P. and Platé, N. A., *Polym. Sci. USSR*, 1977, 1605.
107. Tsukruk, V. V., Shilov, V. V. and Lipatov, Yu, S., *Makromol. Chem.*, 1982, **183**, 2009; *Eur. Polym. J.*, 1983, **19**, 199; *Polym. Commun.*, 1983, **24**, 75, 260.
108. Tsukruk, V. V., Shilov, V. V., Konstantinov, I. I., Lipatov, Yu. S. and Amerik, Yu. B., *Eur. Polym. J.*, 1982, **18**, 1015.
109. Frosini, V., Levita, G., Lupinacci, D. and Magagnini, P. L., *Mol. Cryst. Liq. Cryst.*, 1981, **66**, 21.
110. Shibaev, V. P., Platé, N. A. and Freidzon, Ya S., *J. Polym. Sci., Polym. Chem. Ed.*, 1979, **17**, 1655.
111. Magagnini, P. L., *Makromol. Chem. Suppl.*, 1981, **4**, 223.
112. Newman, B. A., Frosini, V. and Magagnini, P. L., *Am. Chem. Soc. Polym. Prepr.*, 1978, **74**, 71.
113. Osada, Y. and Blumstein, A., *J. Polym. Sci., Polym. Lett. Ed.*, 1977, **15**, 761.
114. Blumstein, A., Osada, Y., Clough, S. B., Hsu, E. C. and Blumstein, R. B., *Am. Chem. Soc. Polym. Prepr.*, 1978, **74**, 56.

115. Blumstein, A., Blumstein, R. B., Clough, S. B. and Hsu, E. C., *Macromolecules*, 1975, **8**, 73.
116. Shibaev, V. P., Talroze, R. V., Karakhanova, F. I. and Platé, N. A., *J. Polym. Sci., Polym. Chem. Ed.*, 1979, **17**, 1671.
117. Paleos, C. M., Margomenou-Léonidopoulou, G., Filippakis, S. E. and Malliaris, A., *J. Polym. Sci., Polym. Chem. Ed.*, 1982, **20**, 2267.
118. Benattar, J. J., Moussa, F. and Lambert, M., *J. de Chimie Phys.*, 1983, **80**, 99.
119. Leadbetter, A. J., in *The Molecular Physics of Liquid Crystals*, Luckhurst, G. R. and Gray, G. W. (Eds), Academic Press, New York, 1979.
120. Doucet, J., in *The Molecular Physics of Liquid Crystals*, Luckhurst, G. R. and Gray, G. W. (Eds), Academic Press, New York, 1979.
121. Blumstein, A., Sivaramakrishnan, K. N., Blumstein, R. B. and Clough, S. B., *Polymer*, 1982, **23**, 47.
122. Roviello, A. and Sirigu, A., (a) *Eur. Polym. J.*, 1979, **15**. 61; (b) *Makromol. Chem.*, 1980, **181**, 1799; (c) *Gazz. Chim. Ital.*, 1980, **110**, 403.
123. Iannelli, P., Roviello, A. and Sirigu, A., *Eur. Polym. J.*, 1982, **18**, 745, 753.
124. Frosini, V., Marchetti, A. and de Petris, S., *Makromol. Chem., Rapid Commun.*, 1982, **3**, 795.
125. Thierry, A., Skoulios, A., Lang, G. and Forestier, S., *Mol. Cryst. Liq. Cryst. Lett.*, 1978, **41**, 125.
126. Usha Deniz, K., Paranjpe, A. S., Amirthalingam, V. and Muralidharan, K. V., in *Liquid Crystals*, Chandrasekhar, S. (Ed.), Heyden, London, 1980.
127. Vainshtein, B. K., *Diffraction of X-rays by Chain Molecules*, Elsevier, Amsterdam, 1966.
128. Bear, R. S. and Hugo, H. F., *Ann. NY Acad. Sci.*, 1951, **53**, 627.
129. Leadbetter, A. J. and Norris, E. K., *Molec. Phys.*, 1979, **38**, 669.

10

OBSERVATIONS ON THE RHEOLOGY OF THERMOTROPIC POLYMER LIQUID CRYSTALS

F. N. COGSWELL

Imperial Chemical Industries plc,
Research and Technology Department, Wilton, UK

INTRODUCTION

Rheology, or the study of deformation and flow, represents the linking stage between polymerization and application: it is concerned with the manufacture of articles and their properties. Liquid crystallinity is a rheological phenomenon; it is observed as a result of a flow process producing persistent anisotropy in the fluid, and that anisotropy determines the property spectrum of the product. Thus an appreciation of rheological phenomena is doubly important in the study of liquid crystal polymers.

The fundamental aspects of the rheology of polymer liquid crystals has been admirably covered by Wissbrun[1,2] including a very detailed review article in 1981. Rather than retrace that ground this paper will make a pragmatic evaluation of the evidence from rheological behaviour and its implications for the exploitation of these materials in an industrial setting, for, today, these materials are poised to become new members of the engineering plastics' fraternity.

In the search for new polymer systems there is a general inverse relationship between ease of processing and service properties. This is not a scientific law, simply an expression of commercial experience—it is easy to make a polymer which cannot be processed and has no useful properties. However, if we redefine processability as ease of shaping and serviceability as retention of shape, that inverse relationship appears a natural pattern obeyed by: molecules of different stiffness or glass transition temperature, molecules of different molecular weight, the addition of fillers, systems having different crystallinity, modification by plasticisers or the addition of orientation. However, Professor Flory's theories[3] have indicated how the combination of high molecular weight, high chain stiffness, and rod-like molecules allow us to achieve materials having a persistent 'liquid crystalline' order combining ease of shaping in the fluid state with very high stiffness in the solid state.

The exploitation of lyotropic liquid crystalline order by Kwolek[4] to process 'Kevlar' fibres established a principle—but one which was retained in the hands of specialized fibre producers. The extension from lyotropic to thermotropic polymers by Jackson and Kuhfuss[5] opened the field to conventional thermoplastics processing. The variability of thermoplastics processing technology takes one into products of complex shapes and more complex flow histories and so an even greater interdependence between flow history, morphology and properties.

THE STATE OF THE ART

First, we can confirm the expectation of easy processability by demonstrating that liquid crystalline polymer melts can be easily moulded into large complex shapes (Fig. 1).

Secondly, the stiffness and strength of such mouldings compare favourably with existing commercial thermoplastics including fibre-reinforced thermoplastics. Further, the properties of such mouldings are highly anisotropic (Table 1) and, as shown by the effect of thickness, clearly depend on the flow history. Finally, even in filling relatively simple shapes, the flow processes are complex so that the microstructure, which is revealed in a fractured moulding, is highly complex (Fig. 2).

Liquid crystal, or self-reinforcing polymers, have the property spectrum to find a position in the field of engineering plastics—plastics designed to carry structural loads. For this field of materials it is not sufficient to

TABLE 1

Modulus ($GN\,m^{-2}$) of Injection Mouldings of Liquid Crystalline Thermoplastic Polyester at 23°C (Typical Values)

Thickness (mm)	*Parallel to flow direction*	*At 45° to flow direction*	*Across flow direction*
0·5	17	8	3
1	16	10	3
2	15	10	4
3	14	7	4

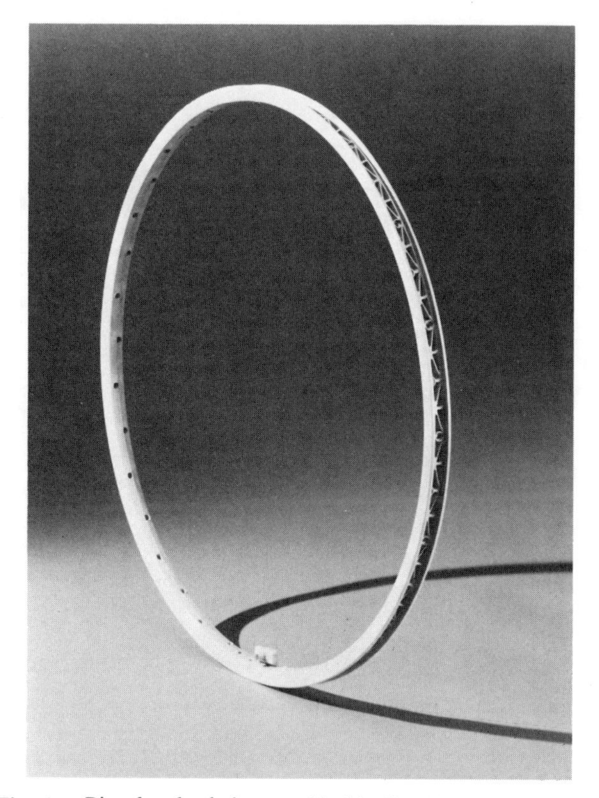

Fig. 1. Bicycle wheel rim moulded in liquid crystal polymer.

demonstrate the ability to make shapes which can have useful properties: we must determine that these properties can be controlled and reproduced.

In the field of conventional engineering thermoplastics we have a detailed understanding of the isotropic state, we appreciate the stress history deployed in a moulding process, we can measure relaxation phenomena and so predict residual orientation, and so we can deduce the property spectrum of a final product. If the time-scale between recognition of liquid crystalline phenomena in melts and its commercial exploitation appears protracted we need only note the observation of Professor J. L. White summing up at a recent conference in Kyoto,[6] that, for liquid crystal polymers, stress history, optical anisotropy and texture are independent variables. In fact, to make the connection from basic material property to performance in the final product, industrial technologists have had to learn a new science.

Fig. 2. Fracture morphology of liquid crystal moulding showing lamella texture.

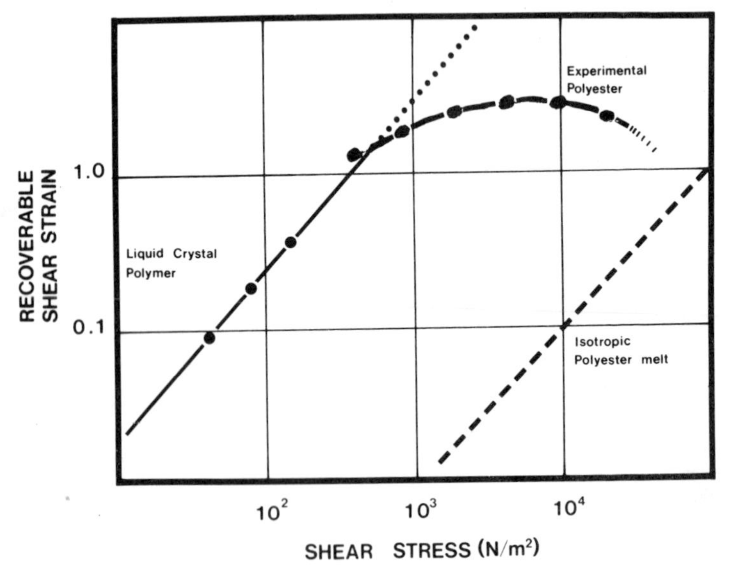

Fig. 3. Elastic response of polyester melts.

RHEOLOGY OF THE NEAR-QUIESCENT STATE

Following the arguments of Wissbrun[1] we may distinguish three deviations in the flow behaviour of liquid crystal polymers in comparison with conventional melts: a high elastic response to small amplitude oscillations but the absence of gross elastic effects such as post-extrusion swelling; a 'three-zone' flow curve incorporating a 'yield stress' a 'pseudo-Newtonian' region followed by 'shear thinning'; and, a strong dependence of rheology on thermomechanical history.

To study the elastic response in the 'near-quiescent' state we used a constant stress cone and plate rheometer to measure the recoverable shear strain after steady flow had been achieved. For an experimental liquid crystal polyester the elastic response at low stress is very large suggesting an elastic modulus of the order 400 N m^{-2}—at least two orders of magnitude lower than a chemically similar polyester melt which does not exhibit liquid crystal phenomena (Fig. 3). However, at a shear stress of about 10^3 N m^{-2}—at which the conventional polyester is just starting to demonstrate significant elastic response—the high elasticity of the liquid crystal polyester collapses.

The idealized 'three region' flow curve has many variants: I prefer to represent a two-region response (Fig. 4). The drawing of lines through experimental data is a subjective process; however this plot is substantiated by observing the response to a preset strain rate in the 'transition' region (Fig. 5). The results of this experiment are qualitatively reproducible if the sample is allowed to rest for two hours between tests, suggesting structural changes during flow. These structural changes are amplified by further experimentation. If, after attaining the steady-state response, the flow is stopped and the stress allowed to relax below 200 N m^{-2} (a matter of a few minutes only) and then shearing is restarted, the stress immediately climbs back to its previous equilibrium value (Fig. 6). However if a much higher shear rate is applied—one well into the 'shear thinning' region—followed by a few minutes cessation of flow to allow the stress to relax below 200 N m^{-2}, and the transition shear rate is then reapplied, the stress initially overshoots to a much higher value before returning to its equilibrium value (Fig. 7).

These data suggest that for one imposed shear rate there is a range of possible rheological response dependent upon the history of the material. It may be significant to note that the 'transition' stress is of the same magnitude as that at which the elastic response collapses.

These observations of rheological response appear to be associated with

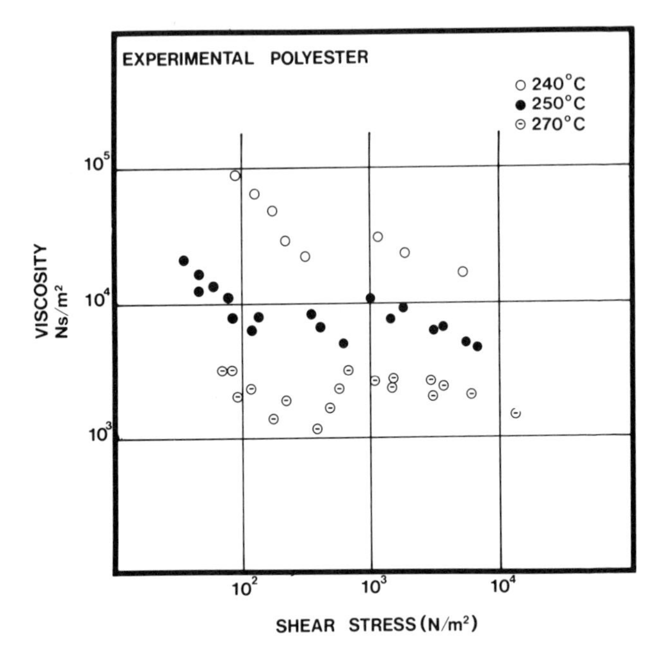

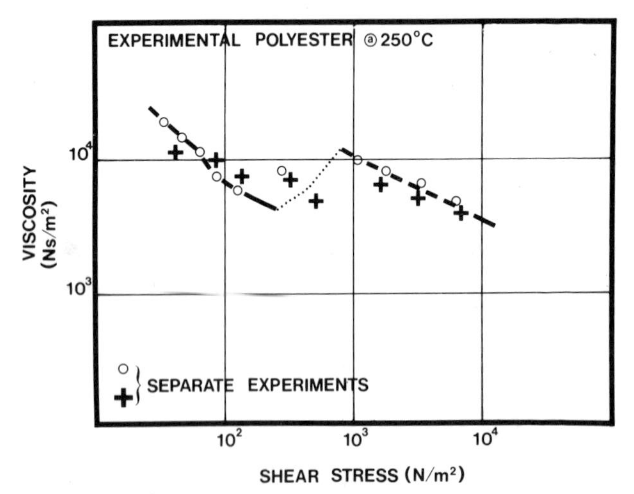

Fig. 4. Viscous flow of liquid crystal polyester melts.

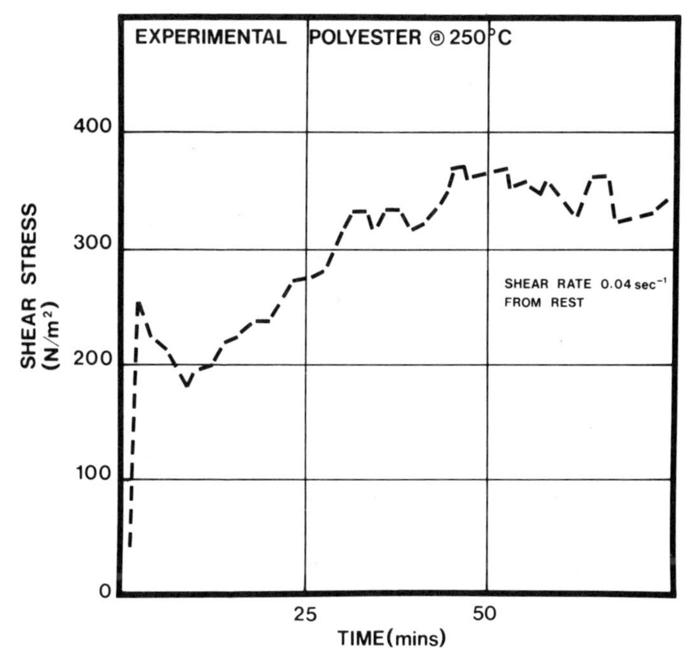

Fig. 5. Stress build-up during steady shear.

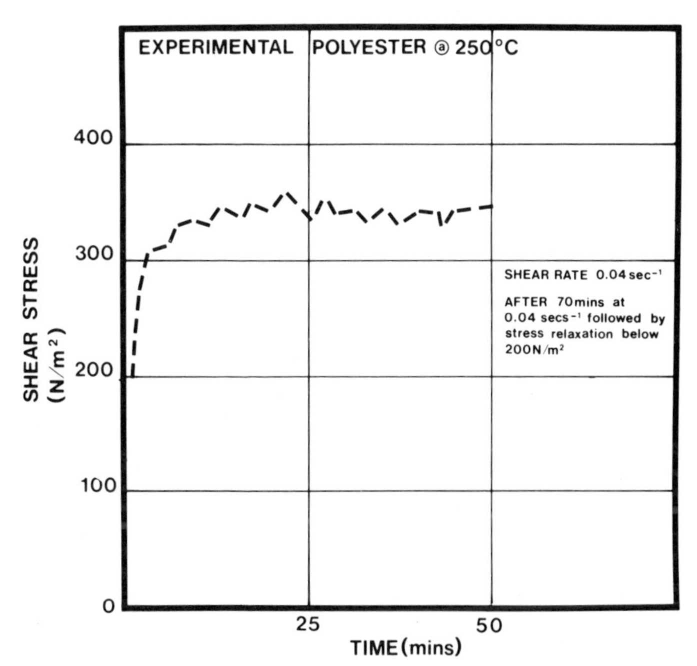

Fig. 6. Stress build-up after interrupted flow.

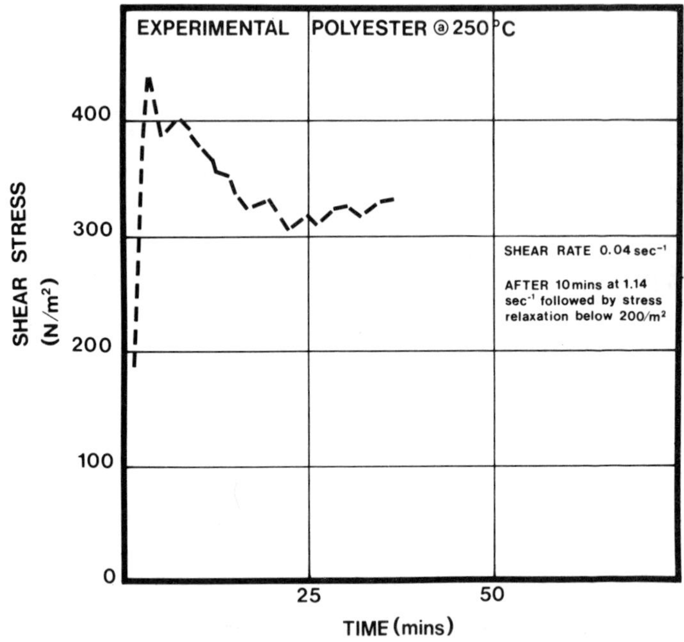

Fig. 7. Stress build-up and decay after intensive shear history.

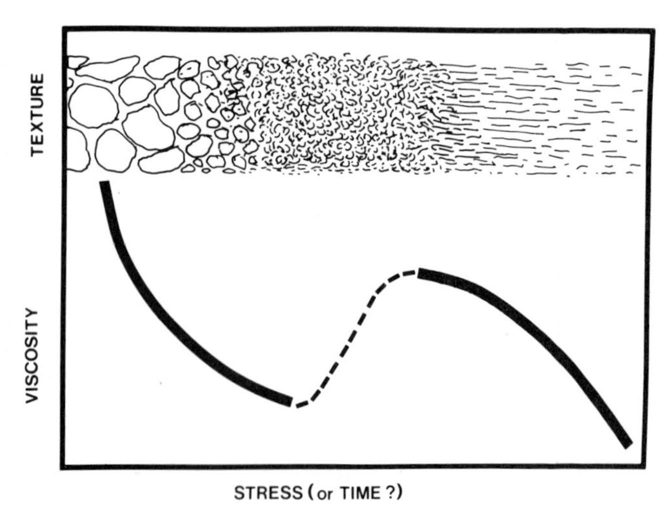

Fig. 8. Relationship between morphology and rheology.

textural changes. On the same family of experimental polyesters Mackley and Graziano[7] observed that as strain rate is increased the microscopic texture of liquid crystal polyester melts changes from a 'droplet' form to a 'vat of seething worms'. It would be reasonable to suppose that the application of a flow field would increase the organization and, thus, the homogeneity of a liquid crystal polymer. The evidence is contrary, and substantiates results presented by Horio for lyotropic systems;[8] it suggests that there is initially a considerable multiplication of nemata possibly followed by a clearing again at very high stress levels.

The evidence from the near-quiescent state suggests an emulsion texture with a yield stress followed by a region of low viscosity, but, that, as the domains are broken down into smaller sizes (having higher surface areas) the resistance to flow increases followed, perhaps, by a final homogenization of the structure and lower viscosity (Fig. 8). These results convey some of the difficulty of defining the rheological state and structure of a liquid crystal polymer in the near-quiescent state. It is that state which is the beginning of its history in a processing operation and it is the state to which it returns at the end of such a flow.

FLOW DURING PROCESSING

The most significant deformations during processing are extensional and simple shear flows at high strain rate. Stretching, or irrotational, flows are

especially potent in inducing high molecular orientation, but note that this set includes not only the obvious free surfaces phenomena like drawing during fibre spinning but also less apparent effects such as converging flow and hidden phenomena like the fountain flow at the advancing front of an injection moulding process.[9]

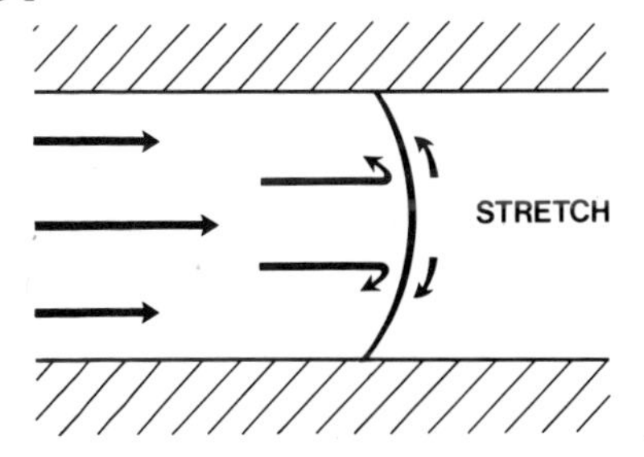

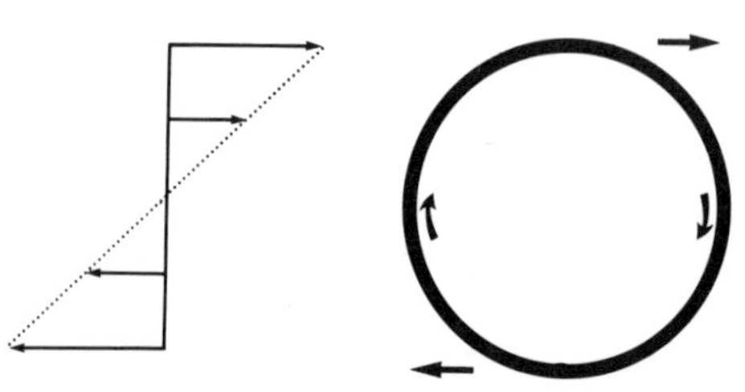

By contrast, so-called 'simple' shear which dominates extrusion or injection moulding possesses a rotational component. Horio[8] surmises that this induces a tumbling of the domains and we have already noted how it appears to lead to a multiplication of nemata. Thus, simple shear, of itself, should not be expected to induce especially high orientation in the melt, but, it has been noted that for the class of simple shear flows produced by relatively moving surfaces the 'homogenizing' influence of very high shear rates can induce persistent order into a melt, or solution, which does not naturally possess it.

Intensive pre-shearing to produce a state of persistent order in a fluid which does not naturally possess it has been shown to be of considerable benefit in easing the processing of both lyotropic and thermotropic systems[10] and identical phenomena are noted for mesophase pitch. Because of the characterization of liquid crystal polymers by a three-zone flow curve including a 'yield stress' it is reasonable to ask if a true *liquid* crystal polymer ever exists at rest. The evidence suggests that rheological history is a critical factor in determining the morphology of the material.

Finally, all thermoplastic processes involve a thermal history and in 'crystalline' polymers the notion of precise temperature transitions must be treated with reservation. As we heat up a 'liquid crystal' polyester we see a phase change from a paste-like material to one which is readily oriented. As with 'solid crystalline' structure the transition temperature is not unique; it may be raised by annealing or depressed by supercooling[1,10] so that the 'fluidity' of a melt at a particular temperature can vary a thousand-fold.[10]

The effect of this rheological complexity is two-fold: it offers the process designer unprecedented scope to employ novel technology, and it is reflected in the mechanical property spectrum of the product. In the long term these two aspects must be considered in parallel: new highly controlled processing technology to achieve enhanced products.

CONCLUSIONS

Thermotropic liquid crystalline polyesters offer an attractive balance of properties as engineering polymers including resistance to chemical attack and to burning, but the most dramatic properties derive from high molecular orientation giving outstanding stiffness per unit weight in a preferred orientation.

A less dramatic 'mechanical' property, associated with intense molecular orientation, is the low coefficient of thermal expansion: fully extended chains have nowhere to go and products made from such resins demonstrate good dimensional tolerance. In a material of extremes it may be this 'zero' effect which will prove a major spur to their commercial development. In an age of high technology, precision is a key statement.

Both the gross mechanical properties and the absence of thermal expansion effects are intimately associated with all levels of structure within the material. Those structures, themselves, depend on the rheological history of the product and understanding them is central to the commercial development of this class of polymer.

ACKNOWLEDGEMENTS

I am especially indebted to my colleagues of the Liquid Crystal Research Team at ICI, in particular Dr B. P. Griffin and Mr W. T. Parkes.

REFERENCES

1. Wissbrun, K. F., *J. Rheol.*, 1981, **25**(6), 619.
2. Wissbrun, K. F. and Griffin, A. C., *J. Polym. Sci., Polym. Phys. Ed.*, 1982, **20**, 1835.
3. Flory, P. J., *Proc. Roy. Soc. London, Ser. A*, 1956, **234**, 73.
4. Kwolek, S. L., Morgan, P. W., Shaefgen, J. R. and Gulrick, C. W., *Macromolecules*, 1977, **10**(6), 1390.
5. Jackson, W. J. and Kuhfuss, H. F., *J. Polym. Sci., Polym. Chem. Ed.*, 1976, **14**, 2043.
6. Japan–US Symposium on Polymer Liquid Crystals, Kyoto, 1983.
7. Graziano, D., Ph.D. Thesis, Cambridge University (1982), supervised by M. R. Mackley, Department of Chemical Engineering.
8. Horio, L., Japan–US Symposium on Polymer Liquid Crystals, Kyoto, 1983.
9. Cogswell, F. N., *Polymer Melt Rheology: A Guide for Industrial Practice*, George Godwin Ltd, London, 1981.
10. Cogswell, F. N., *Br. Polym. J.*, 1980, **12**, 170–3.

11

RHEO-OPTICAL STUDIES OF THE THERMOTROPIC AROMATIC COPOLYESTERS OF POLY(ETHYLENE TEREPHTHALATE) AND *p*-ACETOXYBENZOIC ACID

ANAGNOSTIS E. ZACHARIADES and JOHN A. LOGAN

IBM Research Laboratory, San Jose, California, USA

INTRODUCTION

Many studies have been conducted for the development of polymers with ultra-high modulus and strength. Although most of the work has been on the preparation of highly oriented and extended morphologies using conventional polymers such as high density polyethylene, recently the work has been extended into the processing of rigid polymers—that is polymers with rigid backbone chains.[1-4] Such polymers can be processed in solution, however, under conditions which are far from being attractive. In trying to make a more readily processable rigid polymer, that is, a polymer which will not decompose before melting or dissolve only in highly corrosive solvents, Economy and co-workers[5,6] and Jackson and Kuhfuss[7] synthesized a compositional series of aromatic copolyesters of *p*-hydroxybenzoic acid and biphenyl terephthalate and *p*-acetoxybenzoic acid and poly(ethylene terephthalate).

Attempts to process both types of polymers have been reported previously. Economy *et al.*[8] have used a forging process to prepare biaxially oriented structures of *p*-hydroxybenzoic acid and biphenyl terephthalate and Jackson and co-workers[4] have used injection moulding to obtain high modular specimens with the aromatic copolyesters of *p*-acetoxybenzoic acid and poly(ethylene terephthalate). The high modulus and strength of the injection-moulded specimen correlates with the high degree of chain orientation and extension achieved during the flow of the shear-sensitive thermotropic melt in the mould cavity. In our laboratory, we have investigated both groups of copolyesters for their processability in the melt state and particularly for producing multiaxially oriented morphologies using curvilinear flow conditions under compression.[9,10]

During our studies with the aromatic copolyesters of poly(ethylene terephthalate) with 60 and 80 mole % of *p*-acetoxybenzoic acid, we conducted, concurrently, rheo-optical observations which elucidate their unusual flow behaviour.

EXPERIMENTAL

Materials

The copolyesters of poly(ethylene terephthalate) (PET) with 60 and 80 mole % of *p*-acetoxybenzoic acid (PHBA) were kindly provided by the Tennessee Eastman Company.

Methods of Preparation

The optical and rheo-optical studies were conducted with a Zeiss photomicroscope using polarized light. The rheo-optical studies were conducted on an optical plate–plate rheometer which was described previously.[9] Oriented morphologies of two copolyesters were obtained by torsional shearing under modest compression (≤ 5 atm) at a shear rate of $2\text{–}3 \times 10^2\,\mathrm{s}^{-1}$ (at the periphery of the disc plates). The polymer was prevented from flowing outside the plates by a barrier wall around the plates. Rheological data were obtained with a capillary Instron rheometer using a capillary die ($L/D = 39{\cdot}68$). The measurements were made using well-known methods described by Jerman and Baird.[11]

The internal morphology of the copolyesters was examined with microtomed specimens ($\lesssim 1000$ Å) using a Philips 301 electron microscope equipped with a S(TEM) attachment.

The mechanical properties were determined with ribbon specimen cut along the flow line using a custom-built microtensile tester. The Young's modulus was measured at a strain rate of $1{\cdot}8 \times 10^{-3}\,\mathrm{s}^{-1}$ and the tensile strength is the average value of three tests.

RESULTS

The optical and electron microscopy examination shows that the as-received copolyesters are distinctly different at ambient temperature. The 40/60 (PET/PHBA) copolyester composition appears structureless whereas the 20/80 (PET/PHBA) copolyester composition shows, in addition to a structureless phase, the presence of rather large PHBA

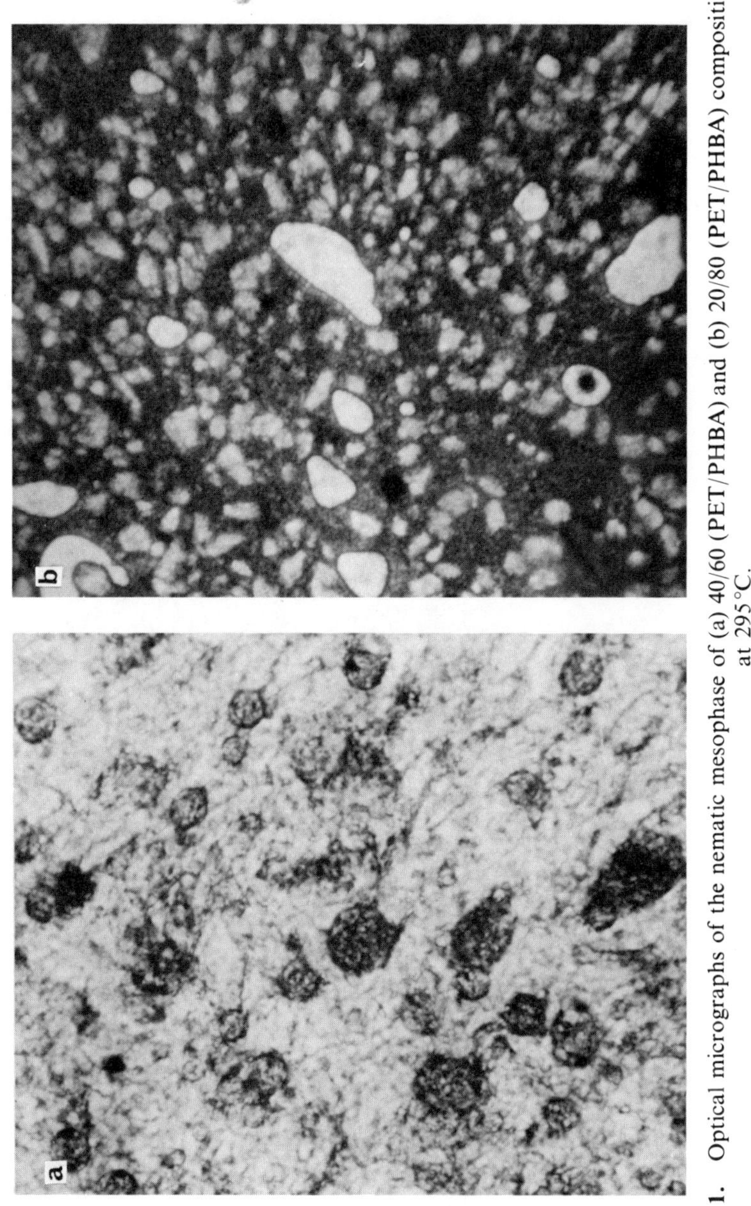

Fig. 1. Optical micrographs of the nematic mesophase of (a) 40/60 (PET/PHBA) and (b) 20/80 (PET/PHBA) composition at 295 °C.

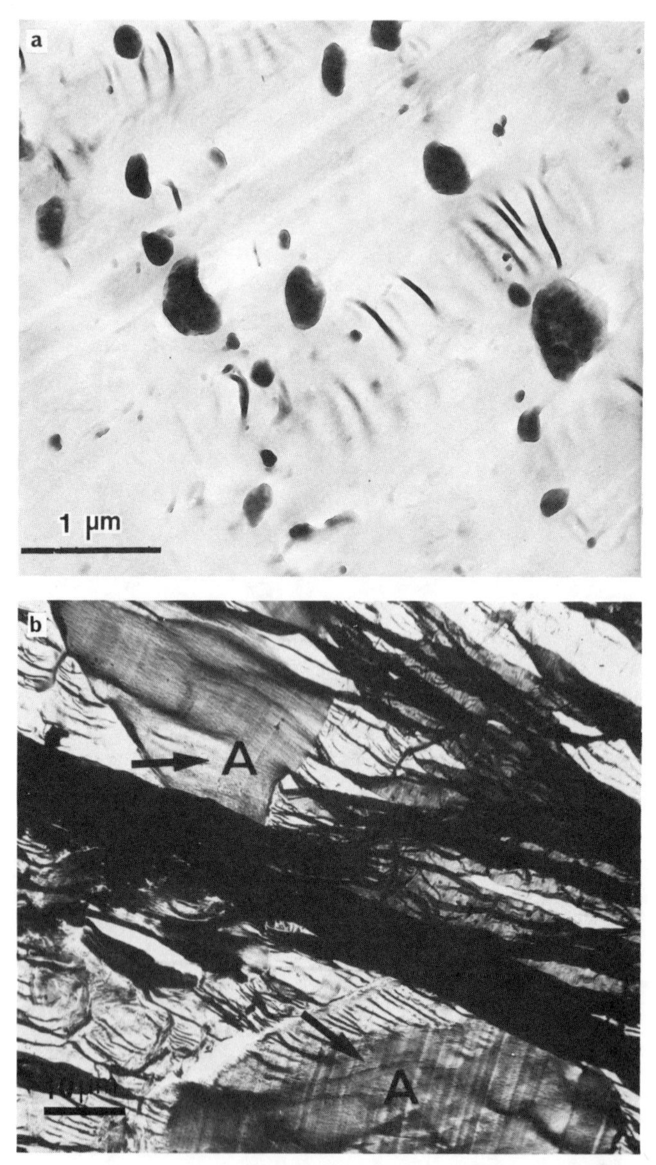

Fig. 2. Transmission electron micrographs of the PHBA domains in the (a) 40/60 (PET/PHBA) and (b) 20/80 (PET/PHBA) copolyester compositions (domains indicated by arrows).

crystalline domains (10–50 μm). Upon heating to 295 °C, both copolyesters formed nematic mesophases which were different in that the mesophase of the 40/60 copolyester composition had small spherical domains (1–2 μm in diameter) included in the phase whereas the mesophase of the 20/80 (PET/PHBA) composition had large PHBA domains which were discernible also in the as-received copolyester at ambient temperature (Fig. 1). The spherical domains in the mesophase of the 40/60 (PET/PHBA) composition disappeared upon heating at ~310 °C, whereas the PHBA domains in the 20/80 (PET/PHBA) composition remained discernible up to 328 °C when they fused in the surrounding mesophase. The transmission electron micrographs in Fig. 2 indicate the presence of domains in both the as-received copolyester compositions. However, the domains in the 40/60 copolyester composition are small (0·1–0·8 μm) and structureless, whereas the domains in the 20/80 composition are large (10–50 μm) and exhibit a high degree of lamellar order. The mesophase of the 40/60 (PET/PHBA)

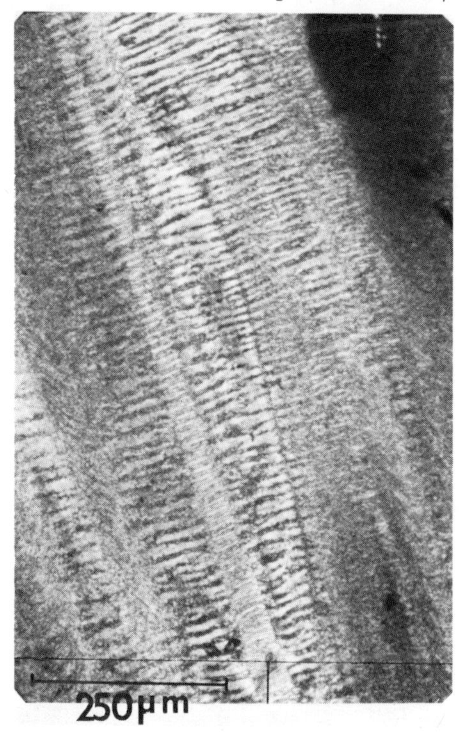

Fig. 3. Optical micrograph of ordered nematic mesophase of 40/60 (PET/PHBA) obtained by shearing at 290 °C and cooling to ambient temperature.

copolyester exhibited a uniformly distributed ordered band structure perpendicular to the flow direction when it was sheared in the plate–plate rheometer at 270–300 °C and subsequently cooled rapidly to ambient temperature (Fig. 3).

A similar ordered structure was obtained also by shearing the mesophase of the 20/80 copolyester composition which was heated at 330 °C but in contrast to the 40/60 composition, it was not uniformly distributed in the sample. The viscosity of the mesophase of the 20/80 copolyester composition at 275 °C was at least one order of magnitude higher than the viscosity of the mesophase of the 40/60 composition at the same temperature and was reduced significantly by heating the polymer at 330 °C, as shown in Fig. 4. The mechanical properties of the oriented films from both copolyester compositions are summarized in Table 1 and suggest

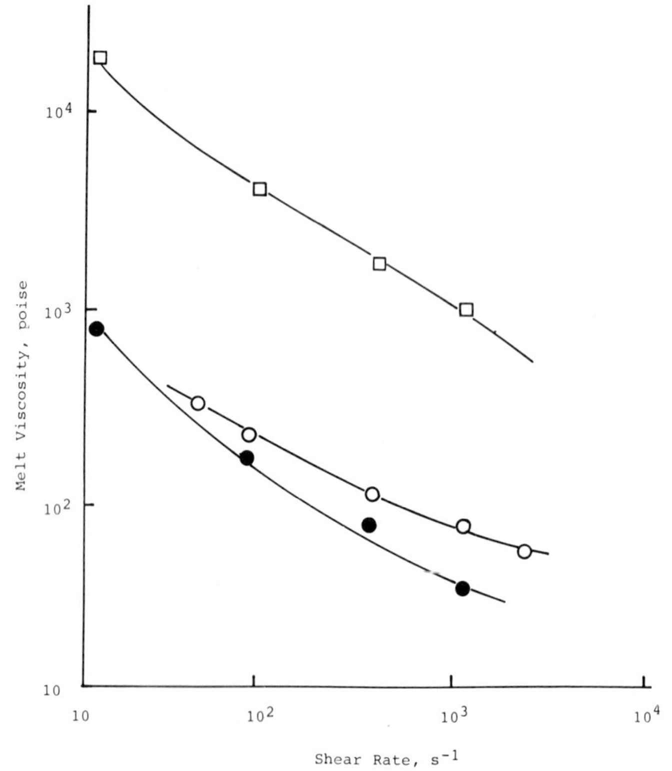

Fig. 4. Viscosity versus shear rate at 275 °C. □, PET/PHBA (20/80), 275 °C; ○, PET/PHBA (20/80), 330 °C; ●, PET/PHBA (40/60), 275 °C.

TABLE 1
Mechanical Properties of Oriented Films of the PET/PHBA Copolyesters

Copolyester sample	*Young's modulus (GPa)*	*Tensile strength (MPa)*	*Elongation at break (%)*
40/60	12	100–110	8
20/80 (prepared from melt <328 °C)	4·5	95–110	5
20/80 (prepared from melt >328 °C)	19	207–226	9

a strong dependence of the modulus and strength on the thermal history of the initial mesophase before shearing.

DISCUSSION

Studies on the melt behaviour of the aromatic copolyesters of poly(ethylene terephthalate) and *p*-acetoxybenzoic acid have been reported previously. Baird and co-workers[11,12] have examined the rheological properties of the 40/60 and 20/80 (PET/PHBA) copolyester compositions over a wide temperature range and observed that the melt viscosity of both copolyester compositions was significantly lower than that of poly(ethylene terephthalate) and dependent on the composition of the copolyester. Wissbrun[13] has also studied the flow properties of liquid-crystalline behaving polymers and observed that the flow curves of the 40/60 (PET/PHBA) copolyester composition are very sensitive to the temperature of the melt and its thermal history. In our studies[14] we observed that the 20/80 (PET/PHBA) copolyester is a block copolymer in which the PHBA blocks aggregate in crystalline domains in contrast to PET/PHBA copolyesters with lower fractions of PHBA which are reported to be random copolymers.[4,7]

The electron micrographs in Fig. 2 indicate that PHBA domains are formed also in the 40/60 (PET/PHBA) composition suggesting that the polymerization may become heterogeneous in the latter stage when the aromatic PHBA component of the copolyester is greater than 50 mole %. However, the PHBA domains in the 40/60 (PET/PHBA) copolyester are small and presumably are composed of short homopolymer segments which are not long enough to crystallize as in the 20/80 (PET/PHBA) copolyester. On the contrary, the crystalline PHBA domains in the 20/80 (PET/PHBA) copolyester are temperature-stable up to 328 °C and therefore act like fillers in the mesophase at temperatures below 328 °C, thus resulting in higher melt viscosities in comparison to the mesophase of the 40/60 (PET/PHBA) copolyester composition at the same temperature (Fig. 4). As the PHBA crystalline aggregates break down to smaller crystals at higher temperatures, the melt viscosity of the mesophase is reduced and when the PHBA domains fuse completely at 330 °C the viscosity approaches the viscosity values of the 40/60 (PET/PHBA) copolyester at 275 °C. It is worth mentioning that a similar relation between the size and structure of the PHBA domains and its concentration in the copolymer has been observed for the aromatic copolyesters of biphenyl terephthalate and *p*-hydroxybenzoic acid.[6]

The development of ordered bands perpendicular to the shear direction in the oriented film of both copolyester compositions is an optical effect observed with liquid-crystalline behaving polymers under mechanical stress conditions and appears to relate to the electro-optical effects observed in nematic liquid crystals. The non-uniform distribution of shear bands in the oriented film of the 20/80 copolyester composition is due to the inhomogeneity of its mesophase which is composed of rich PET and PHBA areas because no shear bands are observed readily in oriented film which is prepared below 310–320 °C, that is when the PHBA domains are not fused.[9]

The mechanical properties appear to be dependent on the thermal history of the initial mesophase before shearing. Thus, the 40/60 (PET/PHBA) copolyester composition when heated to ~280–300 °C, prior to the shearing and cooling experiments, results in oriented and extended chain morphologies with high modulus and strength. Also, the oriented morphologies of the 20/80 (PET/PHBA) copolyester composition, which was heated to 330 °C before shearing, have high modulus and strength. However, oriented morphologies with significantly lower modulus and strength were obtained when the copolyester was heated to <330 °C apparently reflecting the low load bearing of the heterogeneous mesophase in this temperature range.

CONCLUSION

The results of this study suggest that the flow behaviour of the anisotropic melts of the 40/60 and 20/80 (PET/PHBA) copolyesters depends on the structural composition of the copolyester. Although both copolyesters have PHBA domains in their structure, these are too small in the 40/60 copolyester composition to affect noticeably its flow behaviour and the mechanical properties of the oriented morphologies prepared from it. However, the PHBA are large in the 20/80 composition and the effect of the domain size becomes more evident by comparing (a) the flow curves of the copolyester at temperatures below and above the melting point of the domains at 328 °C, and (b) the mechanical properties of the oriented morphologies which are prepared by heating the polymer melt at these temperatures.

REFERENCES

1. Smith, P. and Lemstra, P. J., *J. Mater. Sci.*, 1980, **15**, 505.
2. Kanamoto, T., Tsuruta, A., Tanaka, K., Takeda, M. and Porter, R. S., *Polymer J.*, 1983, **15**, 327.
3. Zachariades, A. E., Mead, W. T. and Porter, R. S., *Chem. Rev.*, 1980, **80**, 351.
4. Cifferi, A. and Ward, I. M. (Eds), *Ultra-High Modulus Polymers*, Applied Science Publishers, Essex, England, 1979.
5. Economy, J., Storms, R. S., Matkovich, V. I., Cottis, S. G. and Nowak, G. E., *J. Polym. Sci., Polym. Chem. Ed.*, 1976, **14**, 2207.
6. Economy, J. and Volksen, W., in *The Strength and Stiffness of Polymers*, Zachariades, A. E. and Porter, R. S. (Eds), Marcel Dekker, New York, 1983, Chapter 8.
7. Jackson, W. J. Jr and Kuhfuss, H. F., *J. Polym. Sci., Polym. Chem. Ed.*, 1976, **14**, 2043.
8. Economy, J., Novak, B. E. and Cottis, S. C., *Sampe, J.*, 1970, **6**, 6.
9. Zachariades, A. E. and Logan, J. A., *Polym. Eng. Sci.*, 1983, **23**, 5, 747.
10. Zachariades, A. E., unpublished data.
11. Jerman, A. E. and Baird, D. G., *J. Rheol.*, 1983, **25**(2), 275.
12. Baird, D. G. and Wilkes, G. L., *ACS Polymer Reprints*, 1981, **22**, 2.
13. Wissbrun, K. F., *Polymer*, 1980, **163**, 163–9.
14. Zachariades, A. E., Economy, J. and Logan, J. A., *J. Appl. Polym. Sci.*, 1982, **27**, 2009.

12

ELECTRON MICROSCOPY OF THERMOTROPIC COPOLYESTERS

A. M. DONALD* and A. H. WINDLE

Department of Metallurgy and Materials Science, University of Cambridge, UK

1. INTRODUCTION

In the study of small-molecule liquid crystalline materials, light microscopy and x-ray diffraction have been the conventional tools for the investigation of microstructures and orientation. These techniques are also appropriate for the examination of molecular organization in liquid crystalline polymers; however, for thermotropic materials it has also proved possible to use transmission electron microscopy (TEM). There are two main advantages to this technique: first, direct correlation of images with diffraction data can be made, and seondly, the spatial resolution is typically two orders of magnitude greater than in light microscopy. (The ultimate resolution for TEM of polymers is limited not by the instrument, but by the beam sensitivity of the specimen, and thus will vary from one material to another.) TEM, however, demands thin specimens ($\sim$100 nm or less in thickness) which can either be obtained by sectioning bulk material or by directly preparing thin films, e.g. by casting from solution, or smearing from the melt. The first method leads to the possible introduction of artefacts such as those due to mechanical damage, whereas the latter means the structures examined may not be typical of the bulk. Notwithstanding these limitations, TEM has shown its power in providing data on many polymer systems which would otherwise be inaccessible (e.g. Refs 1 and 2).

This chapter describes the recent application of TEM to the study of thin films of several thermotropic random copolyesters. The interpretation of the micrographs obtained requires concepts in addition to those previously established for crystalline and amorphous materials, because liquid

* Present address: Cavendish Laboratory, Madingley Road, Cambridge CB3 0HE.

crystalline polymers have attributes which cannot simply be interpreted as intermediate in character between these two limiting states. It has proved necessary to combine a range of TEM techniques: bright and dark field imaging (BF and DF respectively), selected area electron diffraction (SAED) and microdiffraction (MAED) to understand properly the microstructure of thermotropic polymers.

2. EXPERIMENTAL DETAILS

(a) Polymers Studied

The thermotropic polymers used in these studies are all random copolyesters. As has previously been documented for other types of polymers, the presence of aromatic rings and/or conjugated bonds, and the absence of CH_2—CH_2 and ether linkages yield materials most resistant to beam damage (by processes such as chain scission and cross-linking).[3]

The family of thermotropic copolyesters most extensively studied here comprises the following randomly sequenced segments:

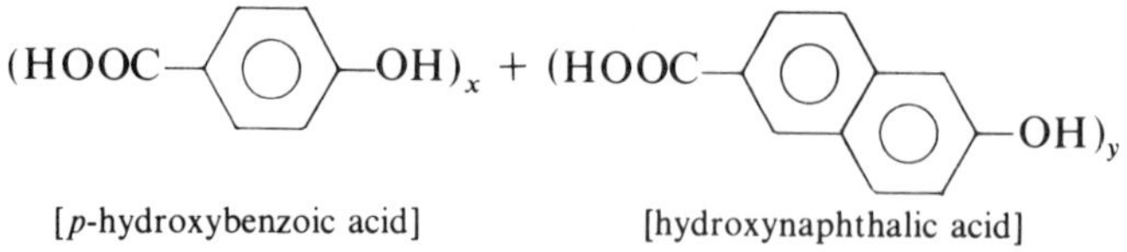

(HOOC—⟨benzene ring⟩—OH)$_x$ + (HOOC—⟨naphthalene ring⟩—OH)$_y$ [B–N]

[*p*-hydroxybenzoic acid] [hydroxynaphthalic acid]

and is designated B–N (x:y). Four different x:y ratios have been examined, with characteristics shown in Table 1. Of these by far the most detailed study has been made of B–N (70:30) which, for convenience, will subsequently be referred to simply as B–N.

A second related polymer, designated N–QT, also exhibited a high beam resistance. It contained the following segments:

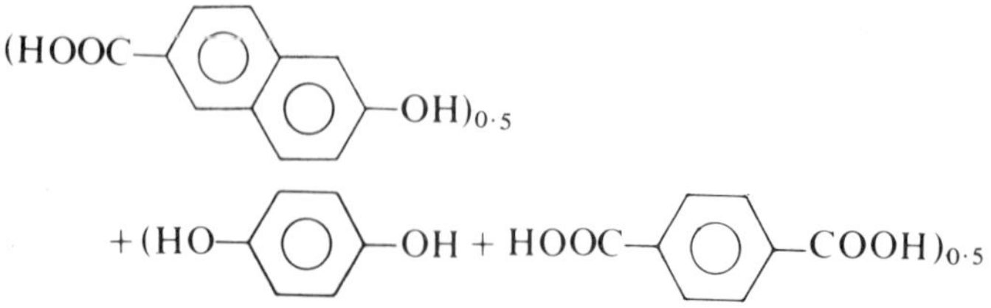

(HOOC—⟨naphthalene ring⟩—OH)$_{0\cdot5}$

+(HO—⟨benzene ring⟩—OH + HOOC—⟨benzene ring⟩—COOH)$_{0\cdot5}$ [N–QT]

In contrast to the above polymers, the copolyester produced by Tennessee Eastman, often referred to as X7G and here designated B–ET, proved rather beam-sensitive. Dark field imaging was not feasible,

TABLE 1
B–N Family of Polymers Examined

$x{:}y$	T_m	IV
70:30	285	5
75:25	290	9·2
58:42	246	5·1
30:70	313	7·8

although it was possible to record a diffraction pattern before significant degradation occurred. This susceptibility can be attributed to the ethylene linkages of the ethylene terephthalate segments, viz:

$$(HOOC{-}C_6H_4{-}OH)_{0\cdot 6}$$

$$+\ (HO{-}(CH_2)_2{-}OH + HOOC{-}C_6H_4{-}COOH)_{0\cdot 4} \quad [\text{B–ET}]$$

Another copolymer studied presented similar problems, so that again only diffraction data and not dark field images could be obtained. It was also a random copolymer, designated C1QT–QG, comprising the units:

$$(HO{-}C_6H_3Cl{-}OH + HOOC{-}C_6H_4{-}COOH)_{0\cdot 5}$$

$$+\ (HO{-}C_6H_4{-}OH$$

$$+\ HOOC{-}C_6H_4{-}O{-}(CH_2)_2{-}O{-}C_6H_4{-}COOH)_{0\cdot 5}$$

$$[\text{ClQT–QG}]$$

(b) Specimen Preparation

As discussed above, for transmission electron microscopy it is necessary to produce specimens in thin film form. In this work the following method has been adopted. A small amount of liquid crystal polymer is placed on a freshly cleaved slice of a rocksalt held at an elevated temperature T_s (typically 10 or 20 °C above the polymer melting point T_m). The polymer is then sheared with a glass slide, to give a thin film, and the crystal and specimen rapidly quenched on a large metal plate which effectively 'freezes-in' the mesophase structure. After quenching the specimen is coated with carbon (to prevent charging problems in the electron microscope), the rocksalt dissolved in distilled water and the polymer film picked up on a copper grid. Some specimens were held at T_s after shearing to anneal the polymer film.

Specimens prepared in this way show a characteristic microstructure due to the process of shear. Thus the structures observed are not identical to those obtained by other techniques such as ultramicrotomy of a bulk sample. Nevertheless, examination of these specimens at the resolution made possible by TEM has yielded much information and provided additional insight into the organization of these semi-rigid polymer molecules.

3. ELECTRON DIFFRACTION AND DIFFRACTION CONTRAST

(a) Selected Area Electron Diffraction (SAED)

A typical SAED pattern obtained from a region 2·5 μm in diameter in an as-sheared specimen of B–N is shown in Fig. 1. The shear axis is vertical and the significant arcing of the reflections indicates a fairly high degree of orientation with this axis. There are no sharp crystalline reflections. A prominent reflection is seen at $s \sim 1{\cdot}4\,\text{Å}^{-1}$ along the shear axis (the 'meridian') and is representative of correlations down the axis of a molecule. Additional weaker reflections are also present on the meridian, together with a very diffuse off-meridional maximum at $s \sim 2{\cdot}5\,\text{Å}^{-1}$ (not apparent in Fig. 1). The equatorial (interchain) scattering consists of a broad arc, with a sharp inner component, and its azimuthal spread gives a measure of the range of orientations present in the specimen. The information provided by SAED is broadly in line with that from x-ray diffraction (see e.g. Ref. 4), and will be discussed further in a subsequent publication.[5]

(b) Dark Field Contrast

The positioning of the objective aperture over a selected part of the equatorial reflection makes it possible to form a dark field image from only those electrons scattered into the aperture. It is therefore possible to identify regions containing molecules correctly oriented for scattering in

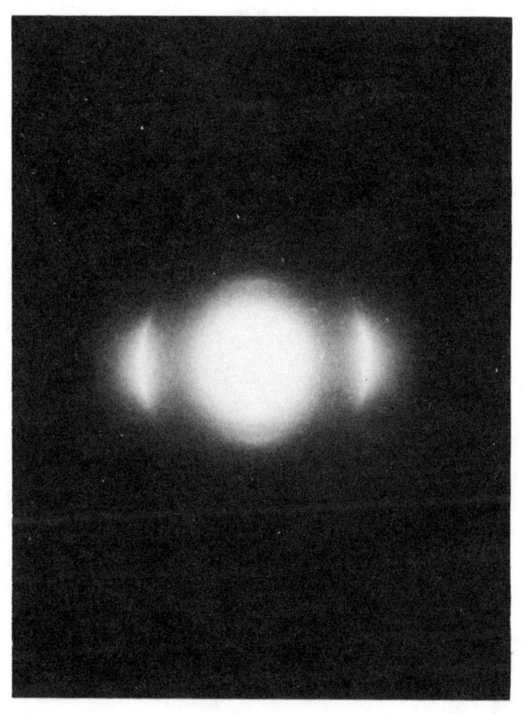

Fig. 1. Selected area electron diffraction pattern from a region 2·5 μm in diameter in an as-sheared specimen of B–N.

this direction. The technique of dark field microscopy thus provides a means for determining the distribution of in-plane molecular orientations.

Figure 2 depicts schematically a diffraction pattern, together with the possible positions of the objective aperture and the nomenclature used to identify the corresponding images. Dark field images cannot usefully be formed in the main meridional reflection, since the orientation-dependent scattering within the arc is convoluted with the molecular transform.[6]

A typical E[W] dark field image of a specimen of B–N, sheared at 300 °C in the direction of the arrow, is shown in Fig. 3(a); Fig. 3(b) shows the image formed in the other wing of the same equatorial reflection. In both micrographs, bands of constant molecular orientation running perpendicular to the direction of shear can be seen. A band which appears bright (strongly scattering) in Fig. 3(a) appears dark in Fig. 3(b), and vice versa. The corresponding bright field image shows virtually no band contrast, which demonstrates that mass thickness variations are not the

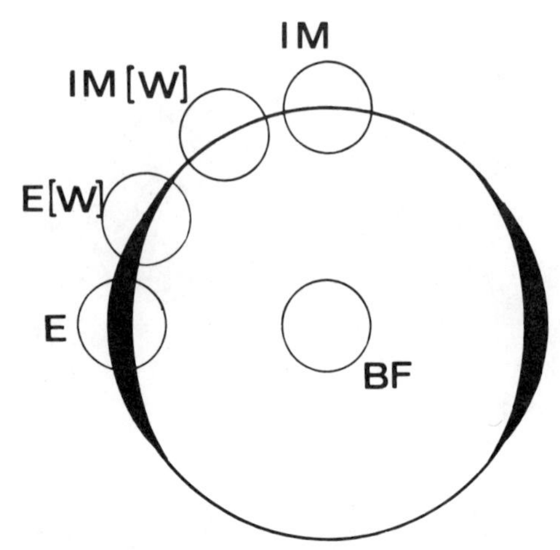

Fig. 2. Schematic representation of an SAED pattern, together with possible positions for the objective aperture. The dark field images so formed will be referred to as: E, equatorial; E[W], equatorial wing; IM, interchain meridian; IM[W], wing of interchain meridian. The bright field image is formed in BF. (After Ref. 14.)

source of the contrast. The bands, where they are sufficiently coarse, can also be seen in the light microscope between crossed polars.[7,8]

Detailed analysis of dark field micrographs such as Figs 3(a) and (b) enables the molecular trajectory underlying the so-called 'banded structure' of as-sheared specimens of B–N to be determined. Figure 4 shows schematically the serpentine trajectory of the molecules obtained from such an analysis.[6] From Fig. 3 it can be seen that the period of the banded structure is 0·8 μm, which is considerably less than the diameter of the diffracting region ($\sim$2·5 μm) from which SAED patterns such as Fig. 1 are usually obtained. This is of importance because it demonstrates that SAED patterns cannot be used in a simple manner to measure the degree of local correlation of molecular orientation, which is often referred to as the order parameter. The advantages of micro-area electron diffraction (MAED) patterns in this context will be discussed in Section 5. It should also be borne in mind that x-ray diffraction patterns, typically obtained from regions a few tenths of a millimetre across, again represent a global average over all molecular parameters. Local order parameters may therefore be much higher than estimated simply from the angular extent of an arc in an x-ray pattern.

Dark field images, for example Fig. 3, can only be obtained for a beam-resistant polymer such as B–N. It is, however, possible to record the

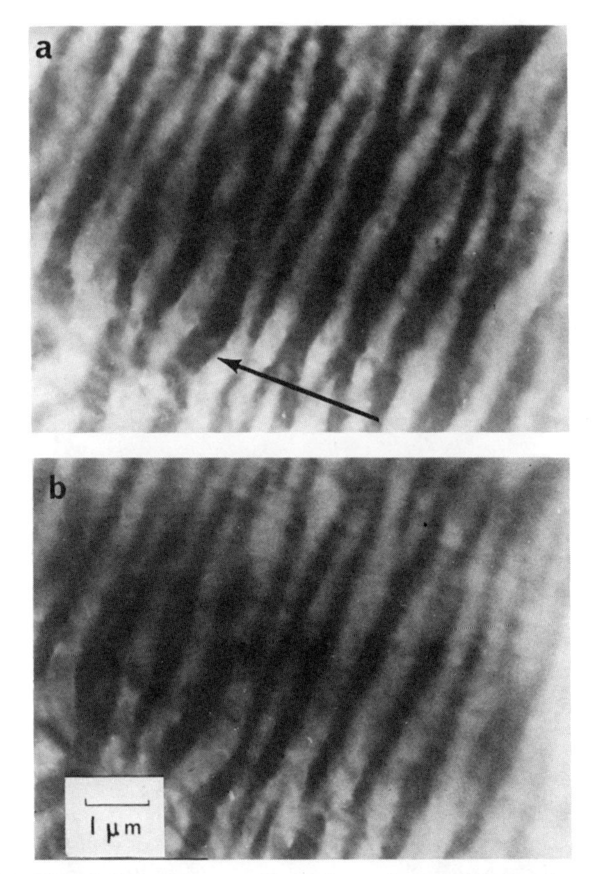

Fig. 3. (a) E[W] dark field image of B–N sheared at 300 °C. The shear direction is along the arrow; (b) the same area as Fig. 3(a); the E[W] dark field image is formed in the other wing of the equatorial reflection. (After Ref. 6.)

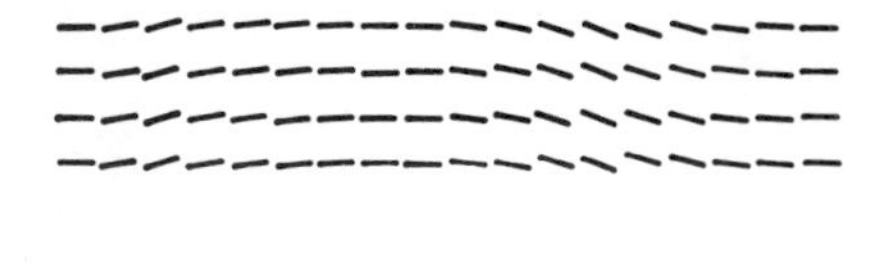

Fig. 4. Schematic representation of the serpentine trajectory of the molecules in the banded structure. (After Ref. 19.)

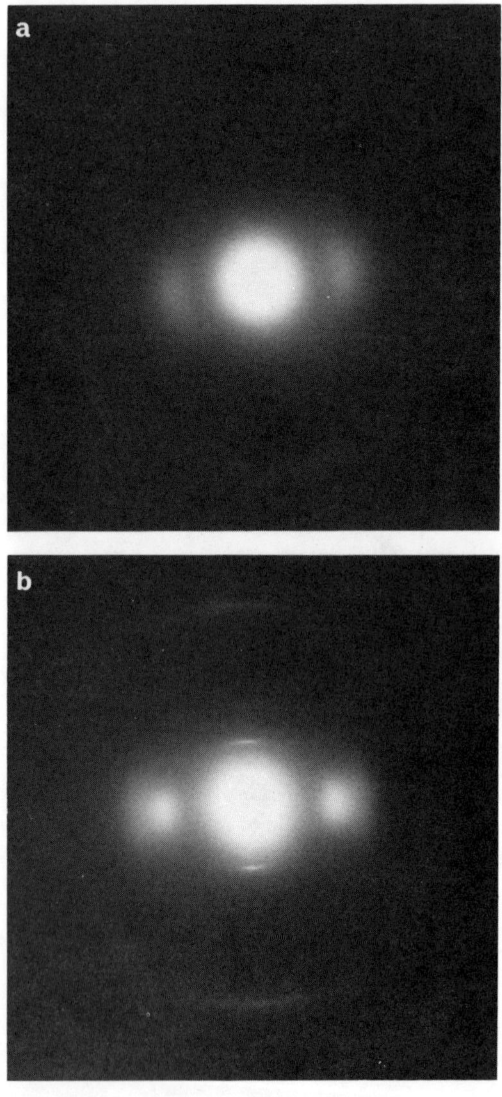

Fig. 5. (a) SAED of B–ET sheared at 300 °C; (b) SAED of ClQT–QG sheared at 300 °C.

diffraction pattern of B–ET and ClQT–QG before they degrade as a result of beam damage. Typical SAED patterns for these two polymers are shown in Fig. 5. They are broadly similar to that of B–N, reflecting the chemical relationship between the polymers. Although dark field microscopy is not possible for these polymers, optical microscopy of as-sheared specimens has shown the presence of bands, of appearance very similar to those seen in B–N.[6] Bands have also been seen in lyotropic systems during and following shear,[9] in spun-fibres of both lyotropic and thermotropic polymers,[10,11] and in thermotropics upon relaxation from shear.[12] It thus seems that the banded structure is common to a wide range of liquid crystalline polymers, and can be thought of as a general response of these materials to particular shear conditions.

(c) Bright Field Contrast

The previous section has described the appearance of the banded structure in dark field images and it has been shown that systematic variations in the orientation of the molecules in the plane of the specimen can give rise to strong contrast effects in E[W] images. However, the bright field (BF) image of such a specimen, in which the predominant variations of molecular orientation are in-plane, only shows contrast due to variations in thickness which do not correlate with the bands.

If a specimen of B–N, which has been sheared at 300 °C, is annealed at the same temperature on the rocksalt substrate, the bright field image shows additional contrast. After only 5 s, a network pattern of light lines appears which is apparent in Fig. 6(a). Longer anneals produce even more striking contrast changes. Figure 6(b) shows the appearance of dark lines, termed ‘veins’, which tend to lie parallel to the direction of prior shear. Close examination of Fig. 6(b) and similar micrographs (e.g. Figs 10(b), 11(b)) shows that within the domains separated by veins, slight contrast variations are also present, so that their centres are systematically lighter than the regions near the veins. The average width of the veins themselves is ~ 30 nm.

As has been demonstrated for the banded structure, in-plane orientation changes lead to contrast in dark field but not in bright field. The appearance of bright field diffraction contrast must be attributed to variation in the out-of-plane component of molecular orientation.[13] The possibility that the generation of abrupt thickness fluctuations during annealing might be the source of the vein contrast can be eliminated by examination of specimens lightly shadowed with platinum at a shallow angle. Figure 7 shows the appearance of such a specimen. The veins are still apparent under the metal

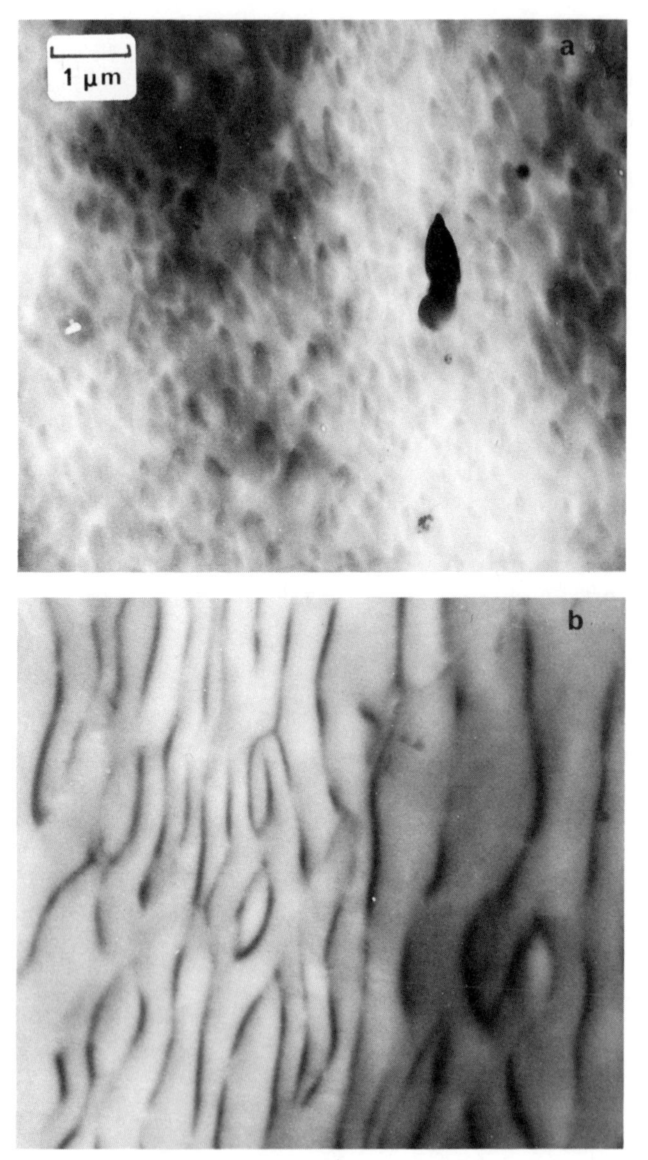

Fig. 6. (a) BF image of B–N sheared and then annealed for 5 s at 300 °C; (b) BF image of B–N sheared and then annealed for 10 min at 300 °C.

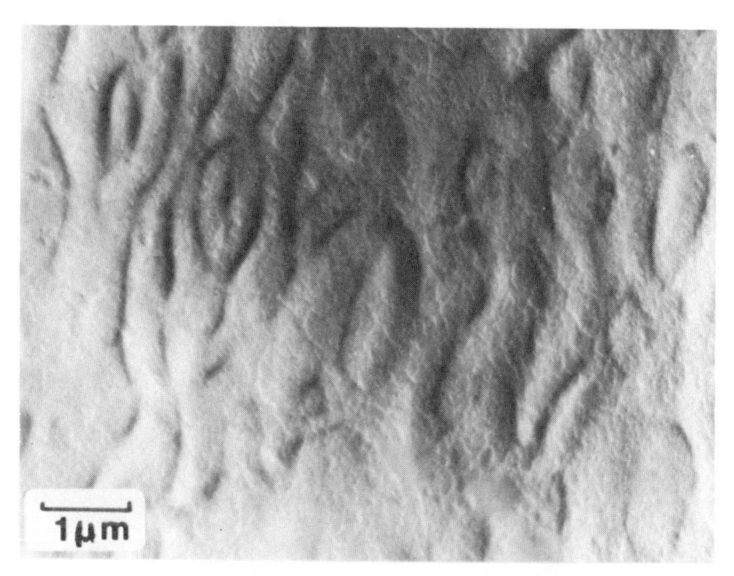

Fig. 7. B–N specimen, annealed for 10 min at 300 °C, and subsequently coated with gold–palladium at a shallow angle. (After Ref. 14.)

coating, but the coverage is fairly uniform, with no shadowing associated with the veins.

As molecules move away from the planar to the homeotropic orientation (perpendicular to the substrate), the appearance of the diffraction pattern will change. Ultimately, if perfect homeotropy were achieved, the whole of the equatorial ring would be excited, and the scattering along the meridian due to intrachain effects would not be recorded. The net effect of homeotropy is therefore that the apparent cross-section for scattering is increased, and the intensity in the bright field image is correspondingly decreased as more electrons are scattered and hence not included in the central, bright field aperture. Local changes in BF contrast, as seen at the veins, therefore indicate the occurrence of corresponding variations in molecular orientation with respect to the specimen plane.

A more quantitative analysis can be performed to examine the relationship between BF intensity and molecular orientation, by deriving an analytical expression for the arc length of the equatorial reflection that is excited for a given molecular orientation.[14] Combined with microdensitometry of the electron image plate to give an absolute value for the magnitude of the contrast changes at the veins, a measure of the out-of-plane component of molecular tilt can be obtained. Such measurements

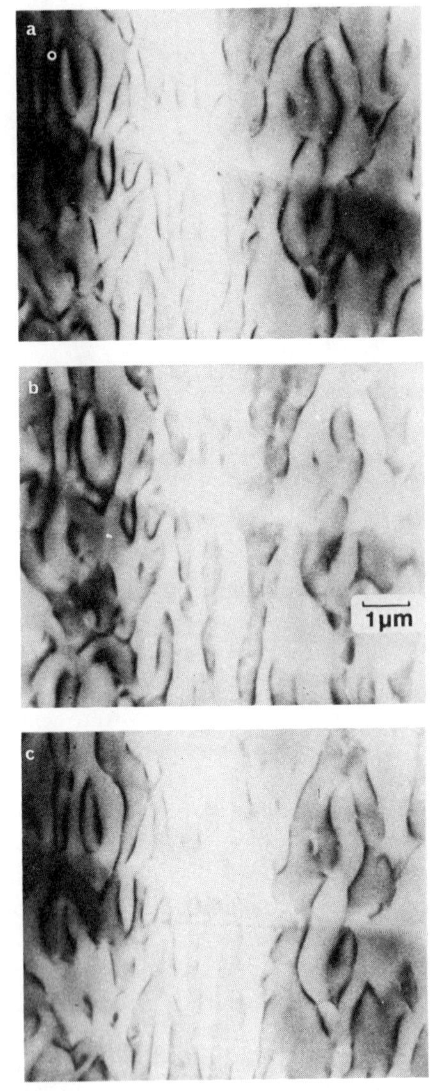

Fig. 8. BF images of B–N, annealed for 10 min at 300 °C, and tilted about an axis perpendicular to the direction of prior shear. Angle of tilt is (a) 0 °, (b) +24 ° and (c) −24 °. (After Ref. 15.)

show that a tilt of at least 45° is present at the veins, but true homeotropy need not occur anywhere to be consistent with the observed contrast.[14]

Observation of bright field images of specimens viewed at normal incidence shows that narrow veins of near-homeotropy are present. The slighter contrast variations that are present within the domains indicate that the out-of-plane component is not uniform. Additional information on the orientation within the domains can be provided by tilting the specimen in the electron microscope. Figure 8 shows the appearance of an annealed specimen of B–N which has been tilted about an axis perpendicular to the shear direction. Viewed at normal incidence, only the veins exhibit strong contrast (Fig. 8(a)). However as the specimen is tilted through relatively large angles, the domains themselves appear in alternating dark and light contrast. Furthermore, a domain which appears bright in Fig. 8(b) (positive sense of tilt) appears dark in Fig. 8(c) (negative sense), and conversely. Thus the sense of tilt out of the specimen plane must change from 'up' in one domain to 'down' in its neighbour, so that the angular sum of molecular plus specimen tilt varies between nearly 90° and approximately zero.

(d) Application to Domain and Wall Analyses

Dark field techniques alone are sufficient to determine the molecular organization in structures where the molecules are essentially confined to the specimen plane. The banded structure is such a structure, for there is no bright field contrast when viewed at normal incidence and only a very limited level of contrast when specimens are tilted through an angle as large as 45°.[15] (The fact that there is BF contrast at all in tilted specimens is significant and indicates the presence of some out-of-plane component, albeit small, associated with the serpentine molecular trajectories.) The use of combined DF and BF techniques (including tilting experiments) is necessary for structures where both in-plane and out-of-plane variations in molecular orientations are present. An example of their use to characterize in detail the domain and wall structures of annealed B–N specimens will be outlined here. Further details are found in Refs 14 and 15.

For a specimen of B–N which has been annealed for only 5 s, strong dark field contrast is observed in E[W] images (Fig. 9(a)). Even after such a short anneal, the structure is very different from the banded structure of as-sheared specimens: domains of constant orientation are no longer elongated perpendicular to the shear direction; on the contrary they tend to lie parallel to it. Only a low level of contrast is observed in the flat BF image (Fig. 9(b)) but upon tilting through 42° about an axis perpendicular to the

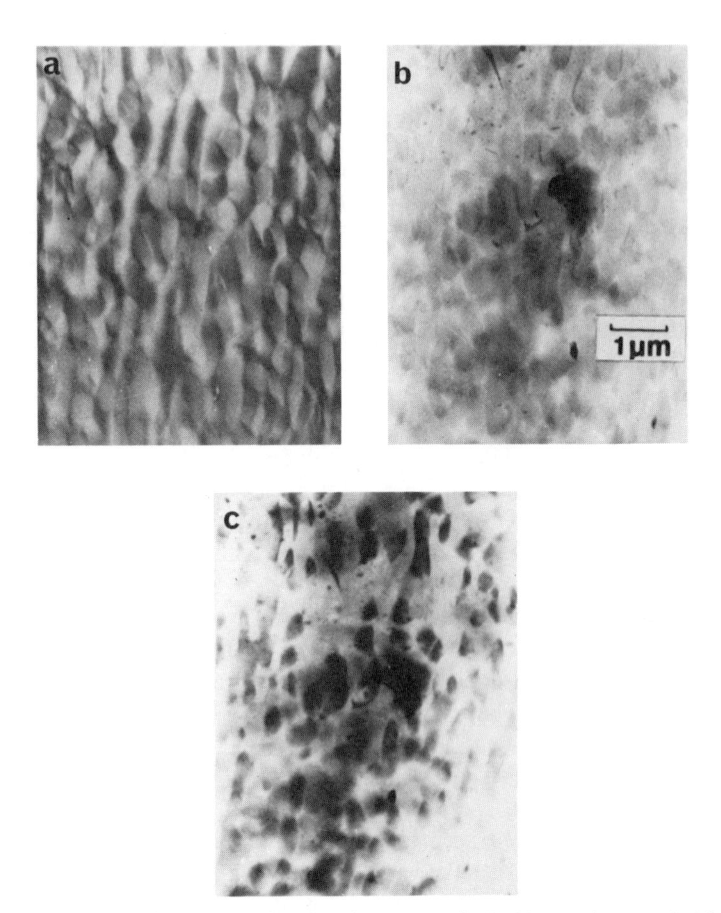

Fig. 9. A series of micrographs of a given area of a B–N sample annealed for 5 s at 300 °C. (a) E[W] dark field; (b) BF, specimen flat; (c) BF, specimen tilted 42° about an axis perpendicular to the shear direction. Shear direction is vertical. (After Ref. 15.)

shear direction domains of dark contrast become visible (Fig. 9(c)). Comparison of Figs 9(a) and 9(c) shows a strong (inverse) correlation, so that domains that appear dark in Fig. 9(c) because they are scattering strongly, are those that are not scattering strongly into the (flat) DF image of Fig. 9(a). The converse is also true. Such a correlation indicates the existence of a coupling between in-plane and out-of-plane orientation changes (see Section 6).

After longer annealing times, two changes occur in DF images. Strong contrast can now be observed in IM (cf. Fig. 2) dark field images, whereas in the (unannealed) banded structure, few electrons appear to be scattered into this region of reciprocal space. Figure 10(a) shows an IM dark field image, which can be compared with the corresponding bright field image of Fig. 10(b). The strongly scattering regions of the IM are seen to correspond to the dark veins of the BF image, consistent with the idea that the veins are regions of near-homeotropy.

E[W] and E images, on the other hand, show lower overall contrast levels than were apparent in specimens annealed for a short time only. The veins may appear either as dark or light; where dark, their width is narrower than in either BF or IM dark field images. Figure 11(a) shows such an E[W] dark field image, to be compared with the corresponding bright field image of Fig. 11(b). Since the BF analysis described in Section 3(c) has shown that the out-of-plane tilts in these specimens are greater than in those annealed for only 5 s, the lower overall contrast levels are to be expected, because the increased excitation of the equatorial ring implies that many in-plane orientations will be able to scatter into any given objective aperture position.

The bright and dark field micrographs taken together show, upon annealing, the development of domains with coupled in-plane and out-of-plane components of orientation. Both components alternate between neighbouring domains and the domains are separated by walls with near-homeotropic orientation. The details of the path of reorientation followed by the molecules as the wall is traversed can most clearly be understood by considering the loci of orientations that will give rise to a given level of contrast in each image. Figure 12 shows the stereographic representation of these loci mapped out for BF, and E and IM dark field images. In this stereogram the centre corresponds to homeotropy, the shear direction is located at the north and south poles, and the circumference corresponds to in-plane orientations.

The appearance of the veins as narrow dark lines in equatorial dark field images indicates that the reorientation path just passes through the lobes of dark contrast; such a reorientation path is shown in Fig. 13 for a wall separating domains with opposite senses of in-plane and out-of-plane tilts. This path is consistent with the requirement of continuity of molecular flux across the wall, which is clearly necessary from a consideration of the macromolecular nature of the polymer.[14]

Finally, the appearance of the light boundaries between domains in specimens annealed for short times must be explained. If dark contrast in

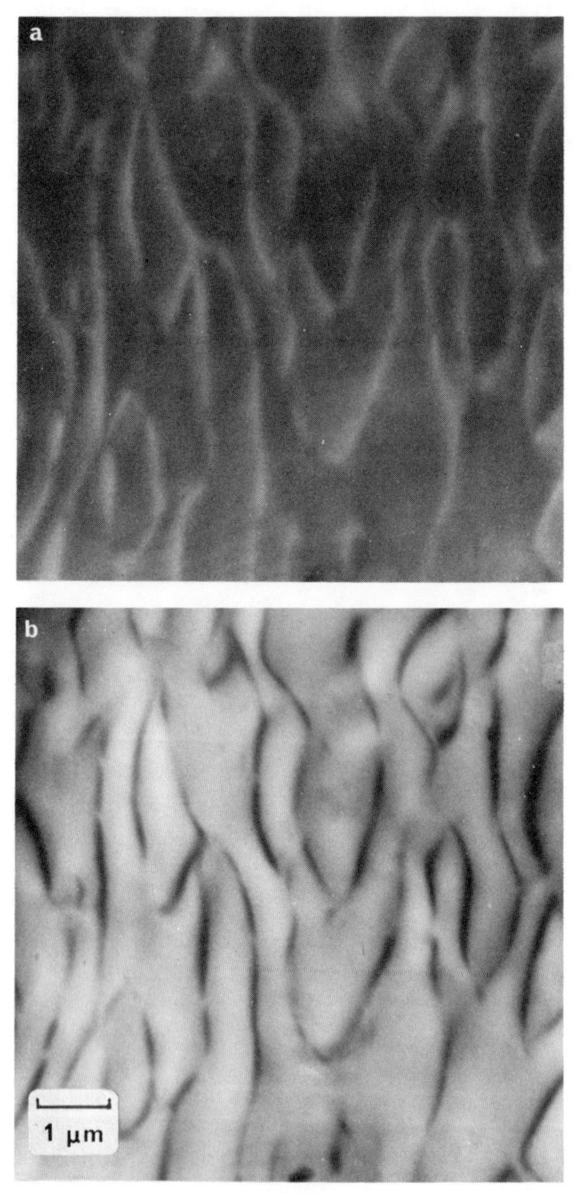

Fig. 10. Micrographs of a specimen of B–N annealed for 10 min at 300 °C. (a) IM dark field; (b) BF image.

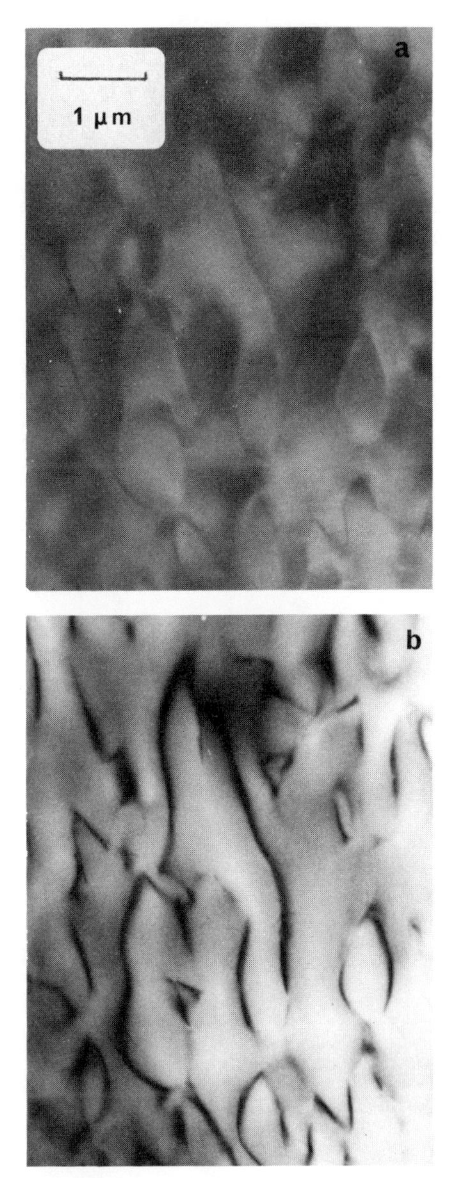

Fig. 11. (a) E[W] dark field image of a specimen of B–N annealed for 10 min at 300 °C; (b) corresponding BF image.

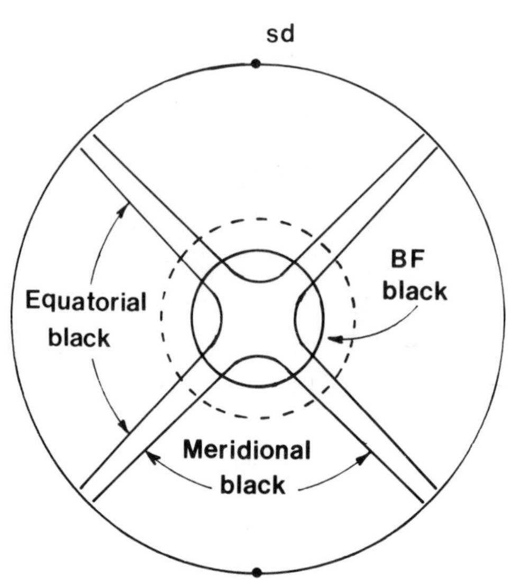

Fig. 12. Stereographic representation of the loci of orientations that will appear dark for bright field and dark field (E and IM) images. (After Ref. 14.)

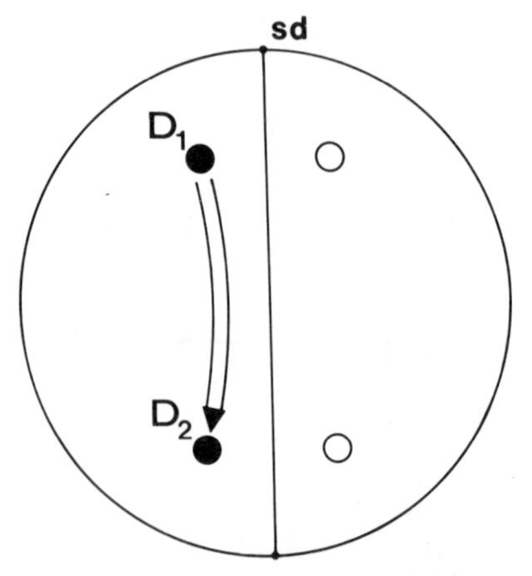

Fig. 13. Reorientation path across a wall separating domains possessing opposite senses of in-plane and out-of-plane orientation. 'sd' is the shear direction, the plane of the wall is the vertical line and the two poles (D_1 and D_2) represent the orientations of the molecular long axes in the domains.

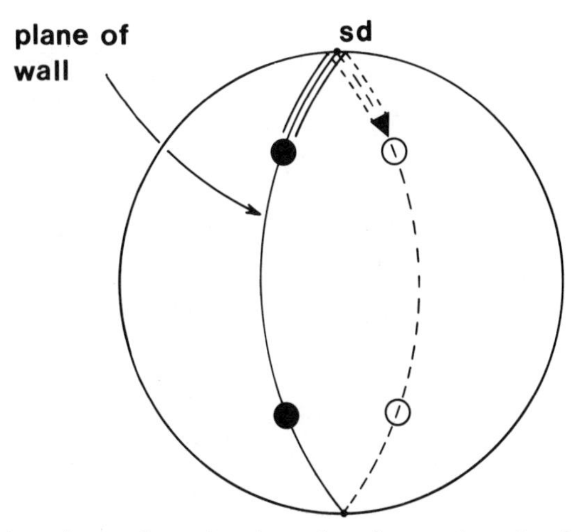

Fig. 14. Reorientation path passing through a planar orientation for specimens annealed for 5 s at 300 °C. (After Ref. 14.)

bright field micrographs corresponds to an increased molecular tilt out of the specimen plane, light contrast must indicate a reduced value for this tilt. Even after 5 s anneals, a substantial development of out-of-plane tilts has occurred within the domains, as is clear from Fig. 9(c). Thus in the early stage of annealing, it seems that the reorientation path passes through the planar orientation rather than near-homeotropy; this path is represented in Fig. 14. The necessary constant flux criterion is also met for this path.

4. SELECTED AREA DIFFRACTION AND MICRODIFFRACTION

X-ray diffraction techniques are commonly used to record diffraction pattern from volumes of bulk specimens $\sim 1\,mm^3$, and structural parameters can be derived from such measurements on oriented liquid crystalline materials (e.g. Refs 16 and 17). Selected area electron diffraction (SAED), in which an aperture limits the diameter of the diffracting region to a few microns, means that a volume of $\sim 2\,\mu m^3$ is sampled. In the TEM there is also the possibility of performing micro-area electron diffraction patterns (MAED) by focusing the electron beam to a spot a few thousands of angstroms in diameter. The resultant pattern is therefore a convergent beam pattern, with consequent loss of angular resolution (i.e. a sharp Bragg

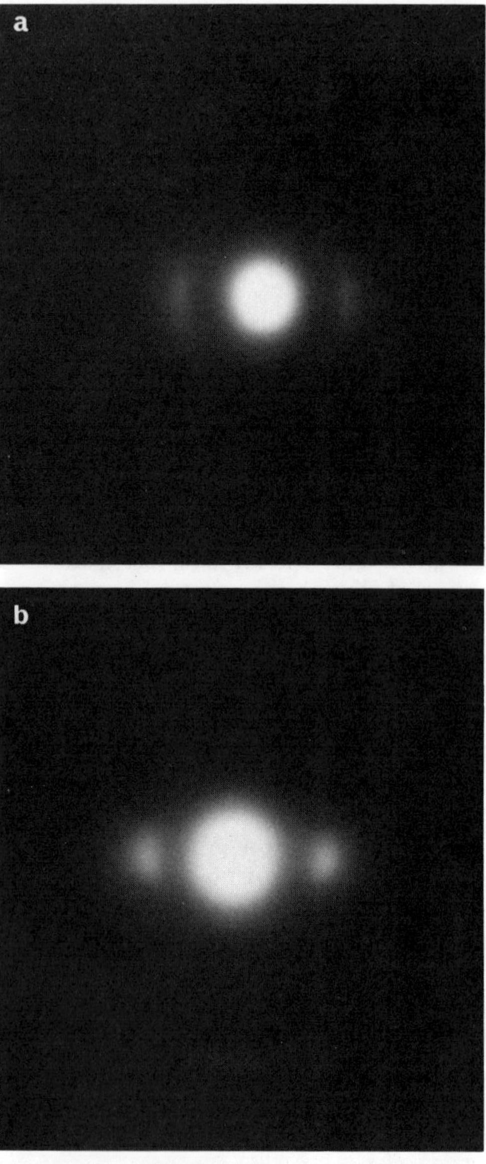

Fig. 15. (a) SAED pattern of B–N (as-sheared) recorded from a region 2·5 μm in diameter; (b) MAED pattern of B–N (as-sheared) recorded from a region 0·2 μm. (A portion of the beam stop is visible.)

spot would be smeared to a disc), but the sampling volume may now be reduced to $\sim 0{\cdot}02\,\mu m^3$. Thus in MAED the diffracting volume can be two orders of magnitude smaller than in SAED, and eleven orders smaller than is usual in x-ray diffraction!

Figure 15 shows SAED and MAED diffraction patterns recorded from a specimen of as-sheared B–N. The increase in apparent orientation in the MAED pattern can be seen. If a series of MAED patterns are recorded as the specimen is traversed, the axis of local orientation can be seen to swing backwards and forwards. The angular spread in orientations can be estimated by measuring the deviation of this local axis from a reference direction, such as can conveniently be introduced by inserting the beam stop of the microscope partially into the path of the electron beam (as is apparent in Fig. 15(b)). Figure 16(a) defines the relevant geometry, and Fig. 16(b) shows the results of such an analysis. The distribution of orientations has a maximum about the shear direction with a spread of $\pm 15°$.

The improvement in apparent orientation, for MAED compared with SAED patterns, and the movement of the local axis of orientation, reflect the spread of orientations present in the banded structure of as-sheared specimens (see Fig. 4). Since the diameter of the diffracting region from which the MAED pattern is obtained is small compared with the period of the banded structure, the MAED pattern (Fig. 15(b)) reflects the micro-orientation within the structure; the SAED pattern, on the other hand, encompasses the total spread of orientations present. In such a situation, the azimuthal spread of the arcs in the SAED pattern cannot be used to

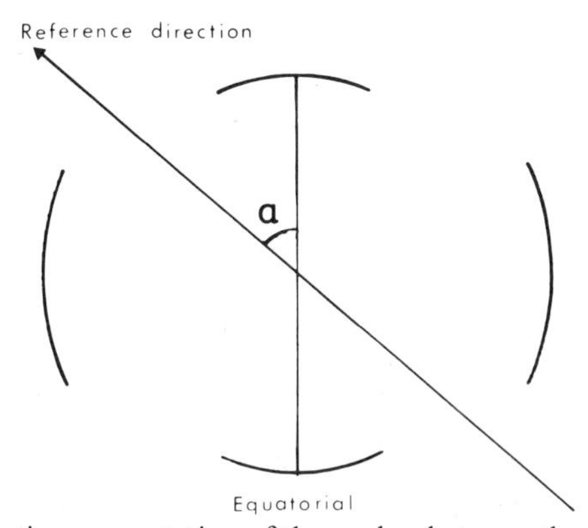

Fig. 16a. Schematic representation of the angle α between the reference direction defined by the beam stop and the line joining the centres of the equatorial reflection. (After Ref. 18.)

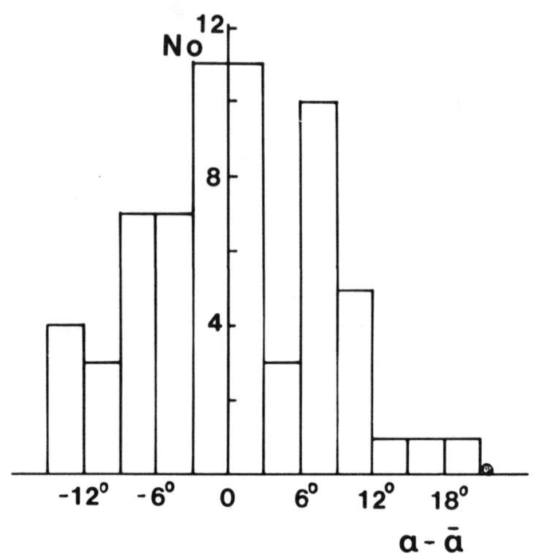

Fig. 16b. Histogram of $(\alpha - \bar{\alpha})$ values for B–N and sheared at 300 °C. $(\alpha - \bar{\alpha}) = 0$ corresponds to molecules lying along the shear direction. (After Ref. 18.)

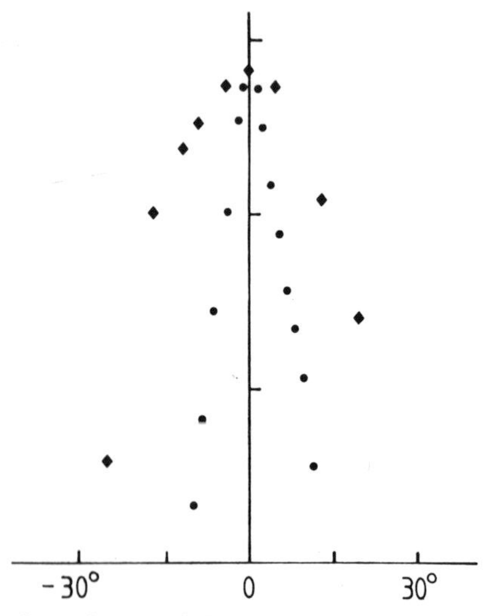

Fig. 16c. Profiles along the arc of the equatorial reflection of a specimen of B–N sheared at 300 °C, measured using a microdensitometer. ◆, SAED pattern; ●, MAED pattern. (After Ref. 18.)

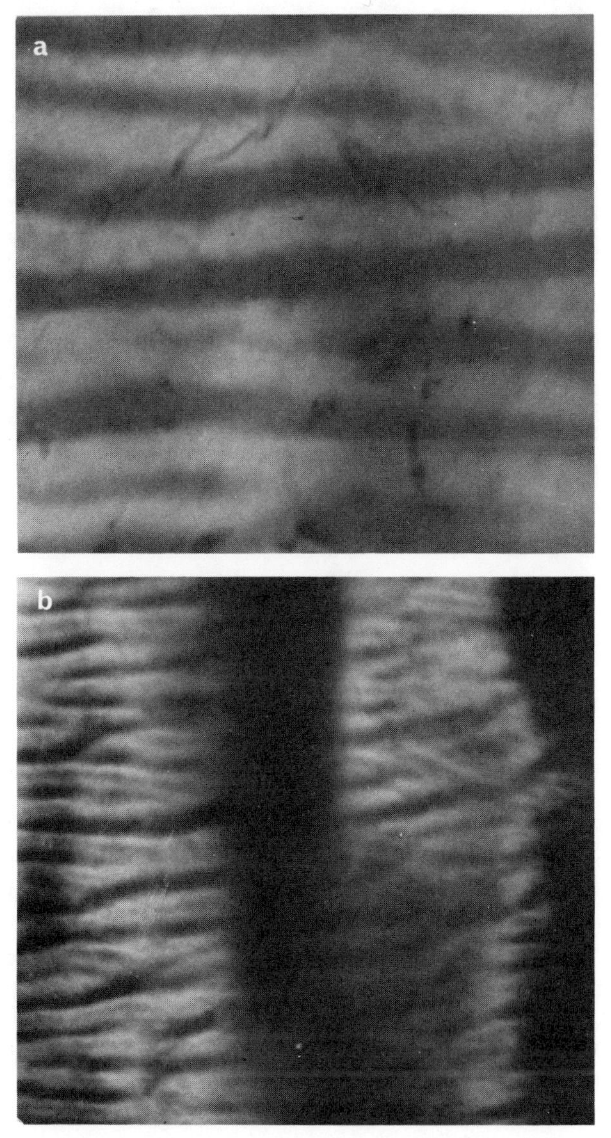

Fig. 17. Dark field micrographs of B–N sheared at (a) 300 °C, (b) 240 °C and (c) 200 °C. The dark field image is formed in E[W].

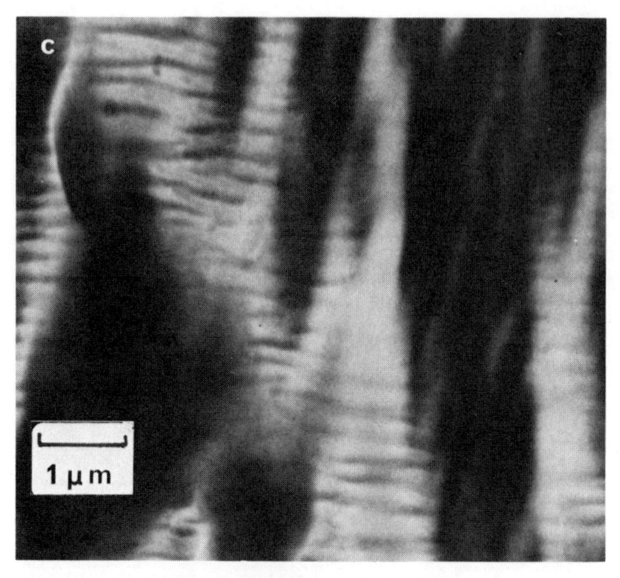

Fig. 17.—*contd.*

estimate the order parameter within the banded structure; this fact also demonstrates that the order parameter cannot be regarded as an intrinsic material property. However, the intensity distribution around the equatorial arc must correspond to the distribution of local orientations present, and hence should be equivalent to the distribution recorded in the histogram of Fig. 16(b). Figure 16(c) shows intensity distributions measured by microdensitometry around the equatorial arc for both SAED and MAED patterns. It can be seen that the azimuthal profile of the MAED pattern is approximately half the width of that of the SAED pattern, whereas it is similar in form and width to the distribution of Fig. 16(b) which was obtained in a different way.

Thus, to obtain a measure of the micro-orientation of a specimen of B–N sheared at 300 °C, SAED techniques are not sufficient and MAED must be used. MAED will be appropriate to probe the micro-orientation of any structure on a scale greater than the achievable size of the diffracting region. If the temperature of shearing, T_s, of B–N is reduced below 300 °C it is found that, although the structure is still formed, its scale decreases with temperature. Figure 17 shows a comparison of banded structures formed at (a) 300 °C, (b) 240 °C and (c) 200 °C. At the lowest temperature, the scale of the structure is 0·2 μm. For this situation, the microdiffraction pattern,

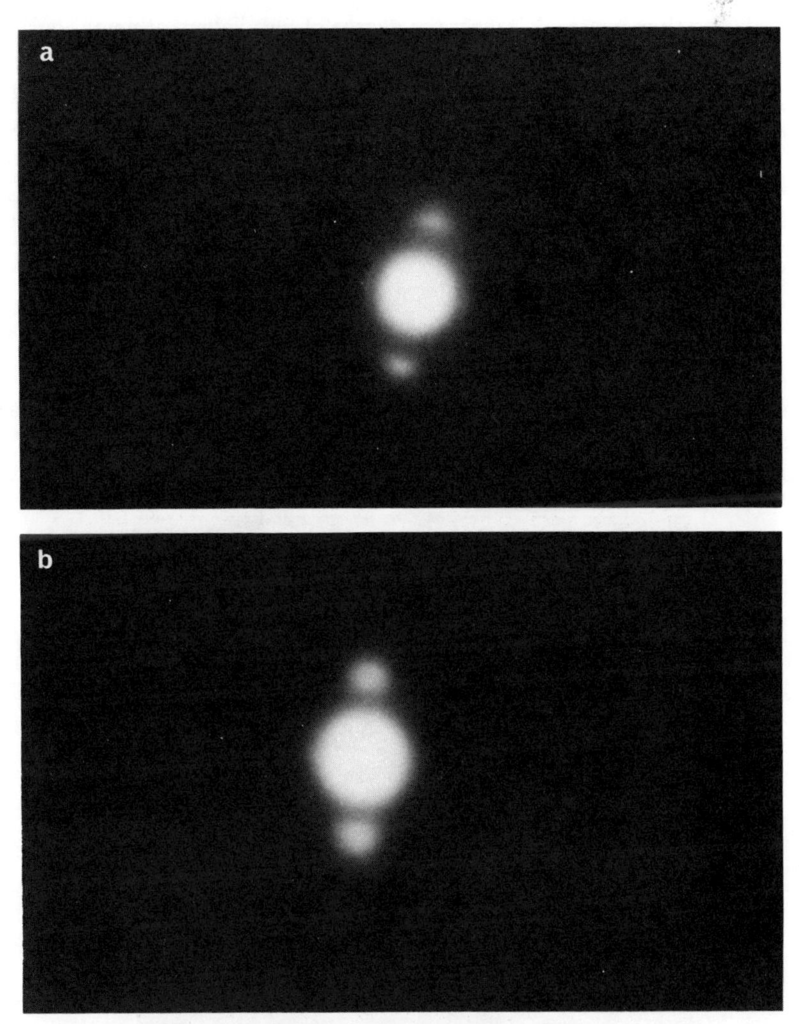

Fig. 18. Diffraction pattern of B–N sheared at 200 °C. (a) SAED pattern and (b) MAED pattern. (After Ref. 18.)

which is recorded from a region of comparable diameter, is indistinguishable from the SAED pattern and is therefore of no special value for such a fine scaled structure. Figure 18 compares the two types of diffraction pattern. In general, as the ratio of the diameter of the diffracting region to the typical length-scale decreases from unity (for the banded structure this implies T_s increasing), the spread in orientation obtained from the MAED analysis will increasingly reflect the degree of parallelism of neighbouring molecules.[18] Theoretically, it should be possible to obtain MAED patterns from regions much smaller than 0·2 μm by using smaller probe diameters as is possible in scanning transmission electron microscopes (STEM). However, for polymers, the ultimate limit of the MAED technique is determined not by the parameters of the electron microscope, but by the beam sensitivity of the polymer and STEM microdiffraction techniques have proved of less use than conventional TEM[19] in this context.

5. ANNEALING EFFECTS ON LOCAL ORDER

Sophisticated analysis of x-ray diffraction patterns of oriented LCPs has been used to analyse the nature of the local packing of neighbouring molecules and to quantify the average orientation in a sample.[16,17] Because

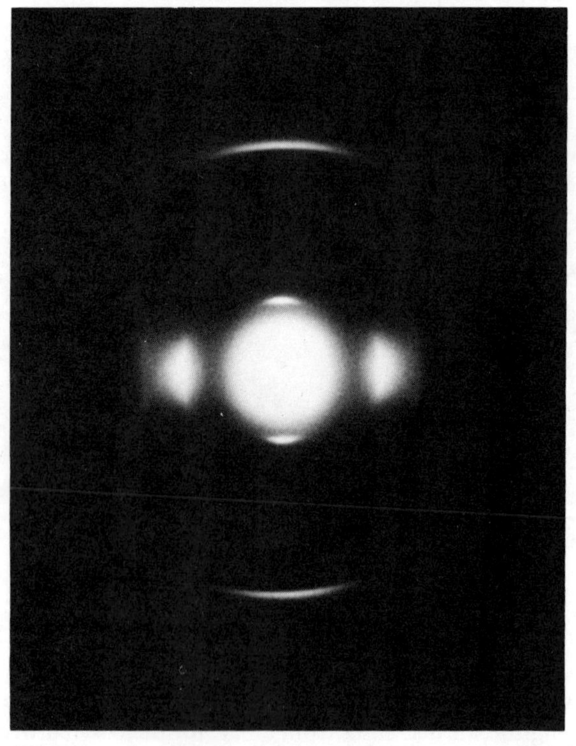

Fig. 19. SAED pattern of B–N sheared at 200 °C and annealed for 20 min.

of problems with background subtraction, electron diffraction is less suitable for such detailed quantitative analysis. However the appearance or disappearance of reflections in the diffraction pattern that may occur upon annealing can readily be detected, and correlated with changes in the underlying structure as revealed by direct imaging[5] which is of course not possible with x-rays.

When B–N is sheared and then annealed at 200 °C, two sharp satellite reflections develop which are positioned either side of the equator and are beyond the outer edge of the main equatorial arc (Fig. 19). No additional (*hkl*) type reflections are seen. Simultaneously with the appearance of these satellites, the equatorial dark field image shows additional scattering entities, overlying the banded structure which, at 200 °C does not transform into domains. Figure 20 shows such a dark field image. Randomly scattered across the specimen are regions 20 nm wide and 100 nm long which appear brighter than the bright bands of the banded structure. Their appearance suggests that some local reorganization of the chains has permitted regions of higher order to develop with a high degree of local chain parallelism. However the absence of *hkl* reflections despite the additional sharp reflections in the equatorial arc, indicates these regions cannot be identified with small, fully developed three-dimensional crystals. The meridional reflections are unchanged by the annealing process, and hence chemical reorganization within the molecule is unlikely. Such improvements in packing seem only able to occur if the original structure is well oriented. A

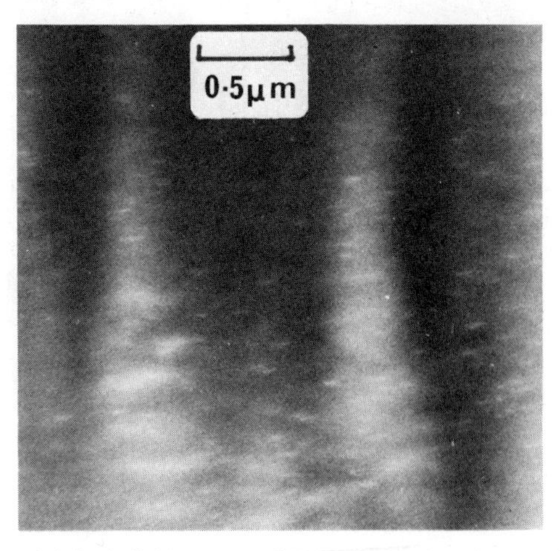

Fig. 20. Equatorial dark field image of B–N sheared and annealed for 20 min at 200 °C.

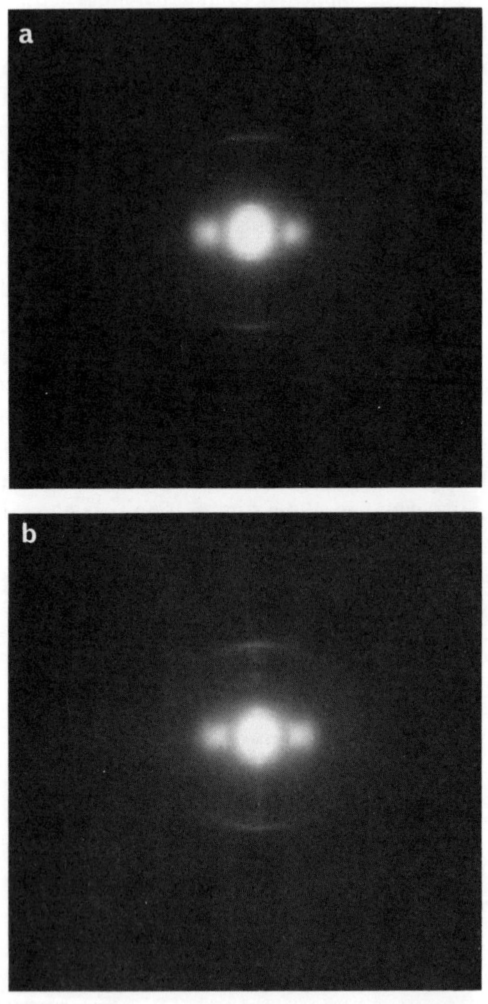

Fig. 21. (a) SAED pattern of N–QT sheared at 300 °C; (b) SAED pattern of N–QT sheared and then annealed for 10 min at 300 °C.

specimen which has been sheared at 300 °C with a broad range of orientations present and then annealed at 200 °C, does not show any improvement in local packing.[5]

In contrast to this behaviour, the response of the polymer N–QT (see Section 2) to annealing is different. When it is sheared at 300 °C, the familiar banded structure is formed, although with less regularity than in B–N. The diffraction pattern is broadly similar to B–N (Fig. 21(a)). Upon annealing for 10 min at 300 °C, however, the diffraction pattern shows dramatic changes along the meridian, with the appearance of a series of additional reflections (Fig. 21(b)). Simultaneously with the appearance of these new reflections the bright field image shows the formation of large lath-like entities lying at approximately $\pm 70°$ to the prior shear direction (Fig. 22). These regions are narrow (< 100 nm), but may be up to several microns in length and are thus large enough to be visible in the polarizing microscope. The change in the meridional reflections demonstrates that alterations to the intramolecular sequence are occurring. The polymer N–QT may therefore be an example of a polymer which exhibits trans-esterification upon annealing,[20,21] and it is possible that the lath-like entities correspond to regions of chemically regular polymer chains. The occurrence of chemical changes in N–QT, appears to preclude any possible tendency towards

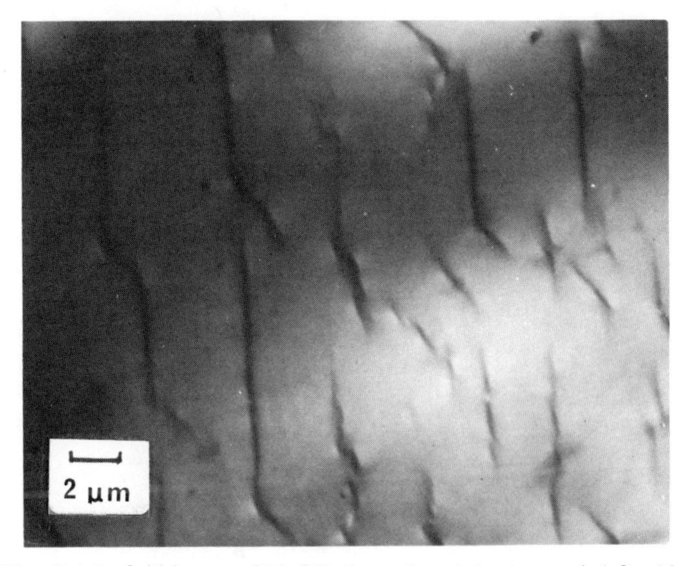

Fig. 22. Bright field image of N–QT sheared and then annealed for 10 min at 300 °C.

homeotropy as occurs in B–N at this temperature, which for both polymers lies 20 °C above their T_m. Thus, although the two polymers show the same response to the initial shear, their subsequent behaviour upon annealing is very different.

6. IMPLICATIONS OF THE MACROMOLECULAR CHAIN NATURE UPON STRUCTURES

A wide variety of structures have been identified in thin films of thermotropic LCPs, and the details have been described above. It is now necessary to consider how these structures (which often seem to have no counterpart in small molecule liquid crystals (SMLCs)) are related to the macromolecular nature of the chains involved.

To describe distortion of the director field of SMLCs, for which the director is equivalent to the long axis of the molecule, three elastic constants are conventionally used (e.g. Ref. 22). Of these K_{22}, the twist

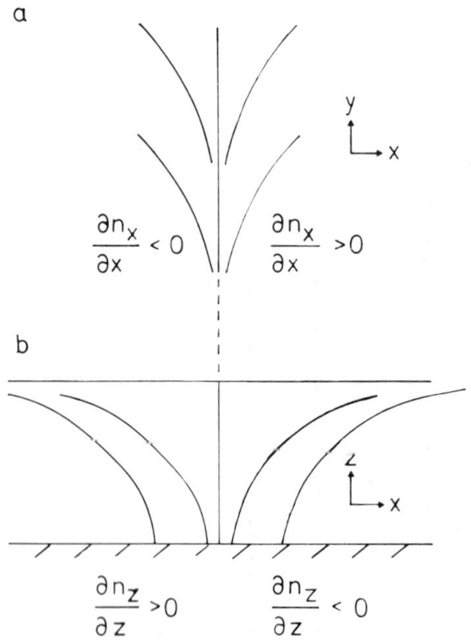

Fig. 23. (a) Proposed in-plane orientation map for domains in B–N annealed for 10 min on a single substrate, (b) possible splay compensation in the third dimension. (After Ref. 15.)

constant, is generally significantly lower than either K_{11} (splay) or K_{33} (bend). For LCPs, because the chains are long with a consequent low density of chain ends, splay distortions are expected to be of high energy.[23,24] However the relative energies of twist and bend distortions are not known: Kleman[25] has suggested that for one particular thermotropic LCP, K_{22} is the lowest Frank constant, whereas DuPre[26] has shown that for the lyotropic polymer poly(γ-benzyl-L-glutamate) in dioxane, $K_{33}$is lower than K_{22}.

If splay distortions are energetically unfavourable, this fact must be reflected in the structures observed. The banded structure contains predominantly bend distortions with some associated splay to maintain constant flux. As the transformation to the domain structure proceeds, further splay distortions occur within the plane of the film. As has been pointed out by Meyer,[24] such distortions can be counteracted by splay distortions of opposite sign through the film thickness. This effect, which has been termed 'splay–splay compensation', provides a rationale for the coupling of in-plane and out-of-plane misorientations observed in the domain structure of B–N.[15] The model is shown schematically in Fig. 23.

In this model it is assumed that there is a tendency towards homeotropy at the rocksalt substrate surface. Annealing of B–N without a substrate,

Fig. 24. B–N annealed for 10 min at 300 °C without a substrate. (After Ref. 15.)

by dissolving the rocksalt and picking up the specimen on a copper grid prior to annealing, shows that the formation of the domain structure is driven by the presence of the substrate: Fig. 24 shows that the banded structure barely changes during a 10 min anneal at 300 °C without a substrate. A similar tendency towards homeotropy for SMLCs at surfaces is well known (e.g. Ref. 27). However homeotropy requires the condensation of chain ends at the surface, and thus must be increasingly difficult as the chain length increases and the chain end density is correspondingly decreased. Hence, although molecular inclination within the domains rapidly increases within the first few seconds of annealing (see Fig. 9(c)),

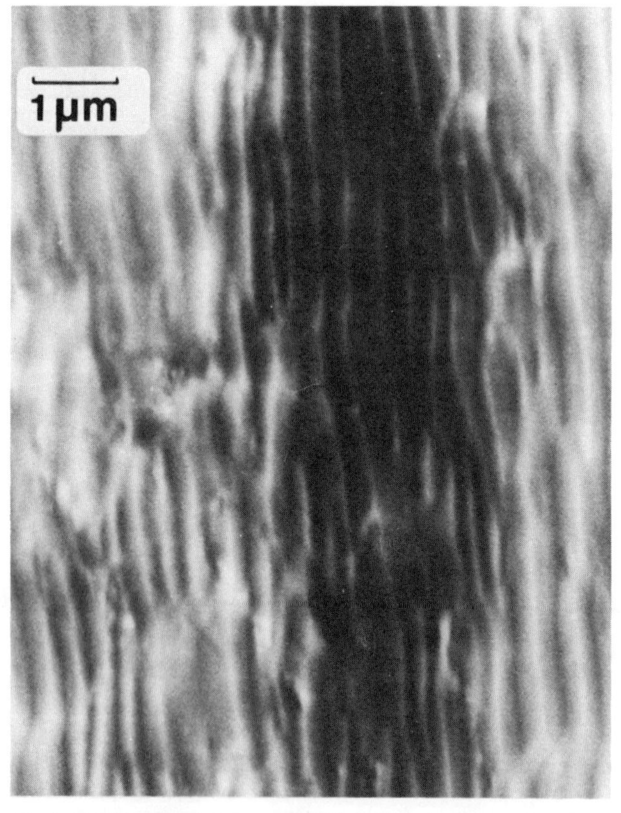

Fig. 25. Dark field image of elongated domains in a sample of B–N (75:25) annealed for 10 min at 320 °C. The original shear direction was vertical. (After Ref. 15.)

thereafter the rate of further increase is less dramatic, as the number of available chain ends drops.

For B–N, however, veins of near-homeotropy do evolve during the longer anneals. As shown in Table 1, the intrinsic viscosity (IV) of this polymer (70:30) is comparatively low, which suggests there may be a significant tail at the low end of the molecular weight distribution. That these short chains may be responsible for the development of the veins is supported by experiments on other polymers of the B–N family, also listed in Table 1.[15] In this context the results on B–N (75:25) and B–N (58:42) are particularly relevant. The former polymer, chemically almost identical to the standard B–N (composition 70:30), has a much higher IV. Upon annealing for 10 min at temperatures up to 320 °C, no veins are observed, but the E[W] dark field image shows the banded structure has transformed to the domain structure, with domains elongated along the prior shear direction (Fig. 25). On the other hand, B–N (58:42), which has an IV of 5·1 (close to that of the standard material), shows the formation of veins as well as domains (Fig. 26).

It therefore seems clear that, although there is an overall tendency towards homeotropy at the surface of the rocksalt substrate, the final

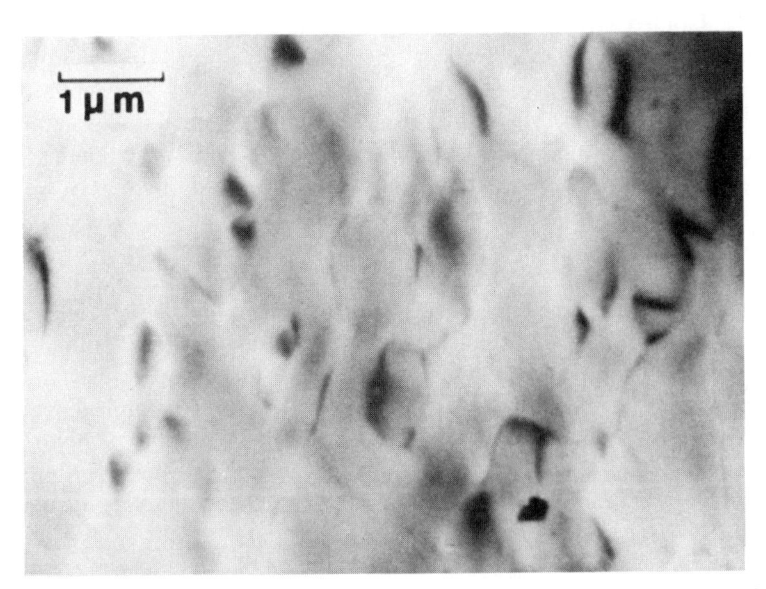

Fig. 26. Bright field image of veins in B–N (58:42) annealed for 10 min at 300 °C. (After Ref. 15.)

molecular inclination is restricted by the comparative scarcity of chain ends in the polymeric material. However, the local segregation of low molecular weight species, where present, permits the formation of narrow regions of near-homeotropy. It is felt that the formation of these segregated regions may have important consequences for the mechanical properties of surface layers of bulk specimens. Thus, although the results presented here are descriptive only of thin film structures, their relevance is far wider.

7. CONCLUSIONS

Transmission electron microscopy is able to identify the orientation of molecules within thin films of those thermotropic polymers which are sufficiently resistant to the necessary intense electron beam. The high resolution possible in TEM enables fine scale structures to be characterized which are invisible in a light microscope. Diffraction data can be recorded from the same area as bright and dark field images. However it must be recognised that the requirement for thin specimens can produce structures not necessarily typical of the bulk, particularly since surface interactions may be of importance.

The banded structure, formed by a wide range of liquid crystalline systems following shear, has been shown to correspond to a serpentine trajectory of the molecules about the shear direction. It is essentially a planar configuration. Microdiffraction techniques show that correlations of molecular orientation on a local scale can be very high, although this fact may not be immediately apparent from SAED or x-ray diffraction patterns.

When the banded structure in B–N is annealed at 300 °C, it rapidly transforms to a non-planar structure if a substrate is present, but in the absence of a substrate it is relatively stable. The development of an out-of-plane component of molecular orientation can be monitored by bright field microscopy, and the correlated changes in the in-plane orientation followed by examination of dark field images. The domains generated during the transformation of the banded structure on annealing are separated by boundary regions of near-homeotropy in samples of sufficiently low molecular weight. In general the tendency towards homeotropy within the domains is limited by the availability of chain ends.

TEM is clearly able to address fundamental questions concerning molecular organization in these novel and important polymers. The future will see the wider use of the technique in this field.

ACKNOWLEDGEMENTS

We should like to thank SERC for funding this research, Professor R. W. K. Honeycombe FRS for the provision of laboratory facilities, Dr B. Griffin of ICI (P & P) and Dr M. Jaffe of Celanese Corporation for the supply of samples, and Dr G. R. Mitchell for useful discussions.

REFERENCES

1. Geil, P. H., *Polymer Single Crystals*, Interscience, New York, 1963.
2. Grubb, D. T., in *Advances in Semicrystalline Polymers*, Bassett, D. T. (Ed.), Academic Press, New York, 1983.
3. Grubb, D. T., *J. Mat. Sci.*, 1974, **9**, 1715.
4. Blackwell, J., Gutierretz, G. A. and Chivers, R. A., *Macromolecules*, in press.
5. Donald, A. M., Mitchell, G. R. and Windle, A. H., to be submitted.
6. Donald, A. M. and Windle, A. H., *J. Mat. Sci.*, 1983, **18**, 1143.
7. Donald, A. M., Viney, C. and Windle, A. H., *Polymer*, 1983, **23**, 155.
8. Viney, C., Donald, A. M. and Windle, A. H., *J. Mat. Sci.*, 1983, **18**, 1136.
9. Kiss, G. and Porter, R. S., *Mol. Cryst. Liq. Cryst.*, 1980, **60**, 267.
10. Dobb, M. G., Johnson, D. J. and Saville, B. P., *J. Polym. Sci., Polym. Phys. Ed.*, 1977, **15**, 2201.
11. Atkins, E. D. T., Fulton, W. S. and Miles, M. J., *5th Int. Conf. Dissolving Pulps* (*Tappi*), Vienna, 1980, p. 208.
12. Graziano, D. J. and Mackley, M. R., *Mol. Cryst. Liq. Cryst.*, in press.
13. Donald, A. M., *J. Mat. Sci. Lett.*, 1984, **3**, 44.
14. Donald, A. M. and Windle, A. H., *Polymer*, in press.
15. Donald, A. M. and Windle, A. H., *J. Mat. Sci.*, 1984, **19**, 2085.
16. Mitchell, G. R. and Windle, A. H., *Polymer*, 1982, **23**, 1269.
17. Mitchell, G. R. and Windle, A. H., *Polymer*, 1983, **24**, 1513.
18. Donald, A. M. and Windle, A. H., *Coll. Polym. Sci.*, 1983, **261**, 793.
19. Donald, A. M., *Phil. Mag.*, 1983, **47A**, L13.
20. Lenz, R. W., Miller, W. R., Pryde, E. H. and Awl, R. A., *J. Polym. Sci.*, 1970, **A-1**, 429.
21. Lenz, R. W., Jin, J. I. and Feichtinger, K. A., *Polymer*, 1983, **24**, 327.
22. de Gennes, P. G., *The Physics of Liquid Crystals*, Oxford University Press, Oxford, 1974.
23. de Gennes, P. G., *Mol. Cryst. Liq. Cryst. Lett.*, 1977, **34**, 177.
24. Meyer, R. B., in *Polymer Liquid Crystals*, Academic Press, New York, 1982, Chapter 6.
25. Kleman, M., Liebert, L. and Strzelecki, L., *Polymer*, 1983, **24**, 295.
26. DuPre, D., private communication.
27. Kleman, M., *Points, Lines and Walls*, Wiley, New York, 1983.

13

NUCLEAR SPIN-LABEL STUDIES OF LIQUID CRYSTAL POLYMERS

K. MUELLER, B. HISGEN,* H. RINGSDORF,*
R. W. LENZ† and G. KOTHE

Institute of Physical Chemistry, University of Stuttgart, Federal Republic of Germany

INTRODUCTION

Nuclear spin-labelling is a reporter group technique. The basic idea underlying this method is to introduce nuclear spins as probes into the system of interest and to deduce from the observed nuclear magnetic resonance (NMR) spectra valuable information about the order and dynamics of the molecular environment. In this chapter we will discuss the principle of the method and some recent applications to liquid crystal polymers.

In the first section the basic NMR theory is developed with particular reference to deuteron spin-labels in anisotropic media. The model takes into account both intermolecular rotational diffusion in an ordering potential as well as intramolecular isomerization. It also considers various double- and multiple-pulse sequences, recently employed in Fourier transform NMR spectroscopy.[1-3]

The theory is then applied to the analysis of temperature and angular dependent ^{2}H-NMR spectra of thermotropic liquid crystal polyesters, specifically deuterated at various positions of the mesogenic unit and aliphatic spacer. Computer simulations provide the orientational distributions and conformations of the polymer chains and the correlation times of the various motions. The results, referring to three different phases, are

* Present address: Institute of Organic Chemistry, University of Mainz, D-6500 Mainz, Federal Republic of Germany.
† Present address: Chemical Engineering Department, University of Massachusetts, Amherst, Massachusetts 01003, USA.

discussed in relation to other studies of the molecular properties of liquid crystal polymers.

THEORY

Generally, time domain experiments are performed in nuclear spin-label studies. The spin system is subject to a sequence of high power rf pulses and the response is then Fourier-transformed in order to obtain a frequency spectrum. According to the sequences, different NMR spectra are obtained. Moreover, significant spectral changes occur when the pulse separations are varied. The same is true for spectra of macroscopically aligned samples of different orientations with respect to the magnetic field. Apparently, variation of typical NMR parameters such as pulse sequence, pulse separation or magnetic field orientation provides a large number of independent experiments.

Analysis of these experiments in terms of molecular order and dynamics requires a comprehensive model. We have developed such a model, based on the density matrix formalism.[1] Before any rf pulse is applied, the spin system is at thermal equilibrium. Application of the first pulse creates a defined initial state. After the pulse the density matrix evolves in time under the influence of the magnetic interactions of the spin-label. Then a second pulse is applied, preparing a new initial condition and so on. After the last pulse the density matrix is Fourier transformed to yield the frequency spectrum.

The action of the different pulses on the density matrix $\boldsymbol{\rho}$ is considered by unitary transformations, employing Wigner rotation matrices.[1,4] Between the pulses the density matrix evolves according to the stochastic Liouville equation:[1]

$$\dot{\boldsymbol{\rho}}_{\mathrm{ABK}} = (i/\hbar)[\boldsymbol{\rho}_{\mathrm{ABK}}, \mathbf{H}_{\mathrm{ABK}}] + (\dot{\boldsymbol{\rho}}_{\mathrm{ABK}})_{\mathrm{rot}} + (\dot{\boldsymbol{\rho}}_{\mathrm{ABK}})_{\mathrm{isom}} \tag{1}$$

Here $\mathbf{H}_{\mathrm{ABK}}$ is the Hamiltonian of the spin-label and $(\dot{\boldsymbol{\rho}}_{\mathrm{ABK}})_{\mathrm{rot}}$, $(\dot{\boldsymbol{\rho}}_{\mathrm{ABK}})_{\mathrm{isom}}$ account for the spin diffusion, caused by the inter- and intramolecular motion of the system. It is assumed that only a finite number of angular positions or sites, denoted by the index ABK, can be occupied during this process.

The spin Hamiltonian of deuterons in C–D bonds, considering Zeeman and quadrupole interactions, may be written as[1]

$$\mathbf{H} = -\gamma\hbar\mathbf{B}\,.\,\mathbf{I} + \mathbf{I}\,.\,\mathbf{Q}\,.\,\mathbf{I} \tag{2}$$

where γ, $\mathbf{B}$, $\mathbf{I}$, $\mathbf{Q}$ are the magnetogyric ratio, the static magnetic field, the nuclear spin operator and the quadrupole coupling tensor, respectively. The equilibrium distribution of the C–D bond vectors is described by an orientational distribution function, depending on internal and external coordinates. The internal part considers different conformations and the external part

$$\begin{aligned} f(\Theta, \Psi) &= N_1 \exp\,[A(\cos\xi \cos\Theta - \sin\xi \sin\Theta \cos\Psi)^2] \\ \cos\xi &= \cos\delta \cos\rho - \sin\delta \cos\varepsilon \sin\rho \\ f(\delta, \varepsilon) &= N_2 \exp\,[B \cos^2\delta] \end{aligned} \tag{3}$$

considers different orientations. Here Θ, Ψ, δ, ε, ρ are Euler angles, relating various molecular and laboratory systems.[1] The coefficient A characterizes the orientation with respect to a local director (micro-order), while the parameter B specifies the orientation of the director axes in a laboratory frame (macro-order). Micro- and macro-order parameters S_{ZZ} and $S_{z'z'}$ are related to the coefficients A and B by mean value integrals:

$$\begin{aligned} S_{ZZ} &= \tfrac{1}{2} N_1 \int_0^{\pi} (3\cos^2\beta - 1)\exp(A\cos^2\beta)\sin\beta\,\mathrm{d}\beta \\ S_{z'z'} &= \tfrac{1}{2} N_2 \int_0^{\pi} (3\cos^2\delta - 1)\exp(B\cos^2\delta)\sin\delta\,\mathrm{d}\delta \end{aligned} \tag{4}$$

The dynamic terms in eqn (1) depend upon the assumptions used to describe the motion. For the intermolecular motion a diffusive process is assumed (rotation through a sequence of small angular steps). In that case intermolecular reorientation can be characterized by two rotational correlation times, $\tau_{R\perp}$ and $\tau_{R\parallel}$. The correlation time for reorientation of the symmetry axis of a molecular diffusion tensor is $\tau_{R\perp}$, while $\tau_{R\parallel}$ refers to rotation about the axis. For the intramolecular motion a random jump process is assumed. Thus, isomerization occurs through jumps between different conformations with an average lifetime τ_J.

EXPERIMENTS AND METHODS

The family of liquid crystal polymers used for this study has the following general structure, in which the Roman numerals, I–III, refer to three different polymers of the same molecular structure but deuterated at different sites in the repeating unit, as indicated in the formula:[5,6]

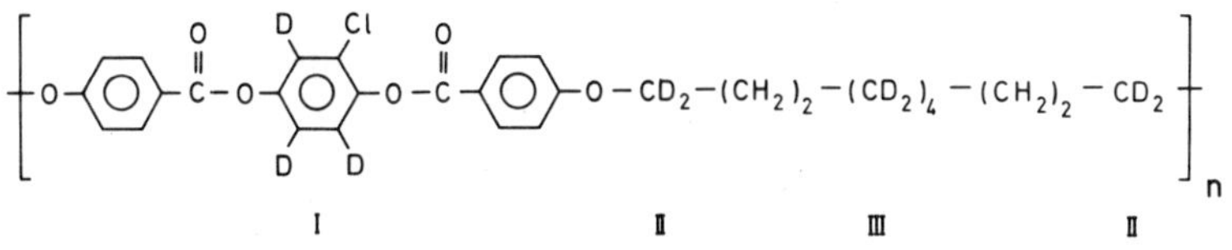

$$\left[-O-C_6H_4-C(=O)-O-C_6(D_3)Cl-O-C(=O)-C_6H_4-O-CD_2-(CH_2)_2-(CD_2)_4-(CH_2)_2-CD_2- \right]_n$$

I (ring D's), II (CD_2 ends), III ($(CD_2)_4$)

They exhibit a glass temperature at 303 K, a melting point at 433 K and a clearing temperature at 553 K, forming a stable nematic melt over the latter temperature range (DSC and polarizing microscopy). The average molecular weight $\bar{M}_n$ of the samples varied between $3000 \leq \bar{M}_n \leq 10\,000$ (vapour pressure osmometry). Deuteron labels were attached either to the central phenyl ring (I) of the mesogenic unit or to various positions (II or III) in the aliphatic spacer, as described elsewhere.[7] A magnetic field of 7·0T was employed to macroscopically align the samples.[8]

The ^{2}H-NMR spectra were measured with a Bruker CXP 300 pulse spectrometer at 46·1 MHz, using inversion recovery ($180^\circ_{x'}$–τ_1–$90^\circ_{x'}$–τ_2–$90^\circ_{y'}$) and quadrupole echo sequences ($90^\circ_{x'}$–τ–$90^\circ_{y'}$). The width for a 90° pulse was 3·5 μs employing a home-built probe, equipped with a goniometer. All spectra were recorded using quadrature detection and phase-alternating sequences. The number of scans varied between 500 and 2000. Angular dependent spectra were obtained by changing the angle between the alignment axis and magnetic field.

A FORTRAN program was employed to analyse the experimental spectra. The program DEUROTJUMP calculates NMR line shapes of $I = 1$ spin systems undergoing inter- and intramolecular motion in an anisotropic medium.[1] The quadrupolar coupling constant, used in the calculations, was $e^2qQ/h = 165$ kHz.

Analysis of the ^{2}H-NMR spectra requires knowledge of the molecular structure and symmetry. To proceed we assume that each molecular conformation can be approximated to a cylindrical rod, the symmetry axis of which is parallel to the long molecular axis (order tensor axis). This approximation is justified by the overall shape of the repeating unit, which is also expected to exhibit axially symmetric rotational diffusion about this axis. In addition, conformational changes, such as *trans–gauche* isomerization or ring flips, can occur. Symmetry considerations suggest that the *gauche* conformations of a particular segment are equally populated.

RESULTS AND DISCUSSION

Typical ^{2}H-NMR spectra of polymer II, specifically deuterated at both ends of the aliphatic spacer are shown in Figs 1–3. They refer to three different temperatures and characterize the anisotropic melt (Fig. 1) and the solid polymer above (Fig. 2) and below the glass transition (Fig. 3), respectively. Figure 1 shows ^{2}H-NMR line shapes in an inversion recovery

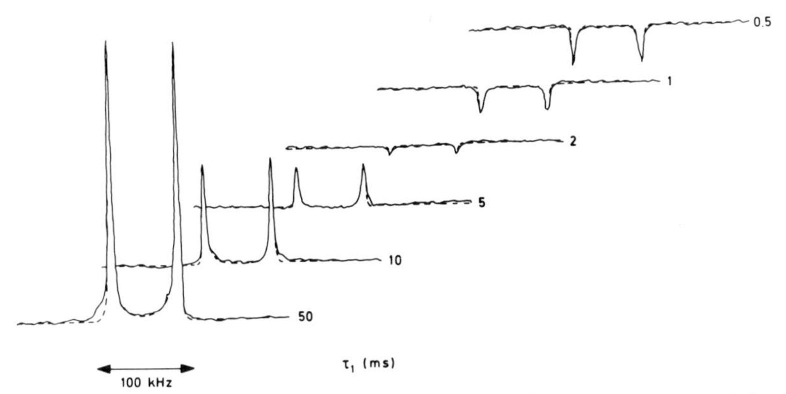

Fig. 1. Experimental (——) and simulated (- - -) ^{2}H-NMR spectra of liquid crystal polyester II in an inversion recovery sequence ($180^\circ_{x'}$–τ_1–$90^\circ_{x'}$–τ_2–$90^\circ_{y'}$) at $T = 443$ K (anisotropic melt) and different pulse separations τ_1. The sample was macroscopically aligned with the director parallel to the magnetic field. The simulations were obtained with $\tau_2 = 20\ \mu$s, $\tau_{R\perp} = 21$ ns, $\tau_J = 2{\cdot}5$ ns, $n_t = 0{\cdot}8$, $S_{ZZ} = 0{\cdot}86$ and $S_{z'z'} = 1{\cdot}0$.

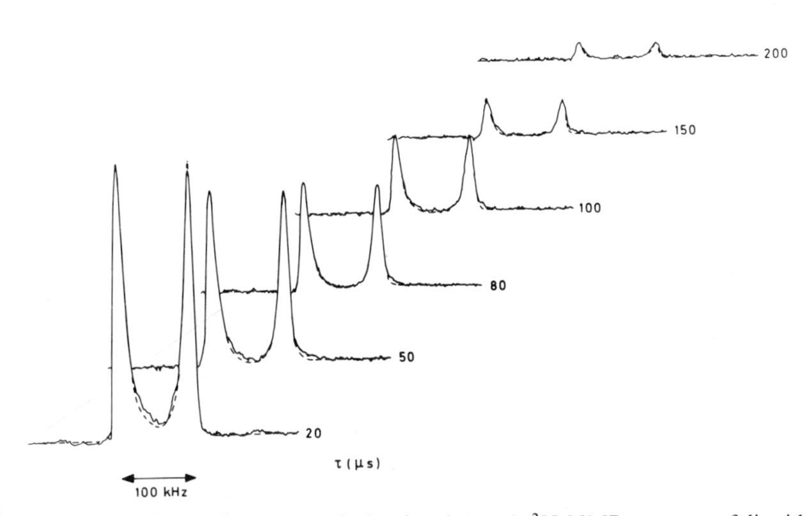

Fig. 2. Experimental (——) and simulated (- - -) ^{2}H-NMR spectra of liquid crystal polyester II in a quadrupole echo sequence ($90^\circ_{x'}$–τ–$90^\circ_{y'}$) at $T = 313$ K (solid state) and different pulse separations τ. The sample was macroscopically aligned with the director parallel to the magnetic field. The simulations were obtained with $\tau_{R\perp} = 7\ \mu$s, $\tau_J = 5\ \mu$s, $n_t = 0{\cdot}85$, $S_{ZZ} = 0{\cdot}89$, $S_{z'z'} = 1{\cdot}0$.

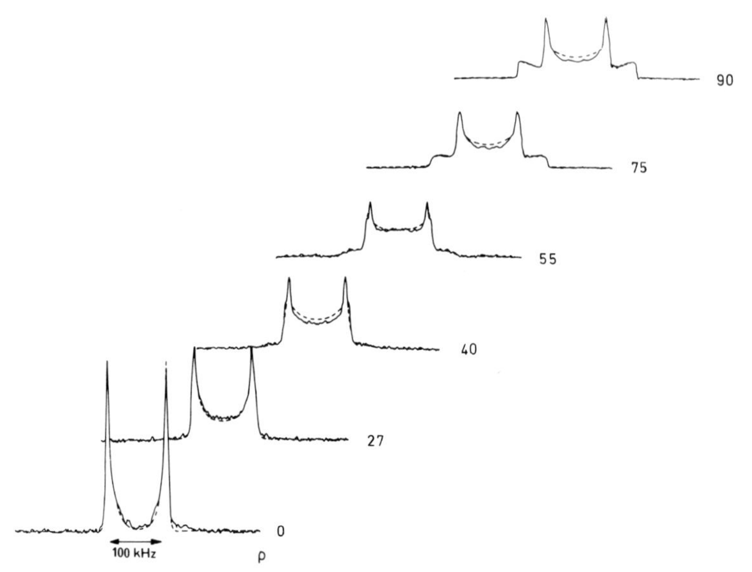

Fig. 3. Experimental (——) and simulated (– – –) ^{2}H-NMR spectra of liquid crystal polyester II in a quadrupole echo sequence ($90^\circ_{x'}$–τ–$90^\circ_{y'}$) at $T = 130$ K (glassy state) and different orientations ρ of the alignment axis and magnetic field. Experimental line shapes were corrected for distortions, due to finite pulse width.[24,25] The simulations were obtained with $\tau = 20\ \mu s$, $\tau_{R\parallel} > 1$ ms, $\tau_{R\perp} > 1$ ms, $\tau_J = 80\ \mu s$, $n_t = 0{\cdot}88$, $S_{ZZ} = 0{\cdot}89$, $S_{z'z'} = 1{\cdot}0$.

sequence at different pulse separations. Drastic spectral changes are observed. They contain valuable information about the dynamics of the system. The same is true for the line shapes in a quadrupole echo sequence (Fig. 2). Figure 3 shows ^{2}H-NMR spectra at different angles ρ between the alignment axis and magnetic field. Again, one recognizes significant changes, which are characteristic of the micro- and macro-order of the system. An iterative fit of several angular and pulse-dependent line shapes for any given temperature provides reliable values for the simulation parameters, i.e. the orientation of the long molecular axis, the micro- and macro-order parameters, the rotational correlation times, and the lifetimes and populations of particular conformations. The dotted lines in Figs 1–3 represent best-fit simulations. They agree favourably with their experimental counterparts.

In Fig. 4 the correlation times for the various motions of polymer II are plotted as a function of $1/T$. They refer to reorientation of the long axis (full circles) and *trans–gauche* isomerization (open squares) of the first spacer

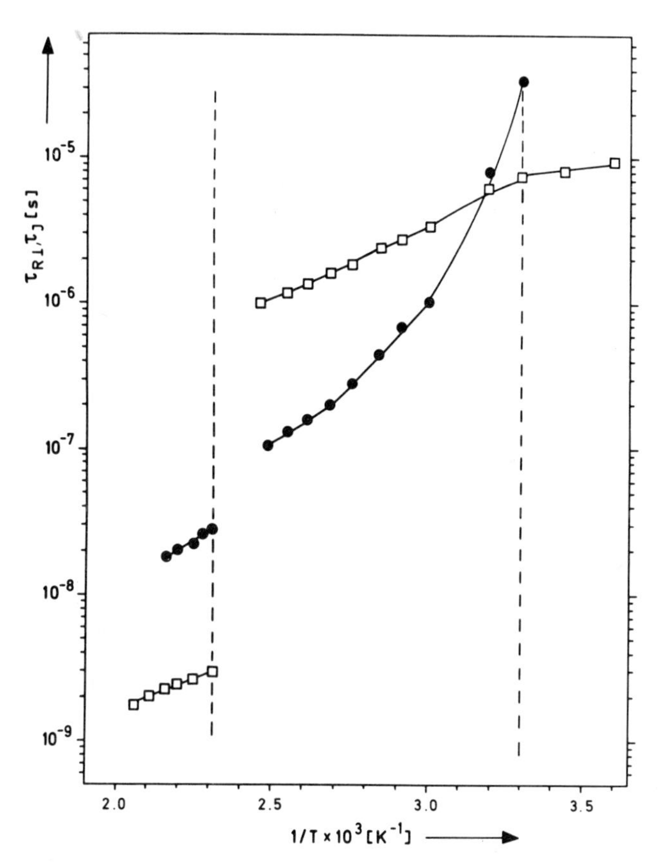

Fig. 4. Arrhenius plot of various correlation times, characterizing the molecular dynamics of liquid crystal polyester II. Full circles refer to intermolecular motion (reorientation of the long axis) and open squares denote intramolecular motion (*trans–gauche* isomerization of the first spacer segment). The dashed lines indicate phase transitions (melting point at $T = 433$ K, glass transition at $T = 303$ K).

segment, respectively. In the melt the correlation time for reorientation of the long axis is of the order of $\tau_{R\perp} = 10^{-8}$ s, while *trans–gauche* isomerization occurs even faster. Apparently these rapid motions are responsible for the unusual rheological behaviour of liquid crystal polymers.[9] At the melting point, within a temperature range of 20 K, all motions decrease abruptly. Lowering the temperature in the solid state first freezes all intermolecular motions. Thus, below the glass transition only intramolecular motions such as *trans–gauche* isomerization can be detected. In fact, isomerization of the aliphatic spacer is observed even at

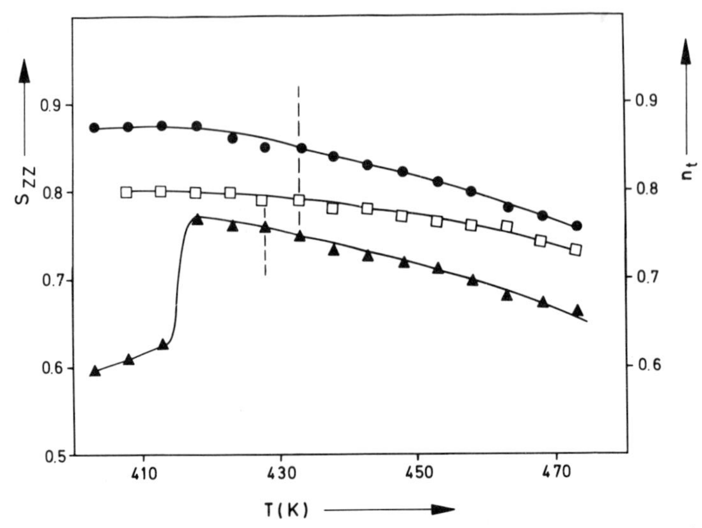

Fig. 5. The temperature dependence of the micro-order parameter S_{ZZ} and *trans* population n_t of liquid crystal polyesters I–III. Full circles (left ordinate) denote S_{ZZ}, while open squares (right ordinate) refer to n_t of the outer and central spacer segments. Full triangles indicate the micro-order parameter S_{ZZ} of the polyester, having only nine segments in the aliphatic spacer. The dashed lines denote the melting points at $T = 433$ K and $T = 429$ K, respectively.

$T = 130$ K with a correlation time of $\tau_J \sim 10^{-4}$ s. The low activation energy of $E_J = 6\,\text{kJ mol}^{-1}$ for this jump process agrees with previous T_1 dispersion measurements on paraffins.[10] Thus, the chain dynamics of liquid crystal polyesters in three different phases, including the glassy state, could be determined.

The micro-order of the liquid crystal polyesters is conveniently discussed in terms of two parameters S_{ZZ} and n_t, characterizing the average orientation of the long molecular axis with respect to the director and the *trans* population of the labelled spacer segment. From the angular-dependent line shapes of polymer I the angle α between the long molecular axis and the *para*-axis of the central ring was determined to be $\alpha = 14°$. In Fig. 5, S_{ZZ} and n_t are plotted as a function of temperature. Full circles denote S_{ZZ}, while open squares refer to n_t of the outer and central spacer segments (polymers II and III). It can be seen that in the nematic melt $S_{ZZ} = 0{\cdot}85$, a value considerably larger than those observed in low molecular weight nematogens. Thus, the polymer chains are highly ordered on a molecular level in agreement with x-ray diffraction,[11,12] ESR,[13] and ^{1}H-NMR studies.[14,15] In addition, this work shows that the chains adopt a highly extended conformation, evidenced by a *trans* population of $n_t \sim 0{\cdot}8$ throughout the entire spacer.

Interestingly, the order parameter S_{ZZ} is retained when the polymer is cooled below the melting point and glass transition, respectively. Thus, in contrast to conventional liquid crystals, nematic order of main-chain polymers can be frozen-in at the glass transition. No change in the order parameter is observed over a period of one year, keeping the sample at room temperature. In all systems studied, S_{ZZ} is independent of the molecular weight of the polymers within the range $3000 \leq \bar{M}_n \leq 10\,000$, according to a plateau effect, observed for liquid crystal side- and main-chain polymers.[16–18]

So far, the discussion has referred to polyesters, having 10 segments in the aliphatic spacer. Reducing the spacer length to 9 methylene groups causes a pronounced change of the micro-order parameter S_{ZZ}. Preliminary results, denoted by full triangles in Fig. 5, indicate a significant reduction of S_{ZZ} in the anisotropic melt and a further decrease at the melting point. This marked even–odd effect, also observed in other thermotropic polymers,[19–22] presents a challenging theoretical problem.

The degree of macroscopic alignment $S_{z'z'}$ of the liquid crystal polymers depends on the orientation method. Because the anisotropic permittivity of the polyester is negative, only a two-dimensional distribution of director axes was achieved using high electric fields. However, a uniform alignment of the domains was obtained using a magnetic field of 7·0T. A detailed analysis of the angular-dependent line shapes yielded a macro-order parameter of $S_{z'z'} = 1{\cdot}0$. Likewise, solid state extrusion of the liquid crystal polyester produces fibres with $S_{z'z'} = 0{\cdot}9$.[13] High modulus and strengths may result from this highly oriented chain configuration.[23] The preparation of fibres by spinning from the anisotropic melt is presently being studied.

ACKNOWLEDGEMENTS

It is a pleasure to thank Dr C. Eisenbach (University of Freiburg) and Dr A. Schneller (University of Massachusetts, Amherst) for advice and help in preparing the specifically deuterated polymers. The authors are also grateful to Dr P. Meier and Dr E. Ohmes (University of Stuttgart) for assistance in the NMR measurements and computations. Financial support of this work by the Deutsche Forschungsgemeinschaft and Fonds der Chemischen Industrie is gratefully acknowledged.

REFERENCES

1. Meier, P., Ohmes, E., Kothe, G., Blume, A., Weidner, J. and Eibl, H.-J., *J. Phys. Chem.*, 1983, **87**, 4904.
2. Lausch, M. and Spiess, H. W., *J. Magn. Reson.*, 1983, **54**, 466.
3. Schwartz, L. J., Meirovitch, E., Ripmeester, J. A. and Freed, J. H., *J. Phys. Chem.*, 1983, **87**, 4453.
4. Wigner, E. P., *Group Theory and Its Application to Quantum Mechanics of Atomic Spectra*, Academic Press, New York, 1959.
5. Jin, J.-I., Antoun, S., Ober, C. and Lenz, R. W., *Brit. Polym. J.*, 1980, **12**, 132.
6. Antoun, S., Lenz, R. W. and Jin, J.-I., *J. Polym. Sci., Polym. Chem. Ed.*, 1981, **19**, 1901.
7. Mueller, K., Eisenbach, C., Hisgen, B., Ringsdorf, H., Schneller, A., Lenz, R. W. and Kothe, G., to be published.
8. Noel, C., Monnerie, L., Achard, M. F., Hardouin, F., Sigaud, G. and Gasparoux, H., *Polymer*, 1981, **22**, 578.
9. Jerman, R. E. and Baird, D. G., *J. Rheol.*, 1981, **25**, 275.
10. Stohrer, M. and Noack, F., *J. Chem. Phys.*, 1977, **67**, 3729.
11. Liebert, L., Strzelecki, L., Van Luyen, D. and Levelut, A. M., *Eur. Polym. J.*, 1981, **17**, 71.
12. Blumstein, A., Vilasagar, S., Ponrathnam, S., Clough, S. B. and Blumstein, R. B., *J. Polym. Sci., Polym. Phys. Ed.*, 1982, **20**, 877.
13. Mueller, K., Wassmer, K.-H., Lenz, R. W. and Kothe, G., *J. Polym. Sci., Polym. Lett. Ed.*, 1983, **21**, 785.
14. Volino, F., Martins, A. F., Blumstein, R. B. and Blumstein, A., *J. Phys. (Lett.)*, 1981, **42**, L305.
15. Martins, A. F., Ferreira, J. B., Volino, F., Blumstein, A. and Blumstein, R. B., *Macromolecules*, 1983, **16**, 279.
16. Finkelmann, H., in *Polymer Liquid Crystals*, Ciferri, A., Krigbaum, W. R. and Meyer, R. B. (Eds), Academic Press, New York, 1982.
17. Wassmer, K.-H., Ohmes, E., Kothe, G., Portugall, M. and Ringsdorf, H., *Macromol. Chem., Rapid Commun.*, 1982, **3**, 281.
18. Blumstein, R. B., Stickless, E. M. and Blumstein, A., *Mol. Cryst. Liq. Cryst. Lett.*, 1982, **82**, 205.
19. Blumstein, A. and Thomas, O., *Macromolecules*, 1982, **15**, 1264.
20. Roviello, A. and Sirigu, A., *Macromol. Chem.*, 1982, **183**, 895.
21. Blumstein, A., Blumstein, R. B., Gauthier, M. M., Thomas, O. and Asrar, J., *Mol. Cryst. Liq. Cryst. Lett.*, 1983, **92**, 87.
22. Griffin, A. C. and Havens, S. J., *J. Polym. Sci., Polym. Phys. Ed.*, 1981, **19**, 951.
23. Schaefgen, J. R., Pletcher, T. C. and Kleinschuster, J. J., Belg. Patent 828.935 (1975) (Du Pont Co.).
24. Hentschel, R. and Spiess, H. W., *J. Magn. Reson.*, 1979, **35**, 157.
25. Bloom, M., Davis, J. H. and Valic, M. I., *Can. J. Phys.*, 1980, **58**, 1510.

14

VISCOSITY AND THE THERMODYNAMIC PROPERTIES OF LIQUID CRYSTALLINE POLYMERS WITH MESOGENIC SIDE GROUPS

J. Springer and F. W. Weigelt

Institut für Technische Chemie der Technischen Universität Berlin, Fachgebiet Makromolekulare Chemie, Berlin, Federal Republic of Germany

INTRODUCTION

Investigations on thermotropic properties of liquid crystalline polymers and on the influence of the chemical structure of the structural units have been reported.[1–3] These polymers were synthesized by linkage of mesogenic side groups with flexible spacers to the polymer main chain.[4] The properties depend not only on the linkage of the mesogenic groups to the polymer main chain[5] and on the chain mobility, but also on parameters such as molecular weight, molecular weight distribution, tacticity, and polymer–polymer interactions. These parameters have been determined and we report here the results[6–8] of our investigations on a poly(methacrylic acid) derivative of the type:

$$\left[-CH_2-\underset{\underset{\displaystyle O{=}C-O-(CH_2)_6-O-\langle O \rangle-\overset{\displaystyle O}{\overset{\|}{C}}-O-\langle O \rangle-OCH_3}{|}}{\overset{\displaystyle CH_3}{\overset{|}{C}}}- \right]_n$$

The polymer was obtained by radical polymerization, initiated with 0·5 mole-% 2,2′-azobisisobutyronitrile (AIBN) in benzene at 333 K, and was fractionated in the classical way using fractional precipitation. The polymer consisted of 30% isotactic, 12% syndiotactic, and 58% heterotactic triads.

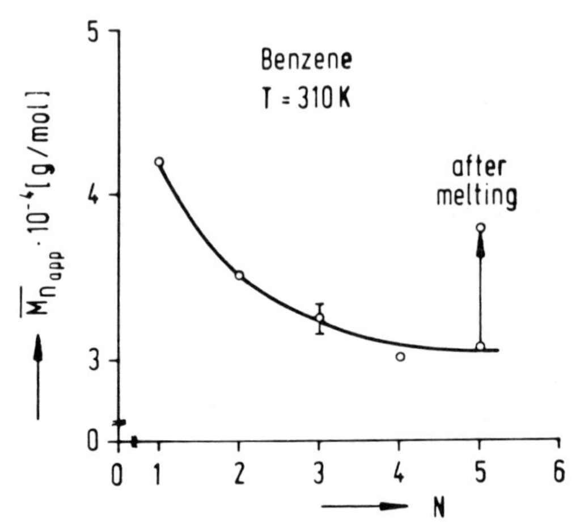

Fig. 1. The apparent molecular weight ($\bar{M}_{n,app}$) as a function of the number of precipitations (N).[6]

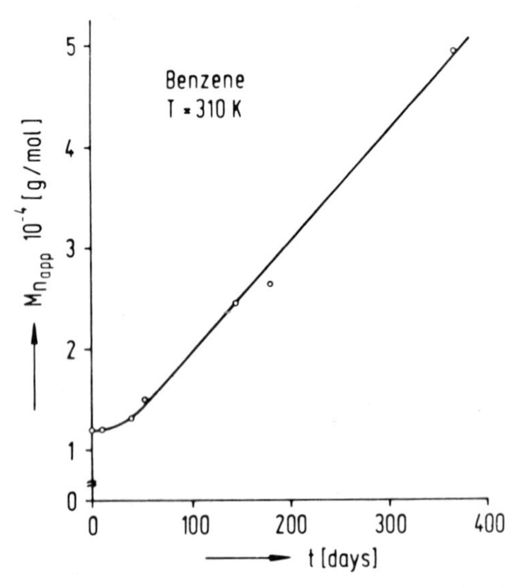

Fig. 2. Influence of storage time (t) on the apparent molecular weight ($M_{n,app}$).

MEASUREMENTS OF MOLECULAR WEIGHT AND SIZE

At the beginning of our investigations only a small amount of polymer was at our disposal. Therefore, for further measurements, the polymer had to be recovered by precipitation. Doing this, we found that against all expectations, the number-average molecular weight (M_n) of an unfractionated sample decreases steadily with increasing number of precipitations (Fig. 1). Of course, this was contrary to our expectations, because the average molecular weight should only have increased due to incomplete precipitation of the low molecular weight parts of the sample. A significant increase of the molecular weight, however, nearly up to the initial value, was measured after melting such samples (Fig. 1).

As shown in Fig. 2, the storage time of the dry polymer has a strong influence on the molecular weight, too, which we have therefore called 'apparent' molecular weight. Both effects and the configuration of the polymeric backbone lead us to the assumption that an aggregation of polymer molecules takes place, similar to the well-known stereo-association of poly(methylmethacrylate) (PMMA).[9]

To test this hypothesis, the solvent dependence of the aggregation process was examined by light scattering measurements, where the formation of larger species (aggregates) could be shown. In accordance

TABLE 1

Light Scattering Data[a] for Unfractionated Polymer[6]

Solvent	*T(K)*	n_1	d*n*/d*c* ($cm^3\ g^{-1}$)	$M_{w,app} \times 10^{-6}$ ($g\ mol^{-1}$)	$A_2 \times 10^5$ ($cm^3\ mol\ g^{-2}$)	$\sqrt{\bar{r}^2}$ (*nm*)
Trichloromethane	298	1·443	0·081	0·92	16·1	53
	308	1·437	0·083	0·88	16·2	51
	318	1·431	0·085	0·87	16·4	50
	328	1·425	0·088	0·87	16·3	48
Benzene	298	1·522	0·079	4·7	2·8	84
	308	1·517	0·082	4·1	3·3	84
	318	1·512	0·087	3·8	4·2	83
	328	1·507	0·090	3·5	4·6	82
THF[b]	298	1·404	0·076	24·1	0·9	115

[a] n_1, Refractive index of solvent; d*n*/d*c*, refractive index increment; $M_{w,app}$, apparent weight-average molecular weight; A_2, second virial coefficient; $\sqrt{\bar{r}^2}$, radius of gyration.

[b] Thanks are due to Dr W. Mächtle (BASF) for this measurement.

with reports on PMMA,[10] solvents enhancing or inhibiting aggregation were chosen. In analogy to these reports trichloromethane is a non-aggregating, benzene a weakly, and THF a strongly aggregating solvent (Table 1). This can be shown at the different apparent molecular weights obtained from the same sample. The mass and the size of the aggregates decrease slightly with increasing temperature. The radii of gyration are relatively small in comparison with the mass of the aggregates. On the other hand, the values of the second osmotic virial coefficients are 'normal'.

VISCOSITY OF DILUTE SOLUTIONS

The light scattering measurements were restricted by the narrow temperature range accessible due to the low boiling points of the chosen solvents. Therefore we had to find a solvent which has a high boiling point and works as a weakly or non-aggregating solvent. Such a solvent is 1,2-dichlorobenzene as is evident from Fig. 3. The reduced viscosity (η_{spec}/c) versus concentration (c) plot of the THF solution is no longer linear, and shows an upward deviation with concentration indicating a strong aggregation.

Because of the linear shape of the curve and the closeness to the trichloromethane curve, 1,2-dichlorobenzene has to be regarded as a weakly or, even, non-aggregating solvent. The polymer solutions in each solvent under investigation exhibit a non-Newtonian flow behaviour. They

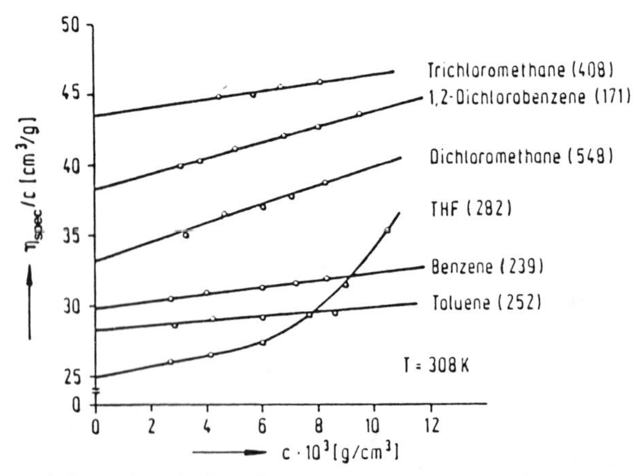

Fig. 3. Plot of the reduced viscosity versus concentration in various solvents. Shear gradient (s^{-1}) of the pure solvent is given in parentheses.[7]

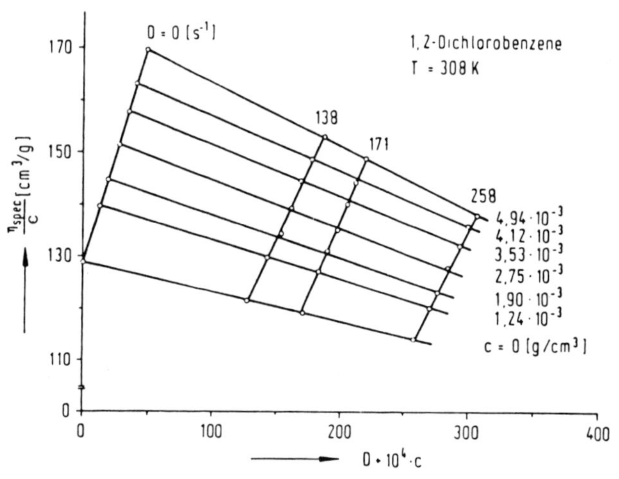

Fig. 4. Double extrapolation plot of reduced viscosity versus concentration and shear gradient.[7]

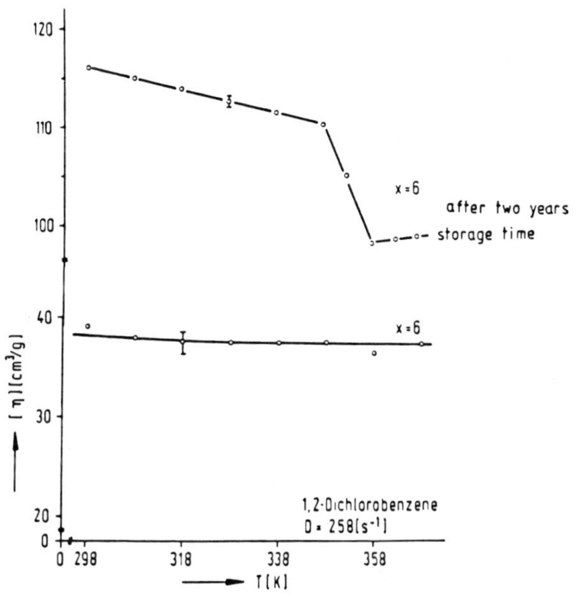

Fig. 5. Intrinsic viscosity as a function of temperature.[7]

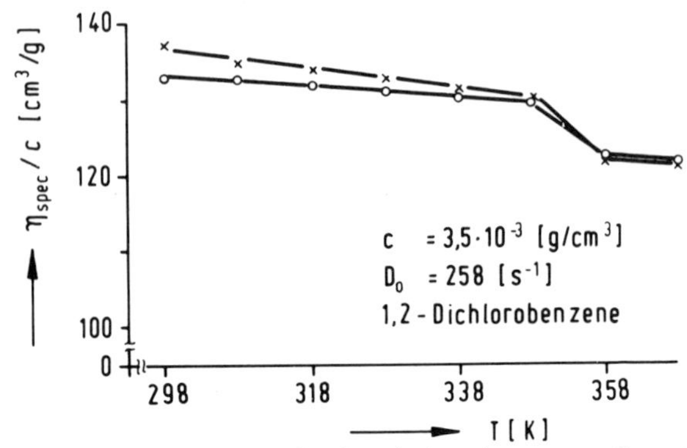

Fig. 6. Temperature dependence of reduced viscosity: ×, cooling; ○, heating.[7]

are structurally viscous. An example of this is shown in Fig. 4 for 1,2-dichlorobenzene as solvent at 308 K. The dependence of the reduced viscosity on the shear gradient D indicates the non-Newtonian behaviour. To obtain the intrinsic viscosity value $[\eta]$, a double extrapolation to zero concentration and to zero shear gradient has to be made. The temperature dependence of a solution of the polymer in 1,2-dichlorobenzene in the temperature interval of 298–368 K is shown in Fig. 5. The lower curve describes the behaviour of a freshly polymerized sample. Its intrinsic viscosity remains constant in the temperature range under investigation. The upper curve shows the behaviour of a stored sample. It shows a relatively small decrease up to 348 K, then the intrinsic viscosity decreases sharply and remains nearly constant above 358 K.

To test whether the decrease in viscosity is due to chain degradation, the reduced viscosity of one solution was measured first with increasing and then with decreasing temperature. As is evident from Fig. 6, the large step at 348 K appears in both directions so that bond scissions in the chain have to be excluded. The differences in the curve shape are probably due to an increase in concentration by solvent evaporation. The step can be attributed only to the decrease in the apparent molecular weight and in order to confirm this effect the temperature dependence of apparent molecular weight as investigated by light scattering.

THERMAL ANALYSIS

If light scattering measurements are carried out between 298 and 368 K changes in the apparent molecular weight of the scattering particles ar

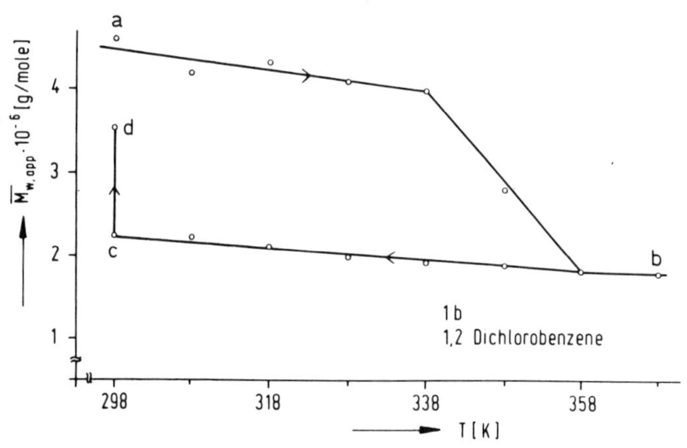

Fig. 7. Temperature dependence of the apparent molecular weight ($\bar{M}_{w,app}$).[8]

observed (Fig. 7), in a manner similar to that of the intrinsic viscosity–temperature relationship (Fig. 5).

When the solutions were cooled after measurement, the scattering particles retained the mass they had at 368 K, in contrast to the reported result that the temperature dependence of the intrinsic viscosity is reversible. To prove whether this disparity was caused by shear stress of the solutions during viscosity measurements, the solutions were subjected, after cooling, to a capillary flow in a viscosimeter before light scattering measurements were performed. As is obvious from Fig. 7, the shear of the solution results in a sudden increase of the apparent molecular weight, e.g. as shown at 298 K (i.e. Fig. 7(d)).

Since the dissolved particles are elongated in shape the shear stress during capillary flow causes them to rotate, taking positions, relative to each other, which are necessary for the formation of aggregates. This hypothesis can be supported by results of calorimetric measurements of solutions. Figure 8 shows the thermograms of the polymer in 1,2-dichlorobenzene ($c = 2{\cdot}5 \times 10^{-3}$ G CM^{-3}) after different pre-treatments. The different thermal pre-treatments correspond to those for the light scattering measurements. Curve (a) was obtained after the light scattering measurement at 298 K (see Fig. 7(a)), curve (b) after heating to 368 K and rapid cooling to 298 K, while curve (c) corresponds to the measurement at 298 K after slow cooling (see Fig. 7(c)). Curve (a) is similar to the DSC diagram of the stored solid sample (Fig. 9, curve 1).

As shown (Fig. 9), three endothermic transitions take place during the heating period. The transition at 332 K (curve 3) can be shown by

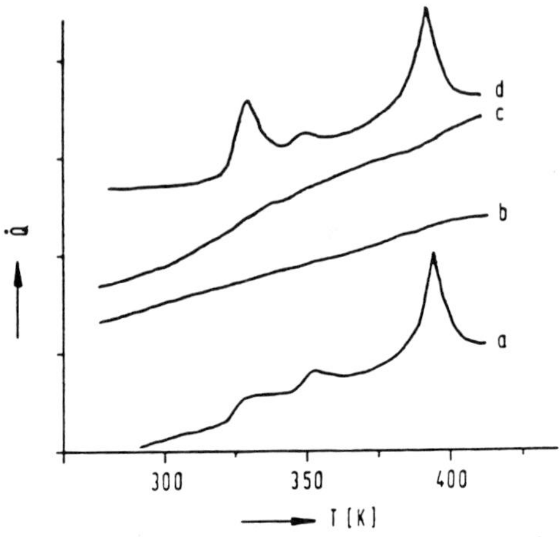

Fig. 8. Thermograms of a polymer solution.[8]

polarization microscopy to be attributed to the glassy–nematic transition. The peak, suggesting a first-order transition (curve 2), results from an overlap of thermodynamic and kinetic processes and is also known to occur in other polymers.[11] The processes considered are the glass transition and the non-equilibrium state arising from relaxation of the chain segments. It is absent during the second heating period (Fig. 9, curve 2) and only the glass transition step at 315 K occurs. The peak at 349 K (see marker, Fig. 9) is not associated with any transition of the liquid crystalline phase. The nematic texture still prevails as was ascertained by microscopy. Besides, the value of the transition enthalpy is much higher than that known for smectic–nematic transitions in other systems.[4] The peak, however, does not appear during the second heating period.

At 382 K, the nematic phase changes into the isotropic melt. To investigate the reasons for the existence and disappearance of the second peak, the sample was annealed at a temperature of 333 K for 72 h and then measured again (Fig. 9, curve 3). The peak reappears at the same temperature. This leads us to believe that it is caused by melting of the crystallites and we assume that crystallization contributes considerably to the formation of aggregates.

In curves (b) and (c) of the solution no endothermic transition can be observed. While in curve (a), the aggregates melt in the solution, they are dissolved in curves (b) and (c). The dissolved particles can, with the same probability, take all spatial positions, but an aggregation is only possible if two or more particles coincidentally lie in a definite direction.

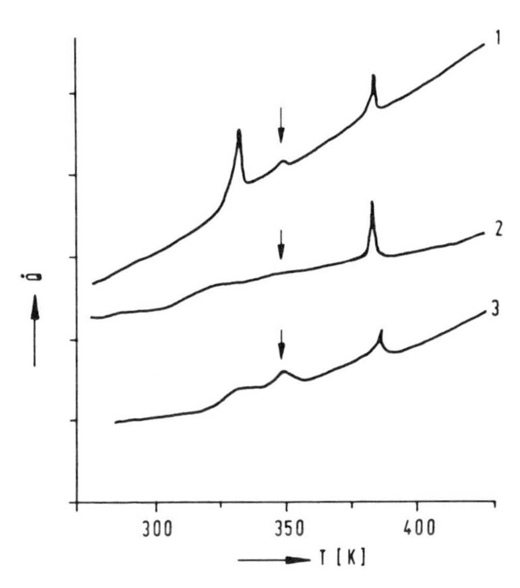

Fig. 9. DSC diagram of a stored sample.[8]

Consequently, the aggregation should take place slowly, but this was not observed during the course of the measurement. Before curve (d) (Fig. 8) was measured, the solution was subjected to shear stress through capillary flow. Here, the particles became oriented and aggregated through crystallization (shear-induced aggregation, i.e. crystallization). The trace of this curve (d) is similar to curve (a).

STRUCTURE OF THE AGGREGATES

The structure of the aggregates was investigated by x-ray analysis. From these results, we conclude that the molecules aggregate in solution by partial crystallization. We propose a structural model of the aggregates, the crystallites, as shown in Fig. 10. According to the position of the diffraction lines the side groups must be ordered orthogonally to the main chain. Due to their space filling, they must be slightly staggered along the main chain, because the phenyl ring in the side group requires a larger diameter for free rotation as we find.

From the width of the diffraction lines the ordered domains are very small, i.e. the lattice is strongly disturbed. The dimensions of the microcrystallites can be estimated with the help of the mean expansion[12] of

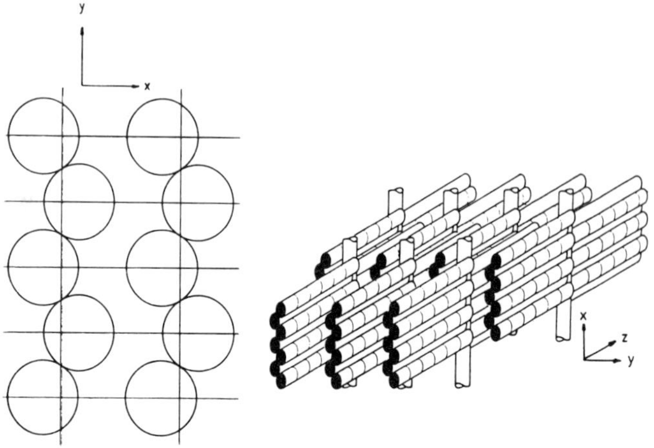

Fig. 10. Proposal for the arrangement of the molecules in microcrystallites.

the diffraction lines. Accordingly, the dimensions of crystallites were 14 units in the z-direction and 5 units in the x- and y-directions. They are encapsuled in an amorphous matrix.

CONCLUSIONS

The aggregation of poly(methacrylic acid) derivatives with mesogenic side groups in solution and in solid takes place by stereo-association, i.e. the formation of microcrystallites. These aggregates are relatively stable and can only be destroyed by melting in solution. The rate of crystallization is enhanced by subjecting the solution to shear stress by capillary flow. A structural model of the aggregates is proposed and discussed.

ACKNOWLEDGEMENT

We should like to thank the Deutsche Forschungsgemeinschaft for financial support.

REFERENCES

1. Shibaev, V. P., Freidzon, Y. S. and Platè, N. A., *Polym. Sci. USSR*, 1978, **20**, 94.

2. Perplies, E., Ringsdorf, H. and Wendorff, J. H., *J. Polym. Sci., Polym. Lett. Ed.*, 1976, **13**, 243.
3. Finkelmann, H. and Day, D., *Makromol. Chem.*, 1979, **180**, 2269.
4. Finkelmann, H., Happ, M., Portugall, M. and Ringsdorf, H., *Makromol. Chem.*, 1978, **179**, 2541.
5. Ringsdorf, H. and Schneller, A., *Brit. Polym. J.*, 1981, **13**(2), 43.
6. Springer, J. and Weigelt, F. W., *Makromol. Chem.*, 1983, **184**, 1489.
7. Springer, J. and Weigelt, F. W., *Makromol. Chem.*, 1983, **184**, 2635.
8. Cackovic, H., Springer, J. and Weigelt, F. W., in preparation.
9. Liquori, A. M., Anzuiono, G., Goiro, V. M., D Alagni, M., de Santis, P. and Savino, M., *Nature*, 1965, **206**, 358.
10. Vorenkamp, J., Bosscher, F. and Challa, G., *Polymer*, 1979, **20**, 59.
11. Bovey, F. A., *Polymer Conformation and Configuration*, Academic Press, New York and London, 1969.
12. Klug, H. P., *X-Ray Diffraction Procedures*, Wiley, New York, 1962.

15

MEASUREMENT OF ORIENTATIONAL ORDER IN LIQUID CRYSTALLINE SAMPLES BY NMR SPECTROSCOPY

J. W. EMSLEY

Department of Chemistry, University of Southampton, UK

INTRODUCTION

Liquid crystalline samples differ from normal liquids in possessing long-range orientational order. For convenience only uniaxial phases such as the nematic and smectic A phases will be considered; the biaxial phases have been extensively studied by NMR and such work has been recently reviewed.[1] For a collection of rigid molecules the extent of orientational ordering is characterized by a singlet orientational distribution function,[2] $f(\Omega)$, which may be expressed as a linear combination of Wigner rotation matrices, $D^L(\Omega)$, where Ω represents the three Euler angles made by axes fixed in the molecule and the mesophase director. Thus

$$f(\Omega) = \sum_{Lmn} f_{Lmn} D^L_{mn}(\Omega) \tag{1}$$

where the expansion coefficients are

$$f_{Lmn} = \bar{D}^{L*}_{mn} \tag{2}$$

where the bar denotes an ensemble average. The f_{Lmn} are also known as order parameters, and an infinite set is required to completely define $f(\Omega)$.

It is possible to obtain $f(\Omega)$ for samples which have been cooled from a liquid crystalline phase into a glass and such experiments are described in Chapter 13. More generally, however, it is not possible to determine $f(\Omega)$, but rather only a limited number of the f_{Lmn}. The order parameters of a particular rank L are related to experimental interactions of the same rank. Most experiments measure partially-averaged components of second-rank interaction tensors along laboratory fixed axes. Thus if $\tilde{A}_{\parallel}$ is such a component along the mesophase director then for a rigid molecule it can be related to components $A_{\alpha\beta}$ in a molecular frame (abc) by

$$\tilde{A}_{\parallel} = A_0 + \tfrac{2}{3}\sum_{\alpha\beta}^{abc} S_{\alpha\beta} A_{\alpha\beta} \tag{3}$$

where A_0 is the isotropic average and $S_{\alpha\beta}$ are elements of the Saupe second-rank ordering matrix.

ORDERING MATRIX AND MOLECULAR SYMMETRY

There are five independent, non-zero elements of S for a general set of axes, hence five independent quantities $\tilde{A}$ are required in order to determine all elements of the order matrix. Having determined all the elements in this way the matrix S, which is real and symmetric, can be diagonalized by a transformation to a set of principal axes (xyz). In the principal frame there are only two independent elements of S; S_{zz}, the largest component, and $S_{xx} - S_{yy}$, which is often termed the biaxiality in S. It is important to realize that S, and hence its biaxiality, are properties of the molecules and should not be confused with the biaxiality of a phase such as the smectic C.

If the molecules have at least two orthogonal mirror planes or a C_n axis with $n > 2$, then xyz can be identified and only two experimental quantities $\tilde{A}_{\parallel}$ are required to completely specify the ordering matrix.[2] It is often assumed that the molecules comprising liquid crystalline phases have $D_{\infty h}$ symmetry ('cylindrically, or axially symmetric') and hence require only one order parameter to describe the transformation between laboratory and molecular reference frames. This is almost invariably wishful thinking since few, if any mesogenic molecules have such symmetry. Moreover, the molecules of interest are also almost invariably not rigid, and usually have several internal degrees of rotational freedom. The description of orientational ordering for such flexible molecules has been described by Emsley and Luckhurst[3] and by Zannoni.[2]

ORIENTATIONAL ORDERING IN FLEXIBLE MOLECULES

To simplify the discussion we will consider a molecule which adopts a finite number of conformations. The partially averaged component along the director of some second-rank interaction is now[3]

$$\tilde{A}_{\parallel} = A_0 + \tfrac{2}{3}\sum_{n} P^n \sum_{\alpha,\beta}^{abc}{}' S^n_{\alpha\beta} A^n_{\alpha\beta} \tag{4}$$

S^n is an ordering matrix for the nth conformation with statistical weight P^n. Note that it is impossible to obtain separately either P^n or $S^n_{\alpha\beta}$ from $\tilde{A}_{\parallel}$; only the products of $P^n S^n_{\alpha\beta}$ can possibly be determined directly. Moreover, for liquid crystals of only moderate complexity there are many conformations of appreciable weight so that there are insufficient experimentally measured values of $\tilde{A}_{\parallel}$ even from NMR to obtain directly the products $P^n S^n_{\alpha\beta}$. It is necessary, therefore, to test theoretical models for predicting the P^n and S^n, such as that formulated by Emsley *et al.*[4] for liquid crystals containing flexible alkyl chains, and which is discussed in Chapter 7 by Luckhurst.

NMR EXPERIMENTS

There are three interactions which may be measured by NMR and which can be used to test theories of orientational ordering via eqn (4). Of these the most attractive is dipolar coupling, D_{ij}, since the components in a molecular frame are determined by geometry alone. For a pair of nuclei in a rigid sub-unit of the molecule the partially-averaged component of D_{ij} along the director is

$$\tilde{D}_{ij} = \frac{-\gamma_i \gamma_j h}{4\pi^2 \langle r_{ij}^3 \rangle} S_{ij} \tag{5}$$

where γ_i and γ_j are the gyromagnetic ratios and r_{ij} is the distance between the two nuclei. The brackets $\langle\ \rangle$ denote an average over vibrational motion. The order parameter S_{ij} is the average

$$S_{ij} = \sum_{n} P^n S^n_{ij} \tag{6}$$

where S^n_{ij} is the local order parameter for the direction of r_{ij} in the nth conformation. If the value of $\langle r^3 \rangle$ is known then S_{ij} is obtained directly from $\tilde{D}_{ij}$; in practice it is usually possible to estimate a value for $\langle r^3 \rangle$ from the known structure of related molecules with a precision, for protons, of perhaps a few per cent.

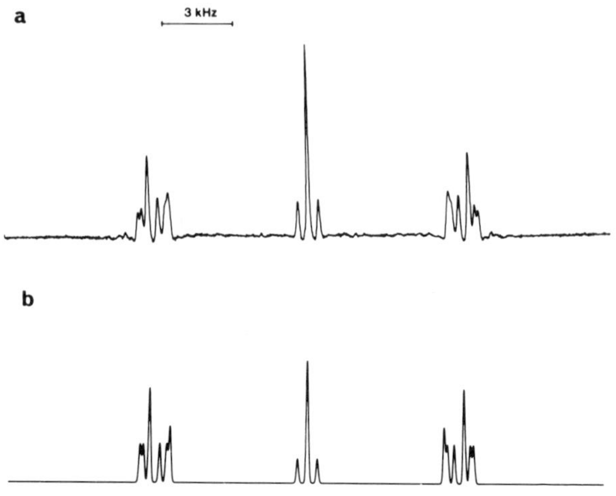

Fig. 1. Proton spin echo spectra of 4-*n*-pentyl-d_{11}-4′-cyanobiphenyl-d_4. (a) Experiment; (b) calculated. From Ref. 8.

When the two nuclei are in rigid sub-units which are moving relative to one another then the measured average dipolar interaction depends on the detailed nature of the conformational distribution.

Proton spectra of liquid crystal samples are too complex to be resolved unless some means of spectral simplification is used. This involves partial deuteriation and removal of the proton–deuteron dipolar coupling either by a double resonance experiment,[5,6] or by using spin echo refocusing.[7,8,9] An example of the spin echo technique is shown in Fig. 1 which is the proton spin echo spectrum of the four protons on 4-*n*-pentyl-d_{11}-4′-cyanobiphenyl-d_4 (5CB-d_{15}),

D D
NC–⟨ring⟩–⟨ring⟩–C_1–C_2–C_3–C_4–C_5
D D

The spectrum yields three distinct dipolar couplings enabling the local order matrix for the protonated ring to be obtained;[6] the biaxiality in this local ordering matrix is of the order of 10 % of the major order parameter.

Even more spectral simplification can be achieved by combining spin echo refocusing with multiple quantum excitation and detection as demonstrated by Sinton and Pines.[10]

The proton work has a great deal of promise for the future but at present

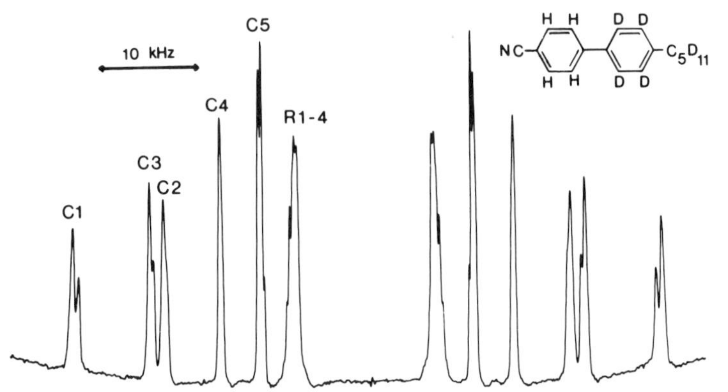

Fig. 2. Deuteron spectrum of 4-*n*-pentyl-d_{11}-4′-cyanobiphenyl-d_4. From Ref. 6.

the number of examples is very small, mainly because of difficulties in obtaining good, decoupled spectra. The quadrupolar interaction is the most widely used in seeking to characterize orientational order. The nucleus used almost exclusively in this way is deuterium, which gives simple spectra, such as that shown in Fig. 2 of 5CB-d_{15}. The quadrupole moment of the deuteron is small and hence spectra are obtained by pulse techniques without undue difficulty. The splitting $\Delta\nu_i$ produced by a deuteron at site i in a liquid crystalline sample yields directly $\tilde{q}_i$ the component of the quadrupolar tensor along the director (assuming that the orientation of the director with respect to the field is known); thus

$$\Delta\nu_i = (\tfrac{3}{2})\tilde{q}_i \tag{7}$$

The value of $\tilde{q}_i$ is related to $q_{\alpha\beta}$, components of the quadrupolar tensor in a molecular frame, by

$$\tilde{q}_i = q^i_{bb}[S^i_{zz}(l^2_{zbi} + \tfrac{1}{3}\eta^i(l^2_{zai} - l^2_{zci})) \\ \times (S^i_{xx} - S^i_{yy})((l^2_{xbi} - l^2_{ybi}) + \tfrac{1}{3}\eta^i(l^2_{xai} - l^2_{xci} + l^2_{yci} - l^2_{yai}))] \tag{8}$$

l_{bi} is the direction cosine of z with b and the xyz axes are fixed in the rigid sub-unit containing the ith deuteron. The asymmetry parameter $\eta^i = (q^i_{aa} - q^i_{cc})/q^i_{bb}$, and the quadrupolar coupling constant q^i_{bb} must be known, as well as the geometrical parameters in order to use eqn (8) to determine the S^i_{zz} and $S^i_{xx} - S^i_{yy}$ from $\tilde{q}_i$. Note that eqn (8) is valid for any choice of abc if xyz are principal axes for the local ordering matrix of the ith rigid unit. Almost invariably, however, in mesogens the principal axes of the local ordering matrices are unknown, but fortunately it is possible to choose abc to be principal axes of q^i in which case eqn (8) is still valid but obviously the order parameters are not the principal components of the local ordering matrix. It is found by experiment that b lies along a C–D bond direction to within a few degrees and that η^i is of the order of 5% and can often be neglected when deriving S^i_{zz}. Choosing xyz to coincide with abc and setting η^i to be zero gives

$$q_i = q^i_{bb} S^i_{bb} \tag{9}$$

It is found that q_{bb} for C–D bonds lie in the ranges:

$$q_{bb} = 168 \pm 5\,\text{kHz for aliphatic hydrocarbons}$$
$$q_{bb} = 185 \pm 5\,\text{kHz for aromatic hydrocarbons}$$

It is possible, therefore, to determine S^i_{bb} values for all the deuteron sites in a molecule to a precision of about $\pm 5\%$.

The third method of determining order parameters uses carbon-13 shielding constants σ_i and for this interaction, which has a finite isotropic average, eqn (3) for a rigid molecule gives

$$\tilde{\sigma}_i = \sigma_i^0 + (\tfrac{2}{3}) \sum_{\alpha\beta}^{abc} S_{\alpha\beta} \sigma_{\alpha\beta} \tag{10}$$

The molecular components $\sigma_{\alpha\beta}$ depend upon the electronic structure and are known with much lower precision than $q_{\alpha\beta}$; moreover, the shielding tensor is affected quite strongly by changes in conformation. This means that it is difficult to use the measured quantity $(\tilde{\sigma}_i - \sigma_i^0)$ to determine local order parameters with good precision and more studies of model systems are necessary in order to assess the usefulness of C-13 experiments.

CONCLUSIONS

The work on low molar mass liquid crystals by NMR spectroscopy has established that the low symmetry and flexibility of the molecules must be taken into account when describing orientational ordering. The neglect of these factors, particularly the flexibility, leads to a serious misinterpretation of the significance of the data. It is necessary to develop models for both the conformational distribution and the ordering matrices for the different conformational states, in order to understand the differences between the local ordering matrices of each rigid sub-unit in a mesogen. These problems are also present when studying liquid crystalline polymers and here the complexity of the molecules means that it is difficult to interpret the variation of measured local ordering parameters. Nevertheless, the analyses should be based on theories of how orientational order varies with conformational state. The view that orientational ordering of molecules in liquid crystalline samples can be described by a single order parameter even when the molecules lack symmetry and are flexible has been shown to be incorrect by NMR experiments. However, the conclusions are generally applicable and mean that the order parameters determined by techniques such as the refraction of light must be interpreted with caution. They can be used as crude monitors of the degree of ordering of the molecules in a sample but they give some unknown combination of the population and ordering of individual molecular conformations.

REFERENCES

1. Doane, J. W., in *N.M.R. of Liquid Crystals*, Emsley, J. W. (Ed.), Reidel, Dordrecht, 1984.
2. Zannoni, C., in *Molecular Physics of Liquid Crystals*, Luckhurst, G. R. and Gray, G. W. (Eds), Academic Press, New York, 1979.
3. Emsley, J. W. and Luckhurst, G. R., *Mol. Phys.*, 1980, **40**, 19.
4. Emsley, J. W., Luckhurst, G. R. and Stockley, C. P., *Proc. Roy. Soc., London*, 1982, **A381**, 117.
5. Emsley, J. W., Luckhurst, G. R., Gray, G. W. and Mosley, A., *Mol. Phys.*, 1978, **35**, 1499.
6. Emsley, J. W., Luckhurst, G. R. and Stockley, C. P., *Mol. Phys.*, 1981, **44**, 565.
7. Emsley, J. W. and Turner, D. L., *J. Chem. Soc. Faraday Trans. II*, 1981, 1493.
8. Avent, A. G., Emsley, J. W. and Turner, D. L., *J. Magn. Reson.*, 1983, **52**, 57.
9. Avent, A. G., Emsley, J. W. and Luckhurst, G. R., *Mol. Phys.*, 1983, **49**, 737.
10. Sinton, S. and Pines, A., *Chem. Phys. Lett.*, 1980, **76**, 263.

16

PHASE BEHAVIOUR OF DYE-CONTAINING LIQUID CRYSTALLINE COPOLYMERS AND THEIR MIXTURES WITH LOW MOLECULAR WEIGHT LIQUID CRYSTALS*

H. RINGSDORF and H.-W. SCHMIDT

Institut für Organische Chemie, Universität Mainz, Federal Republic of Germany

and

G. BAUR and R. KIEFER

Fraunhofer Institut für Angewandte Festkörperphysik (IAF), Freiburg im Breisgau, Federal Republic of Germany

INTRODUCTION

Liquid crystalline side-chain polymers can be obtained by connecting the mesogenic groups and the polymer chain via a flexible spacer. The liquid crystalline polymers combine the well-known properties of low molecular weight liquid crystals with those of polymers.[1]

Pleochroic dyes dissolved in low molecular weight nematic liquid crystals have been investigated intensively and are used in guest–host displays.[2] Dye-containing copolymers with mesogenic units in the side chain have recently been discussed.[3–5] They may open interesting application possibilities in the field of display technology. Liquid crystalline dye-containing copolymers (Fig. 1) can be obtained by copolymerization of mesogenic monomers and dye monomers.

By the covalent fixation of dye and mesogenic group to the same polymer chain it is possible to obtain systems with high and temperature-independent dye concentrations. This is the main difference from low molecular weight guest–host systems in which the solubility of the dye

* This contribution is an extended abstract of a paper presented at the ESF Workshop on Liquid Crystal Polymer Systems.

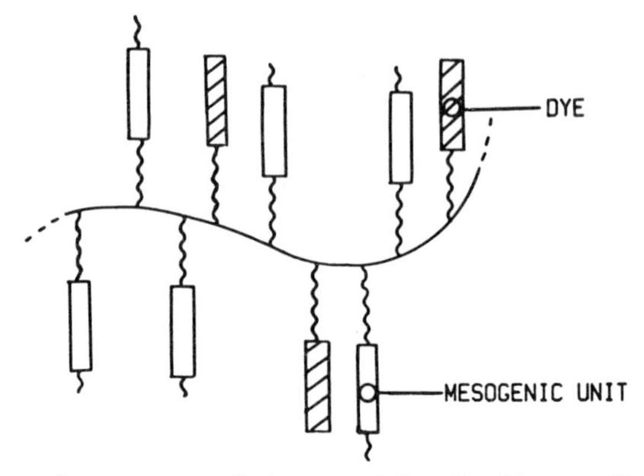

Fig. 1. Schematic structure of dye-containing liquid crystalline side-chain copolymers.

in the nematic host phase is frequently insufficient and temperature dependent.[6,7]

In this chapter we describe the synthesis and the phase behaviour of liquid crystalline anthraquinone copolymers with different mesogenic units and azo dye-containing copolymers of the general structure:

$$\left[CH_2 - CH - COO + CH_2 +_6 N \right]_n \quad \text{(O, } NH_2\text{, O; O, } NH_2\text{, O)}$$

$$\left[CH_2 - CH - COO + CH_2 +_6 \square \right]_m$$

$\square$ =

$-O-C_6H_4-COO-C_6H_4-CN$

$-O-\overset{O}{\overset{\|}{C}}-C_6H_4-\langle H \rangle-C_3H_7$

$-O-C_6H_4-C_6H_4-CN$

$$\left[CH - COO + CH_2 +_6 O - C_6H_4 - COO - C_6H_4 - CN \right.$$
$$\left. CH_2 \right]_n$$
$$\left[CH - COO + CH_2 +_6 O - C_6H_4 - N{=}N - C_6H_4 - CN \right.$$
$$\left. CH_2 \right]_m$$

In addition, mixtures of these copolymers with low molecular weight liquid crystals were investigated and compared with the corresponding low molecular weight guest–host system. The order parameters of the anthraquinone dye in the mixtures were determined.

EXPERIMENTAL

Monomers

The mesogenic acrylate monomers were prepared by procedures described in the literature (phenylbenzoate-,[8] phenylcyclohexane-,[9] biphenyl-derivatives,[10] azo dye monomer[11]). The anthraquinone monomer was synthesized by Dr Etzbach, BASF, Ludwigshafen.

Copolymers

The random copolymers were obtained by solution polymerization at 70 °C in dioxane or toluene with AIBN as initiator. They were purified by GPC with PVA (Merck) and THF as eluent. The copolymer compositions were determined by UV measurements and elemental analysis.

Mixtures

The mixtures of the copolymers with low molecular weight liquid crystals were prepared in various weight ratios and dissolved in THF. The solvent was then evaporated and the mixtures dried *in vacuo*.

Characterization

The thermal behaviour of the copolymers and the mixture was investigated by differential scanning calorimetry using a DSC-2C (Perkin–Elmer). The optical characterization was made using a polarizing microscope POL-BK II (Leitz) with a Mettler FP2 hot stage. The solubility of the dye in the low molecular weight host and the degrees of orientation of the mixtures were determined at the IAF in Freiburg.

RESULTS AND DISCUSSION

Phase Behaviour of the Dye-Containing Copolymers

The phenylbenzoate copolymers with the anthraquinone dye form nematic phases which decrease with increasing content of the dye monomer as shown in Fig. 2. The copolymer with an anthraquinone monomer content of about 39 % by weight is no longer liquid crystalline.

This phase behaviour points to the fact that the anthraquinone dye has

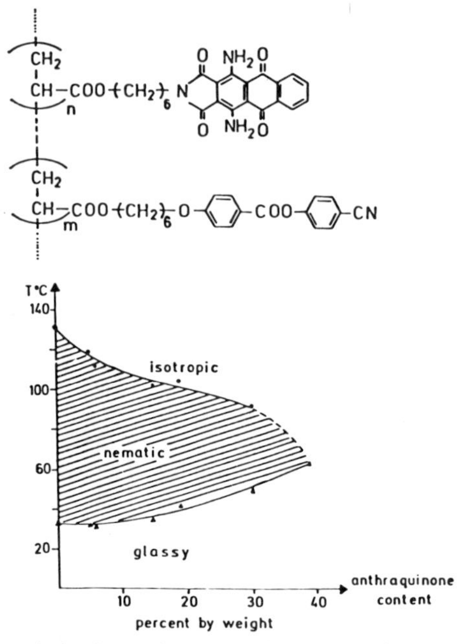

Fig. 2. Phase behaviour of anthraquinone-containing copolymers.

TABLE 1

Dye Contents and Phase Behaviour of Anthraquinone-Containing Liquid Crystalline Copolymers

–[ ]	*Dye content* (*wt.* %)	*Phase behaviour*[a] (°C)
$-O-C_6H_4-COO-C_6H_4-CN$	15	g 34 n 111i
$-O-C(=O)-C_6H_4-C_6H_{10}(H)-C_3H_7$	8	g −1 n 27i
$-O-C_6H_4-C_6H_4-CN$	11	g 39 n 118i

[a] Phase transition temperatures (°C) measured by DSC. Phases are: g, Glassy; n, nematic; i, isotropic.

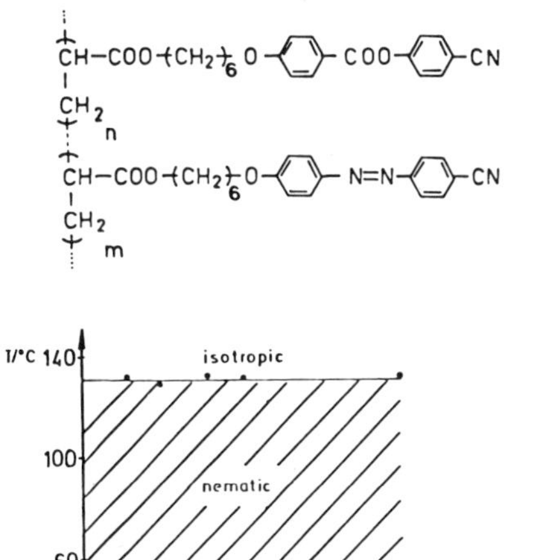

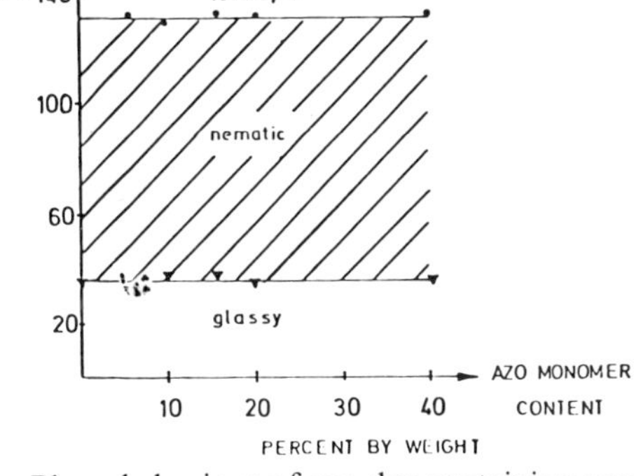

Fig. 3. Phase behaviour of azo dye containing copolymers.

only slight mesogenic properties and disturbs the formation of the nematic phase. On the other hand, the glass transitions increase with anthraquinone content. This is in contrast to comparable experiments with azo dye containing copolymers of the same type[3,11] as shown in Fig. 3. In this case the dye is a mesogenic system itself. Thus, the phase transitions of these copolymers are independent of the copolymer composition.

For the phenylcyclohexane–anthraquinone copolymers and the biphenyl–anthraquinone copolymers examples are given in Table 1 and compared with a phenylbenzoate–anthraquinone copolymer.

The influence of the mesogenic comonomer is obvious: the biphenyl- and the phenylbenzoate-derivatives show a comparable phase behaviour while the phenylcyclohexane copolymer is nematic at room temperature.

An ordered nematic orientation of the copolymers can be frozen-in below the glass transition yielding films with dichroic properties.

Mixtures of the Anthraquinone-Containing Copolymers with Low Molecular Weight Liquid Crystals

Liquid crystalline side-chain polymers are miscible with low molecular weight liquid crystals.[12–14] These mixtures show improved properties

TABLE 2
Mixtures of the Anthraquinone Monomer, the Phenylbenzoate–Anthraquinone Copolymers with the Binary Phenylbenzoate Host Mixture

Dye components		*Mixtures*		
Type	*Dye content (wt-%)*	*Dye content (wt-%)*	*Copolymer content (wt-%)*	*Clearing point (°C)*
Monomer	100	(0·46[a])	—	57
Copolymer 1	15	1·5	10	59
Copolymer 2	19	1·5	8	58
Copolymer 3	30	1·5	5	57

[a] Maximum solubility at 31 °C.

(phase behaviour, surface and electric field orientation) compared to the pure polymers.

Miscibility studies of the nematic dye-containing copolymers with suitable low molecular weight liquid crystals yield homogeneous nematic mixtures. A binary mixture of 4-cyanophenyl-4-heptylbenzoate and 4-cyanophenyl-4-pentylbenzoate is miscible with the phenylbenzoate–anthraquinone copolymers.

HOST: Binary nematic mixture of

$CH_3(CH_2)_6$–C₆H₄–COO–C₆H₄–CN 60 mol-%

$CH_3(CH_2)_4$–C₆H₄–COO–C₆H₄–CN 40 mol-%

The dye content in these mixtures is higher than the maximum solubility of the anthraquinone monomer in the same nematic phase as shown in Table 2.

By using copolymers with different dye compositions it is possible to reduce the copolymer content in the mixture while retaining a constant dye concentration. It should be noted that the value of 1·5% (by weight) is not the maximum dye content for these mixtures. In low molecular weight guest–host systems derivatives of this type of anthraquinone dye are less soluble than 1% (by weight) in several nematic hosts.[6]

Order Parameter of the Anthraquinone Dye in the Mixtures

The order parameters of the dye (S_D) were determined from the optical absorption spectra parallel and perpendicular to the homogeneous orientation of the mixtures as shown in Fig. 4.[2]

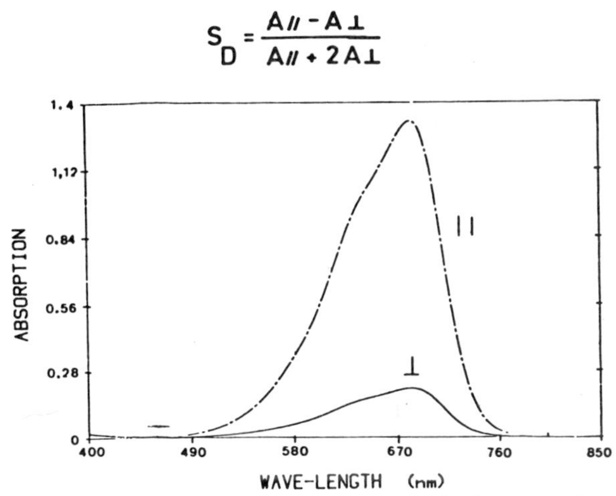

Fig. 4. Absorption spectra parallel (‖) and perpendicular (⊥) of copolymer 2 (8 wt-%) in the binary phenylbenzoate host at 23 °C ($A_{\parallel}$, absorption parallel; $A_{\perp}$, absorption perpendicular).

The measurements show that the order parameters of the dissolved anthraquinone monomer are higher ($S_D = 0{\cdot}61$) at low temperatures (30·9 °C) than those of a mixture of copolymer 2 ($S_D = 0{\cdot}58$) in the same low molecular weight host. It is interesting to note, however, that the temperature dependence of the order parameter of the polymer system is lower. At 47·3 °C both systems show the same order parameter ($S_D = 0{\cdot}51$). The temperature dependence of the order parameters for both systems is shown in Fig. 5.

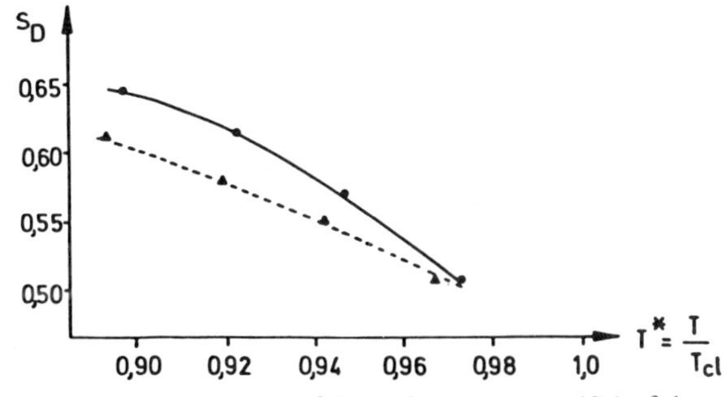

Fig. 5. Temperature dependence of the order parameter (S_D) of the anthraquinone monomer and copolymer 2 in the binary nematic phenylbenzoate mixture. ———, Anthraquinone monomer in the binary nematic host; – – – – –, anthraquinone copolymer 2 (19 wt-%) in the binary nematic host.

REFERENCES

1. Ciferri, A., Krigbaum, W. R. and Meyer, R. B., *Polymer Liquid Crystals*, Academic Press, New York, 1982.
2. Cox, R. J., *Mol. Cryst. Liq. Cryst.*, 1979, **55**, 1.
3. Ringsdorf, H., Schmidt, H.-W. and Schneller, A., *12. Freiburger Arbeitstagung Flüssigkristalle*, March–April, 1982.
4. Finkelmann, H., Benthack, H. and Rehage, G., *J. Chimie Phys.*, 1983, **80**, 163.
5. Ringsdorf, H., Schmidt, H. W., Baur, G. and Kiefer, R., *Polymer Preprints*, 1983, **24**(2), 306.
6. Aftergut, S. and Cole, H. S., *Mol. Cryst. Liq. Cryst.*, 1981, **78**, 271.
7. Cognard, J. and Hieu Phan, T., *Mol. Cryst. Liq. Cryst.*, 1981, **68**, 207.
8. Portugall, M., Ringsdorf, H. and Zentel, R., *Makromol. Chem.*, 1982, **183**, 2311.
9. Kreuder, W., Master Thesis, Mainz, 1982.
10. Shibaev, V. P., Kostromin, S. G. and Platé, N. A., *Eur. Polym. J.*, 1982, **18**, 651.
11. Ringsdorf, H. and Schmidt, H.-W., *Makromol. Chem.*, 1984, **185**, 1327.
12. Cser, F., Nyitrai, K., Hardy, G. and Varga, J., *J. Polym. Sci., Polym. Symp.*, 1981, **69**, 91.
13. Ringsdorf, H., Schmidt, H.-W. and Schneller, A., *Makromol. Chem., Rapid Commun.*, 1982, **3**, 745.
14. Finkelmann, H., Kock, H.-J. and Rehage, G., *Mol. Cryst. Liq. Cryst.*, 1982, **89**, 23.

17

DIELECTRIC RELAXATION MEASUREMENTS AND X-RAY INVESTIGATIONS OF LIQUID CRYSTALLINE SIDE-CHAIN POLYMERS

R. ZENTEL, G. STROBL and H. RINGSDORF

Fachbereich Chemie, Universität Mainz, Federal Republic of Germany

In the last few years several liquid crystalline side-chain polymers have been synthesized,[1-4] which combine the properties of low molecular weight liquid crystals with those of polymers.[5-9]

In order to get more information concerning the molecular arrangement of the mesogenic groups and the dynamics of these systems, a series of liquid crystalline polymers of the general form

$$\left[\begin{array}{c} CH_2 \\ | \\ R_1-C-COO-(CH_2)_n-O-C_6H_4-COO-C_6H_4-R_2 \end{array}\right]_x$$

were synthesized and investigated (see Table 1).

These polymers were studied by x-ray diffraction of fibre samples and by frequency and temperature dependent dielectric measurements. For similar investigations on other polymers see Ref. 8.

In the dielectric relaxation spectrum up to five different kinds of motion were found. There are up to *three local motions* (β_1, β_2 and γ-relaxation), which are active below and above the static glass transition temperature. The β_1-relaxation is found in all investigated samples independent of the spacer length, the polymer chain and the end group R_2, and is attributed to internal relaxations of the mesogenic groups. The β_2-relaxation is only found for long spacers with six methylene groups and therefore assigned to relaxations within this sequence. The γ-relaxation is observed in polymers, which possess a more extended end group (R_2: n-OC_4H_9) capable of performing additional internal relaxations.

TABLE 1
Chemical Structure and Phase Behaviour of the Polymers under Investigation

Number	R_1	*n*	R_2	M_n	*Phase transition*[a] (°C)
1	H	2	OCH_3	39 000	g 62 n 116 i (4)
2	H	6	OCH_3	43 000	g 35 s_A 97 n 123 i (4)
3	H	2	CN	18 000	g 75 n 110 i (4)
4	H	6	CN	20 000	g 33 n 133 i (4)
5	CH_3	2	OCH_3	>100 000	g 97 n 120 i (1, 4)
6	CH_3	6	OCH_3	>100 000	g 47 n 111 i (1, 4)
7	CH_3	2	OC_4H_9	>100 000	g 137 s_A 163 i
8	CH_3	6	OC_4H_9	>100 000	g 39 s_A 109 n 114 i
9	Cl	2	OC_4H_9	>100 000	g 132 s_C 174 i
10	Cl	6	OC_4H_9	>100 000	g 39 s_C 109 n 116 i

[a] g, Glass transition temperature; n, nematic phase; s_A, smectic A phase; s_C, smectic C phase.

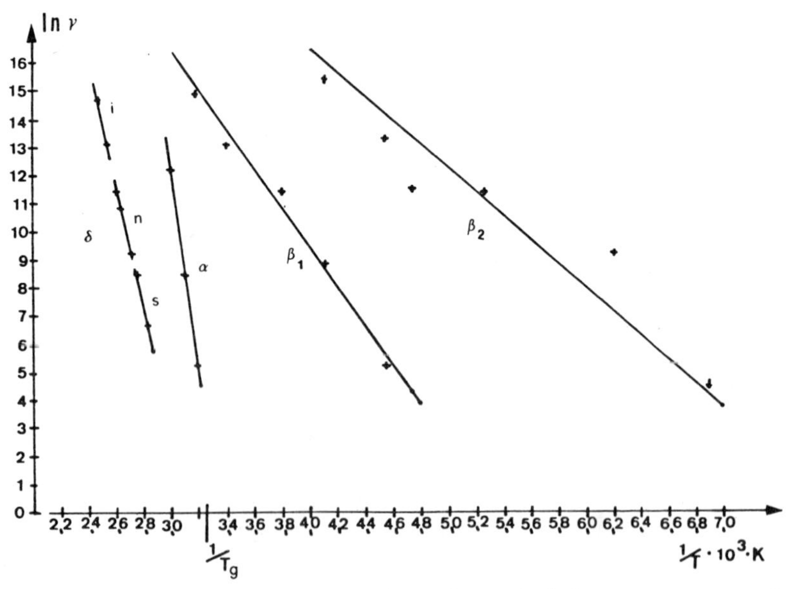

Fig. 1. Dependence of $\ln \gamma$ on $1/T$ for polymer **2** (i, isotropic; n, nematic; s, smectic).

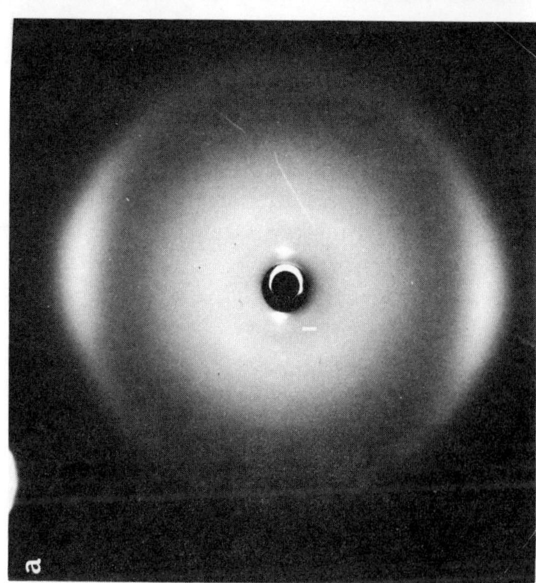

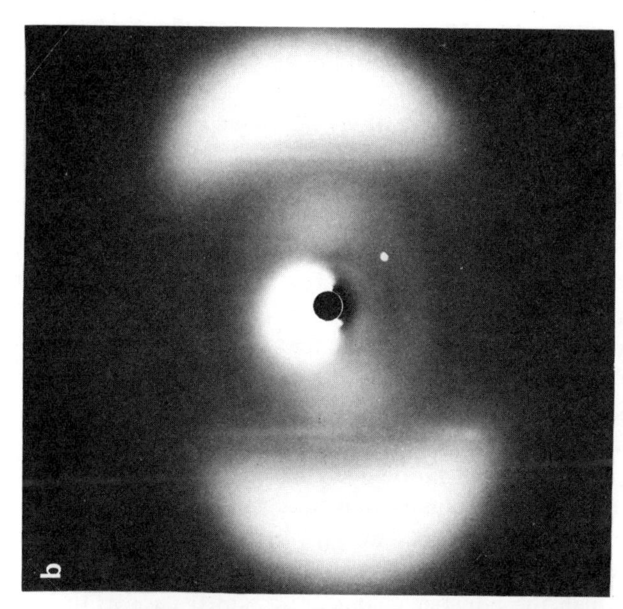

Fig. 2. X-ray fibre pattern of polymers in the (a) smectic phase (e.g. polymer **8**, independent of the spacer length) and of polymers with long spacer groups in the nematic phase (except the small angle reflection) and (b) x-ray fibre pattern of polymers with short spacer groups in the nematic phase (e.g. polymer **1**). The fibre axis is in the vertical direction.

At the *glass transition* (α-relaxation) the centres of gravity of the mesogenic groups and the polymer chain become mobile. Dipole reorientations associated with this motion contribute to the dielectric relaxation.

Additionally there is a *reorientation of the long axes* of the mesogenic groups about the polymer chain (δ-relaxation) which can be seen independently of the liquid crystalline phase (smectic, nematic, isotropic).

For liquid crystalline polyacrylates (R_1: H, polymers **1–4**) all these relaxations can be observed separately. Figure 1 shows the temperature dependence of the frequencies of the different relaxation processes using an Arrhenius plot.

If the polymer chain is altered to a polymethacrylic acid (R_1: CH_3, polymers **5–8**) or a polychloracrylic acid chain (R_1: Cl, polymer **10**) the local relaxations remain essentially unchanged. However, instead of the separated α and δ-relaxations only a mixed relaxation is found.

X-ray diffraction patterns of oriented fibres indicate a different arrangement of the mesogenic groups in polymers with short ($n = 2$) and long ($n = 6$) spacer groups (see Fig. 2). The x-ray pattern for polymers with long spacer groups (see Fig. 2(a)) agrees well with the proposed patterns for a structure in which the polymer chains are oriented parallel to the fibre axes, while the long axes of the mesogenic groups are arranged parallel to each other and perpendicular to the fibre axes. Their short-range order is as for low molecular weight liquid crystals.

The x-ray patterns of the polymers with short spacer groups (see Fig. 2(b)) can be interpreted in two ways:

—As a result of a different orientation behaviour on drawing, the long axes of the mesogenic groups are parallel to the fibre axes.

—The orientation of the mesogenic groups is still perpendicular to the fibre axes and the polymer chain, but the positional correlation of the mesogenic groups has changed.

REFERENCES

1. Finkelmann, H., Ringsdorf, H. and Wendorff, J. H., *Makromol. Chem.*, 1978, **179**, 273; Ringsdorf, H. and Schneller, A., *Makromol. Chem., Rapid Commun.*, 1982, **3**, 557.
2. Shibaev, V. P., Platè, N. A. and Freidzon, Ya. S., *J. Polym. Sci., Polym. Chem. Ed.*, 1979, **17**, 1655; Shibaev, V. P., Kostromin, S. G. and Platè, N. A., *Eur. Polym. J.*, 1982, **18**, 651.

3. Finkelmann, H. and Rehage, G., *Makromol. Chem., Rapid Commun.*, 1980, **1**, 31; Finkelmann, H., Kock, H. J. and Rehage, G., *Makromol. Chem., Rapid Commun.*, 1981, **2**, 317.
4. Portugall, M., Ringsdorf, H. and Zentel, R., *Makromol. Chem.*, 1982, **183**, 2311.
5. Wassmer, K. H., Ohmes, E., Kothe, G., Portugall, M. and Ringsdorf, H., *Makromol. Chem., Rapid Commun.*, 1982, **3**, 281.
6. Geib, H., Hisgen, B., Pschorn, U., Ringsdorf, H. and Spiess, H. W., *J. Am. Chem. Soc.*, 1982, **104**, 917.
7. Finkelmann, H. and Rehage, G., *Makromol. Chem., Rapid Commun.*, 1980, **1**, 733; *ibid.*, 1982, **3**, 859.
8. Kresse, H., Kostromin, S. and Shibaev, V. P., *Makromol. Chem., Rapid Commun.*, 1982, **3**, 509; Talroze, R. V., Sinitzyn, V. V., Shibaev, V. P. and Platè, N. A., *Mol. Cryst. Liq. Cryst.*, 1982, **80**, 211; Kostromin, S. G., Talroze, R. V., Shibaev, V. P. and Platè, N. A., *Makromol. Chem., Rapid Commun.*, 1982, **3**, 803.
9. Ringsdorf, H. and Zentel, R., *Makromol. Chem.*, 1982, **183**, 1245.

18

STRUCTURAL INVESTIGATIONS ON LIQUID CRYSTALLINE SIDE-CHAIN POLYMERS

P. ZUGENMAIER and J. MÜGGE

Institut für Physikalische Chemie der TU Clausthal, Clausthal-Zellerfeld, Federal Republic of Germany

INTRODUCTION

Small molecules with geometrical form-anisotropy and high polarizability may exhibit, besides the well-known crystalline and isotropic (liquid) phases, one or more liquid crystalline phases. If these molecules are attached to a polymer backbone, i.e. a polysiloxane chain by a flexible spacer, liquid crystalline phases are still present. The temperature range in which the liquid crystalline phases are stable may widen and liquid crystalline phases may even appear when the small molecular species exhibit only a crystalline phase. Studies with liquid crystalline side-chain polymers show that optical properties and the orientation function remain almost the same[1,2] as compared to the small molecular species.

The structure for different liquid crystalline phases of small molecules, that is the packing of the molecules, is in principle well established and depicted in many textbooks dealing with the liquid crystalline state. Structural models for liquid crystalline side-chain polymers in different liquid crystalline phases have also been proposed analogous to the packing of small molecules[3] but were challenged by other research workers.[4,5,6] The difficulty in establishing a unique structural model for liquid crystalline side-chain polymers is the poor quality of x-ray data. X-ray analysis is the preferred method for structural investigations on a molecular level. It only shows a few reflections in the liquid crystalline phase and cannot be interpreted in a unique way. The alignment of samples in an external field, which yields fibre x-ray diagrams, has not led to much improvement as regards the idea of packing of polymers. However, it can lead to information about the orientation of the polymer and establishes the idea of small domains within the liquid crystalline phase.[7]

In previous papers[6,8] a method was proposed to overcome the difficulties in evaluating structural models for liquid crystalline phases. If the crystalline phase of the side-chain polymer is available and the changes between the crystalline and liquid crystalline phases are known, the packing of the liquid crystalline phase may be determined. The accessibility of the polymer crystal structure by x-ray methods combined with conformation and packing analysis has become a standard method in polymer science.[9-11]

TABLE 1

Chemical Constitution of the C-5 and C-6 Polymers[a]

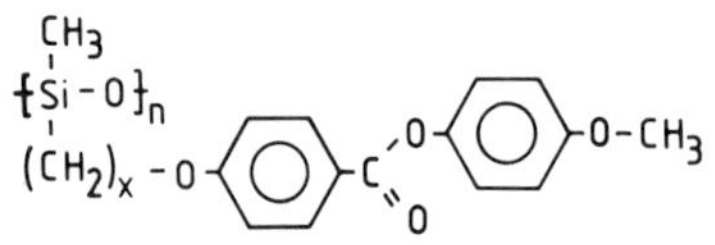

$-[Si(CH_3)(-(CH_2)_x-O-C_6H_4-C(=O)-O-C_6H_4-O-CH_3)-O]_n-$

x = number of methylene groups	polymer
5	C-5
6	C-6

[a] Reprinted from Ref. 6 by permission of Hüthig & Wepf Verlag, Basle.

This method has been applied to the smectic phase of two very similar liquid crystalline side-chain polysiloxanes.[6,7,12] They differ, as is shown in Table 1, only in the length of the alkyl spacer, connecting the side chain to the main chain. The two polymers are termed, according to this spacer length, C-5 and C-6. Both polymers exist in a crystalline and a smectic phase and can serve as model compounds for packing considerations of the smectic phase.

The degree of polymerization of the atactic polysiloxanes is about 100. The x-ray investigations were carried out with differently oriented specimens:

(i) mechanically drawn (with tweezers) fibres out of the liquid crystalline phase;
(ii) oriented fibres in a magnetic field.

The resulting x-ray patterns will be discussed with regard to the orientation of main and side chains, the domain structure and, qualitatively, the packing of the molecules in the crystalline and smectic phases.

X-RAY INVESTIGATIONS

Well-oriented samples of C-5 and C-6 were produced by drawing fibres out of the melt with a pair of tweezers. The crystallinity was improved by annealing.[6,12,13] The fibre x-ray patterns for crystalline, well-oriented samples of C-5 and C-6 are shown in Fig. 1. If the samples of C-5 and C-6 are crystallized and annealed in a magnetic field, the fibre x-ray diagrams of Fig. 2 are obtained. The fibre x-ray patterns of differently oriented C-5 polymer (Figs 1(a) and 2(a)) clearly show layer lines. The layer lines are missing on comparable diagrams of C-6 (Figs 1(b) and 2(b)). A uniaxial orientation (fibre axis) is observed for C-5 and C-6 along the drawing direction in Fig. 1 (vertical line) and along the magnetic field in Fig. 2 (horizontal line). This has been proved by a rotation around the fibre axis which produces the same x-ray fibre pattern or by placing the x-ray beam parallel to the drawing direction or magnetic field which gives a pattern with rings, indicating random orientation of the crystallites in a plane perpendicular to the fibre axis (Fig. 3).

Fibres of the smectic phase can be produced by drawing samples out of the liquid crystalline phase or by alignment in a magnetic field with subsequent quenching of the fibres.[6,12] The corresponding x-ray fibre patterns are represented in Fig. 4 for C-5 and C-6 polymers. The fibre axes are vertical for drawn fibres and horizontal for samples aligned in a magnetic field. The orientation is exactly the same as observed for the crystalline samples (Figs 1 and 2). The domains of the smectic phases are

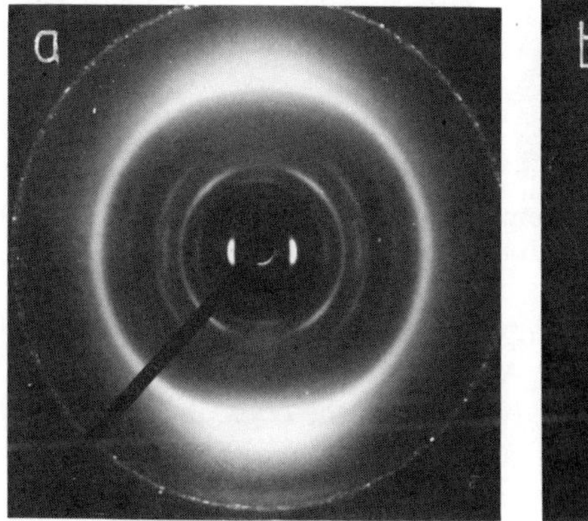

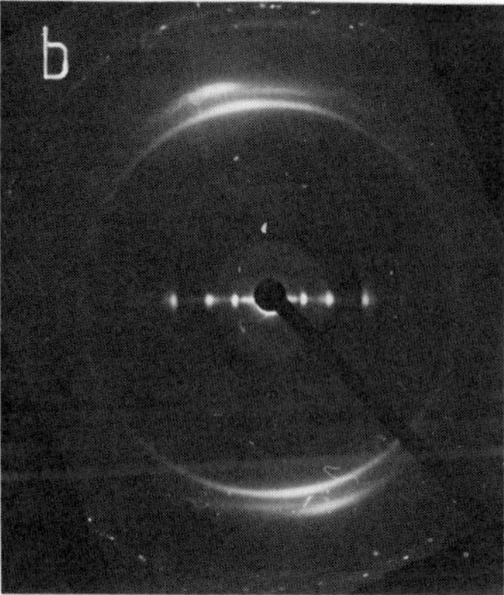

Fig. 1. Crystalline x-ray fibre patterns from drawn samples of (a) C-5 and (b) C-6. Fibre axes and drawing direction are vertical.

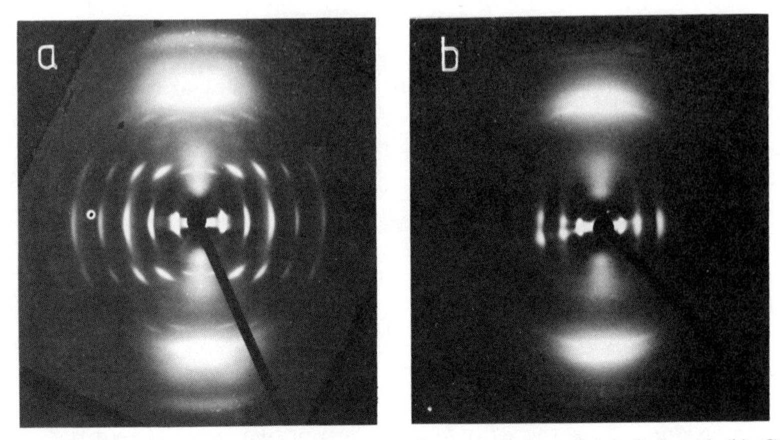

Fig. 2. Crystalline x-ray fibre patterns from samples of (a) C-5 and (b) C-6 aligned and crystallized in a magnetic field. Fibre axes and magnetic field direction are horizontal; the x-ray beam is perpendicular to the magnetic field direction.

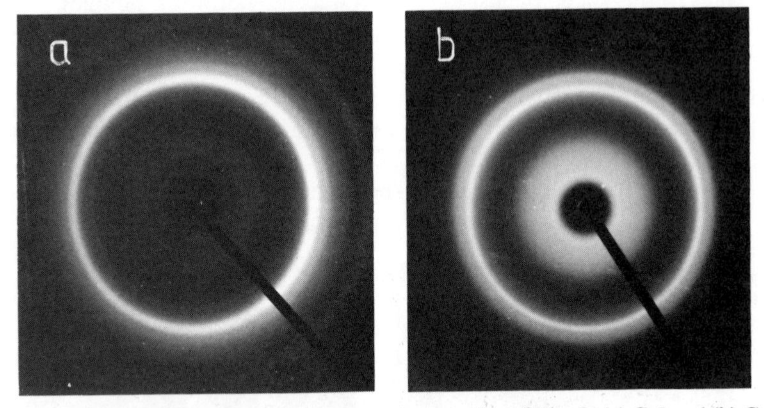

Fig. 3. Crystalline x-ray diagrams from the samples of Fig. 2: (a) C-5 and (b) C-6, with fibre axes parallel to the magnetic field direction.

distributed at random perpendicular to the fibre axis which is demonstrated with the x-ray diagram of Fig. 5. A different x-ray fibre pattern for the crystalline C-6 polymer is obtained (Fig. 6) if the crystallization process proceeds from the oriented smectic sample (Fig. 4(b)) without any magnetic field applied. The horizontal line remains a single line, while for the crystalline C-6 pattern, crystallized in a magnetic field, Fig. 2(b), two lines in the form of an X are observed.

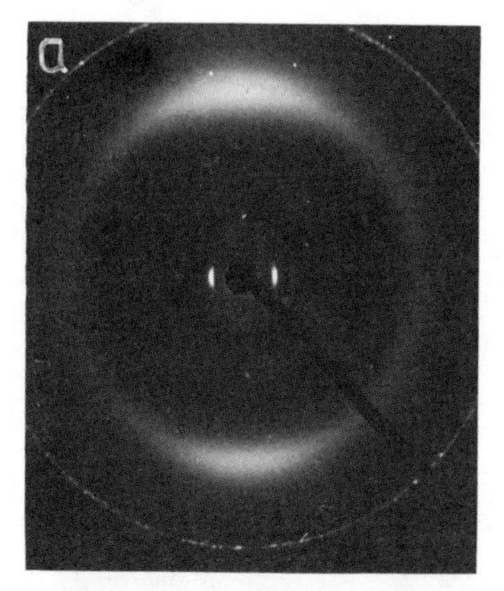

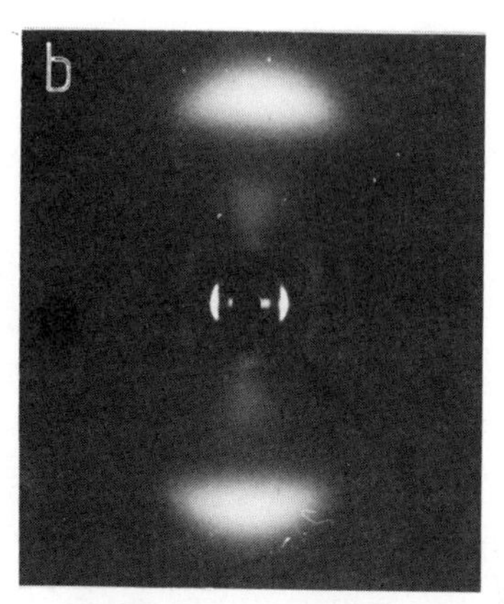

Fig. 4. Smectic x-ray fibre pattern from (a) a drawn sample of C-5 and (b) a C-6 sample aligned in a magnetic field. The fibre axis is vertical for (a) and horizontal for (b). The x-ray beam is perpendicular to the drawing or magnetic field direction.

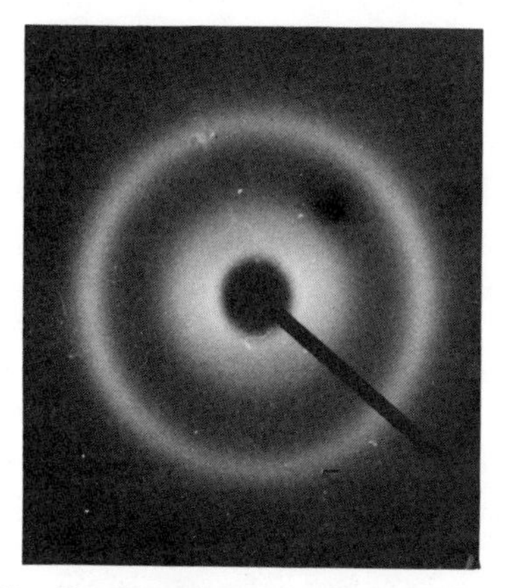

Fig. 5. Smectic x-ray diagram from sample C-6 of Fig. 4(b) with fibre axis parallel to the x-ray beam.

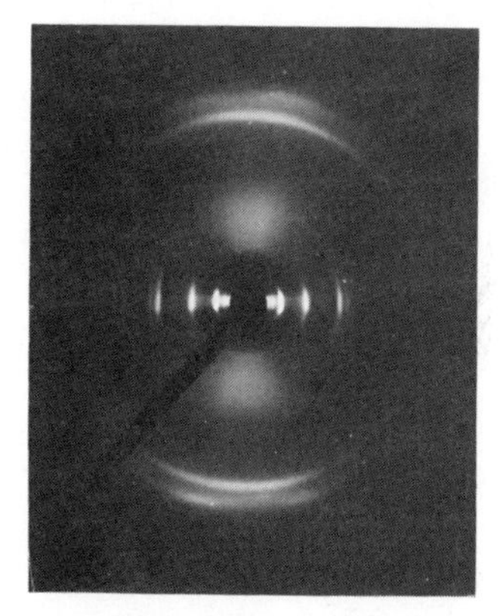

Fig. 6. Crystalline x-ray fibre pattern from the oriented smectic sample C-6 of Fig. 4(b) crystallized without external field.

RESULTS AND DISCUSSION

Packing Considerations

A careful evaluation of the x-ray diagrams of the C-5 and C-6 polymers leads to basic structural models in the crystalline and smectic phase,[6,7,12] which are the same for differently oriented samples.

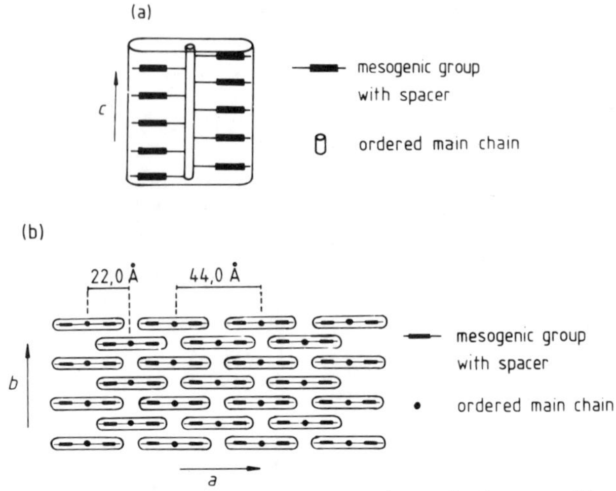

Fig. 7. Schematic representation of the C-5 polymer in the crystalline state. (a) Projection of a single chain along its axis; (b) projection of a domain into the *a*, *b* plane. Reproduced from Ref. 6 by permission of Hüthig & Wepf Verlag, Basle.

A schematic representation of the packing for the crystalline C-5 polymer is represented in Fig. 7. The periodicity in the *a*-direction results from the equally spaced reflections along the horizontal direction of the fibre x-ray patterns of Figs 1(a) and 2(a). It seems that the correlation of chains along the *b*-direction is very weak and, therefore, this dimension can scarcely be detected in the x-ray diagram. The periodicity along *c* is caused by the packing of the side chains. Systematically extinct reflections suggest the kind of packing presented in Fig. 7. The distance between two adjacent parallel running chains in *a* is approximately two times the length of a side chain.

The packing of the crystalline C-6 polymer represented in projection into the *a*,*b* plane (Fig. 8) is comparable with the C-5 polymer. A strict extinction rule for the reflections along *a* does not exist and small changes in the ideal packing arrangement of Fig. 8 are to be expected. The side chains are arranged at a certain tilt angle from the main chain and form staples. The polymer chains are not correlated along *c*, and therefore, layer lines are not observed on the x-ray diagram of Fig. 1(b). Reflections on the x-ray patterns of crystalline C-5 and C-6 polymers are detected, which to some extent signal an ordered main chain along *c*.

A schematic representation of the smectic phase of C-5 and C-6 is reproduced in Fig. 9. The size of the periodicity in *a* remains the same as for

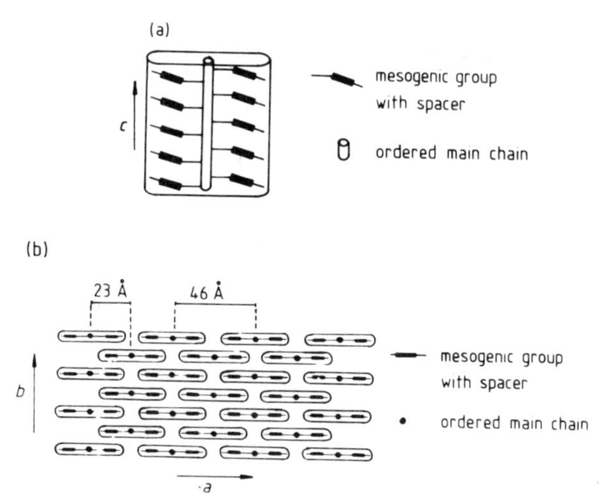

Fig. 8. Schematic representation of the C-6 polymer in the crystalline state. (a) Projection of a single chain along its axis; (b) projection of a domain into the *a*, *b* plane. Reproduced from Ref. 7 by permission of Hüthig & Wepf Verlag, Basle.

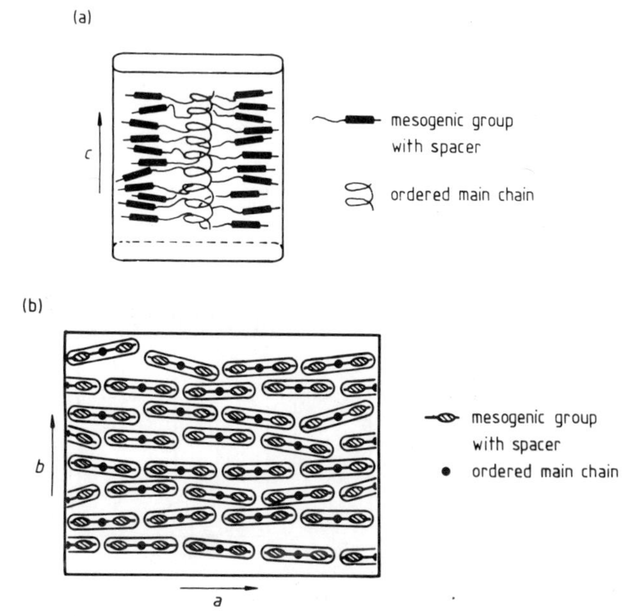

Fig. 9. Schematic representation of the C-5 or C-6 polymer in the smectic state. (a) Projection of a single chain along its axis; (b) projection of a domain into the *a*, *b* plane. Reproduced from Ref. 6 by permission of Hüthig & Wepf Verlag, Basle.

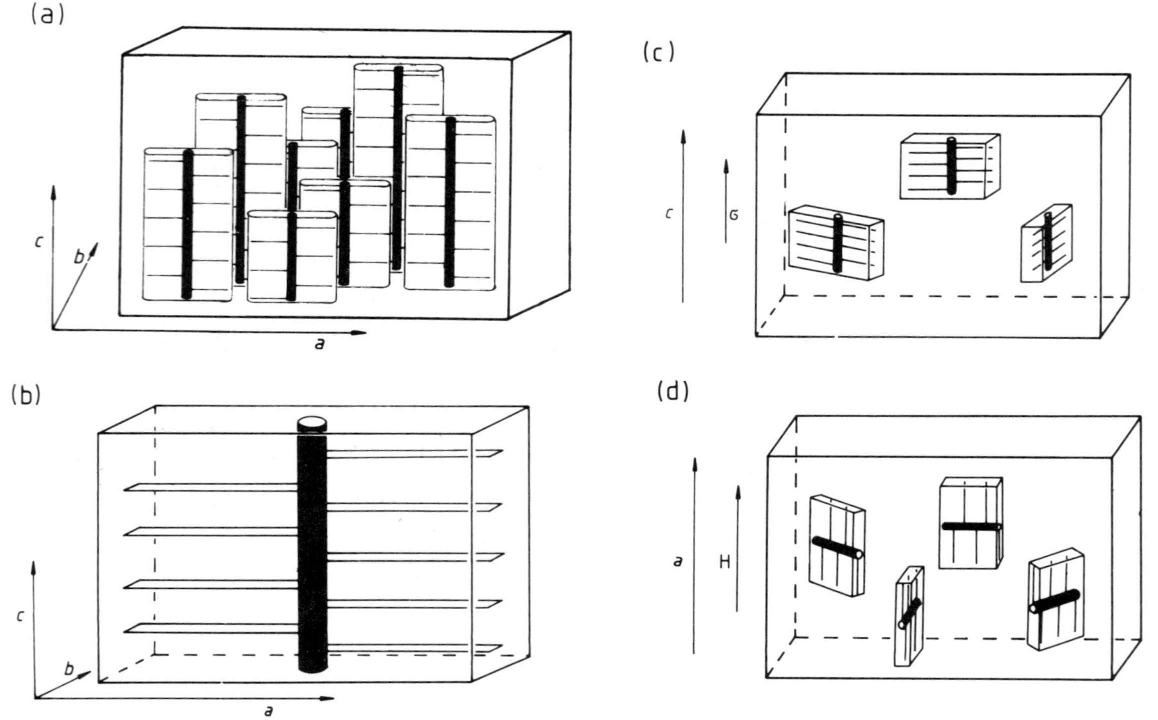

Fig. 10. Schematic representation of the orientation of crystallites and smectic domains for C-5. (a) Orientation of chains within a domain; (b) schematic drawing of the orientation of main and side chains as represented in (a); (c) orientation of domains for a drawn sample; (d) orientation of domains for a sample crystallized in a magnetic field.

the crystalline phases but the packing order is much poorer leading to fewer reflections in the *a*-direction of the x-ray pattern (Fig. 4). A correlation between different parallel chains along the main chain direction is missing resulting in absent layer lines in the smectic x-ray diagrams. The side chains lie perpendicular to the main chain. Their packing is less ordered compared with the crystalline state, and the order of the polymer backbone is also reduced. Therefore, the reflections in Fig. 4 caused by the packing of the side chains are very broad.

Orientation of the Crystalline and Smectic Domains

The x-ray patterns of Figs 1–6 result from scattered electro-magnetic waves on more or less ordered domains, called crystallites or smectic domains, which are differently oriented in space. The orientations of these domains are schematically depicted with respect to external fields for C-5 in Fig. 10. A main-chain orientation is observed in all the different domains of the crystalline or smectic sample, if a drawn sample is considered. All main-chain axes lie parallel to the drawing direction which is also the fibre axis *c*. The dimensions *a* and *b* of the domains are distributed at random perpendicular to *c*. If the samples are aligned in a magnetic field, a side-chain orientation is observed (Fig. 10(d)). All side chains are oriented parallel to the magnetic field which is now the fibre axis *a*. The domains are distributed with *b* and *c* at random in a plane perpendicular to *a*.

This orientation distribution also holds for smectic C-6. A more detailed consideration is necessary for the crystalline C-6 because of the tilted side chains. The main-chain orientation of C-6 by mechanical drawing is still in principle represented by Fig. 10(c) with tilted side chains, however. If the side chains are oriented along *a* in a magnetic field during crystallization, different orientations of the main chains in space result and are depicted schematically in Fig. 11. According to this drawing, the main-chain direction is no longer perpendicular to the magnetic field. Therefore the regular spacing in *a* from the lamellar type of packing of the polymer chains causes an X-shaped splitting of reflections in Fig. 2(b). The reflections resulting from the packing of the side chains of a single C-6 polymeric chain are placed on the vertical line through the origin of the x-ray pattern of Fig. 2(b). Those of the ordered main chain split in X-form with the vertical line as the centre line.

Crystallization of C-6 from the oriented smectic phase without an external field leads to a different x-ray pattern (Fig. 6). The crystallization proceeds from a smectic phase with an orientation shown in Fig. 10(d). The main chains remain fixed in space while the side chains are tilted. Therefore,

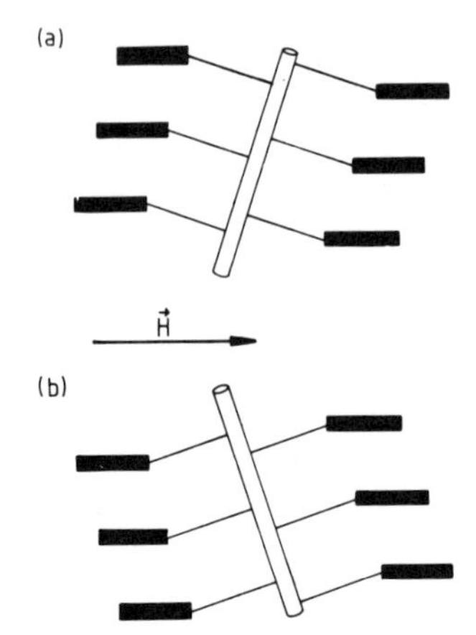

Fig. 11. Schematic drawing of different polymer chain orientations for C-6 crystallized in a magnetic field.

an X-shaped splitting of reflections occurs in the side-chain region of the x-ray diagram which is the vertical line. Reflections of the ordered backbone chain are unaffected by this splitting and are still placed on the vertical line. The lamellar type packing of polymer chains along the *a* dimension remains the same in space as in the smectic phase and is represented by an unchanged horizontal line in the pattern of Fig. 6.

CONCLUSIONS

The evaluation of fibre x-ray patterns of the same samples of polysiloxanes with mesogenic side groups in the crystalline and smectic phase is an excellent tool to develop structural models for both phases. The same orientational behaviour for the domains of the crystalline and smectic phases is observed when the external fields are applied. Only small changes are present in packing order and arrangements when the structures within the crystalline and smectic domains are compared: in the crystalline state a lamellar type of packing occurs extremely well spaced; in the smectic state, this order is considerably reduced, leading to mobile smectic layers known from models of small molecular compounds. The main chains all lie parallel within the domains of both phases with the side chains placed perpendicularly or at a tilt angle. Only in the crystalline state of C-5 a strong correlation between different chains along the chain axes is detected, whereas in the crystalline C-6 such a correlation is absent. The difference between these two polymers is a change in alkyl spacer length of one methylene group.

Although different orientations of crystalline or smectic domains can be produced by different alignment procedures, the packing within the domains is the same. This is also valid for a sample with random orientation of domains for which the reflections on a Debye–Scherrer x-ray diagram exhibit the same spacings as observed for oriented samples.

Main and side chains are partially decoupled by the flexible spacer. The orientation functions of the side and main chains are in general not identical. They can be evaluated by considering the arcing of the corresponding reflection in the fibre x-ray patterns.

The packing model for crystalline and smectic polysiloxanes with mesogenic side groups has been qualitatively evaluated in this x-ray investigation. Further studies are necessary for a quantitative description and more detailed structural models.

ACKNOWLEDGEMENT

This work was supported by the Deutsche Forschungsgemeinschaft.

REFERENCES

1. Finkelmann, H., Benthack, H. and Rehage, G., *J. Chim. Phys.*, 1983, **80**, 163.
2. Stevens, H., Rehage, G. and Finkelmann, H., *Macromolecules*, 1984, **17**, 851.
3. Strzelecki, L. and Liebert, L., *Bull. Soc. Chim. Fr.*, 1973, 603.
4. Platé, N. A. and Shibaev, V. P., *J. Polym. Sci., Polym. Symp.*, 1980, **67**, 1.
5. Cser, F., *J. Phys. Colloq.*, 1979, **40** (C3), 459.
6. Zugenmaier, P. and Mügge, J., *Makromol. Chem., Rapid Commun.*, 1984, **5**, 11.
7. Zugenmaier, P., *Makromol. Chem., Suppl.*, 1984, **6**, 31.
8. Zugenmaier, P. and Mügge, J., Report: 23. Sitzung des Dechema Arbeitsausschusses 'Polyreaktionen', Frankfurt, 20 January 1982; Mügge, J., Diplomarbeit, Institut für Physikalische Chemie der TU Clausthal, D-3392 Clausthal-Zellerfeld, 1982.
9. Zugenmaier, P. and Sarko, A., *Biopolymers*, 1976, **15**, 2121.
10. Smith, P. J. C. and Arnott, S., *Acta Cryst.*, 1978, **A34**, 3.
11. Tadokoro, H., *Structure of Crystalline Polymers*, John Wiley & Sons, New York, 1979.
12. Mügge, J. and Zugenmaier, P., to be published.
13. Frenzel, J. and Rehage, G., *Makromol. Chem.*, 1983, **184**, 1685.

19

MACROMOLECULAR ORDER AND CONFORMATION IN THE SOLID AND NEMATIC PHASES OF SEMI-RIGID POLYMERS AND POLYMER–MONOMER MIXTURES—NMR STUDY

A. F. MARTINS

Faculdade de Ciências e Tecnologia, UNL, and Centro de Física da Matéria Condensada, INIC, Lisbon, Portugal

F. VOLINO

CNRS and Département de Recherche Fondamentale, Centre d'Etudes Nucléaires de Grenoble, France

and

R. B. BLUMSTEIN

Polymer Science Program, Department of Chemistry, University of Lowell, Massachusetts, USA

1. INTRODUCTION

The observation of liquid crystalline mesophases in concentrated solutions or in melts of certain natural and synthetic polymers, and the exploitation of this property for the development of high modulus materials, is perhaps one of the most interesting recent developments in the field of polymer science. Molecular and macroscopic properties of various types of liquid crystalline polymers have been extensively studied in the last few years, see for example Ref. 1.

The characterization of quasi-equilibrium macromolecular order and conformation in these systems, as well as the dynamics of cooperative chain orientation in samples subjected to external forces are of paramount importance for the understanding of their macroscopic properties. In this work we pursue previous efforts along these lines[2,3] and present the main results of a proton and deuterium nuclear magnetic resonance (NMR) study of molecular order and chain conformation, and their temperature dependence, in the solid and thermotropic nematic phases of semi-rigid polymers, oligomers, and polymer–monomer mixtures.

Details of this study are reported here for the polyester DDA9:

$$-\!\!\left(CO-(CH_2)_{10}-CO-O-C_6H_3(CH_3)-N{=}N(\rightarrow O)-C_6H_3(CH_3)-O\right)\!\!-_x$$

for its model compound 9DDA9:

$$CH_3O-C_6H_3(CH_3)-N(\rightarrow O){=}N-C_6H_3(CH_3)-O-CO-(CH_2)_{10}-CO-O-C_6H_3(CH_3)-N{=}N(\rightarrow O)-C_6H_3(CH_3)-OCH_3$$

and for mixtures of DDA9 with the low molecular weight ('monomer') nematic PAA:

$$CH_3O-C_6H_4-N{=}N(\rightarrow O)-C_6H_4-OCH_3$$

The reason for considering PAA as a physical model for the 'monomer', rather than structures such as

$$CH_3COO-C_6H_3(CH_3)-N{=}N(\rightarrow O)-C_6H_3(CH_3)-OOCCH_3$$

or

$$CH_3O-C_6H_3(CH_3)-N(\rightarrow O){=}N-C_6H_3(CH_3)-O-CO-(CH_2)_{10}-CH_3$$

is that these structures do not display liquid crystalline behaviour.[4] On the other hand, as we shall see below, the sequencing of the rigid–flexible

TABLE 1
Physical Properties[a] of Samples Investigated

Sample	$\bar{M}_n$ *or M*	$\overline{DP}$ ($\bar{x}$)	T_g (°C)	T_{KN} (°C)	T_{NI} (°C)	S_{NI} (J/kg°C)
DDA9-L	4000	9	15	99	135·5	39·6
DDA9-H	20000	42	9	119	165	36·5
9DDA9	738·8	—	—	(32)[b]	(101)[b]	22·8
PAA	258·3	—	—	118; (90)	135	5·4

[a] $\bar{M}_n$, number average molecular weight; $\overline{DP}$, average degree of polymerization; S_{NI}, entropy of the NI transition; T_g, glass transition temperature; T_{KN}, solid–nematic transition temperature; T_{NI}, nematic–isotropic transition temperature.
[b] Monotropic nematic.
Values in parentheses, on cooling.

units in 9DDA9 (rigid–flexible–rigid, i.e. a flexible chain with both ends attached to mesogenic rods) is the meaningful sequence for modelling the situation of the flexible moiety of the polymer—as contrasted to the reverse sequencing (flexible–rigid–flexible) usually found in standard low molecular weight nematics, where the flexible chains have one end free. The relevant physical properties of the samples used in this work are displayed in Table 1.

The remainder of this paper is organized as follows. In Section 2 we briefly refer to the techniques used in this work. Section 3 is a survey of the results of our NMR studies on phase transitions, degree of order, heterophase behaviour, and conformation of flexible spacers in DDA9. Section 4 is devoted to a polymer–monomer nematic mixture (DDA9 + PAA); in particular we give the degrees of order of each component of the mixture. In Section 5 we discuss macromolecular organization in the solid state of DDA9 just below the solid–nematic transition in comparison with the unoriented nematic state of the high molecular weight polymer. Section 6 concludes this paper with a few remarks on the preceding results and future directions of research on this subject matter.

2. NOTE ON THE EXPERIMENTAL TECHNIQUES

The synthesis and characterization of the samples used in this work were reported elsewhere.[4–6] Proton NMR experiments were performed on a

Bruker CXP-100 spectrometer working either at 49·9 MHz or at 75 MHz, and deuterium NMR on a Bruker WM-250 spectrometer working at 38·4 MHz. The samples were degassed and sealed under vacuum. This operation involved several melting–solidification cycles in the absence of any orienting field. The NMR data reported here were always recorded at decreasing temperature, except those quoted from Ref. 3 which were recorded at increasing temperature. The samples were equilibrated at ~20 °C above T_{NI} for about 40 min and then cooled down slowly by steps of 2–3 °C, with 5–10 min residence time at each temperature before recording the spectra. The temperature of the samples was controlled to within ±0·3 °C with a Bruker B-UT 1000 temperature-regulating system, and the estimated error in the absolute values was less than 1 °C. Absolute values of the degree of order S were deduced from dipolar splittings as in Ref. 3.

3. NMR OF HOMOGENEOUSLY ORIENTED NEMATIC MELTS OF POLYMER AND MODEL COMPOUNDS

The first systematic investigation of semi-rigid liquid crystalline polymers by proton nuclear magnetic resonance (NMR) was reported in Ref. 3, and this technique proved to be very powerful in demonstrating the main features of macromolecular order and conformation in both the solid and nematic states of those polymers. In particular, for DDA9-L, homogeneous orientation of the macromolecules in the nematic phase was demonstrated, for a sample with number average molecular weight $\bar{M}_n \simeq 4000$, as well as unusually high values of the nematic degree of order (ranging from 0·69 at the nematic–isotropic transition to 0·84 at the solid–nematic transition), and comparable ordering of the mesogenic moiety and of the flexible —CO—$(CH_2)_{10}$—CO— spacer. These last features are not found with conventional (non-polymeric) nematics.

The results of Ref. 3 are complemented and extended here and in the following sections on the basis of new proton and deuterium NMR data obtained with normal and partially deuterated samples of the polymer DDA9-L and its model compound 9DDA9.

3.1. Phase Transitions, Order Parameter and Heterophase Behaviour

The transition temperatures T_{KN} and T_{NI} indicated in Table 1 may be clearly defined with reasonable accuracy (say ±1 °C or better) by NMR methods, as shown by Figs 4–6 in Ref. 3. The observed transition

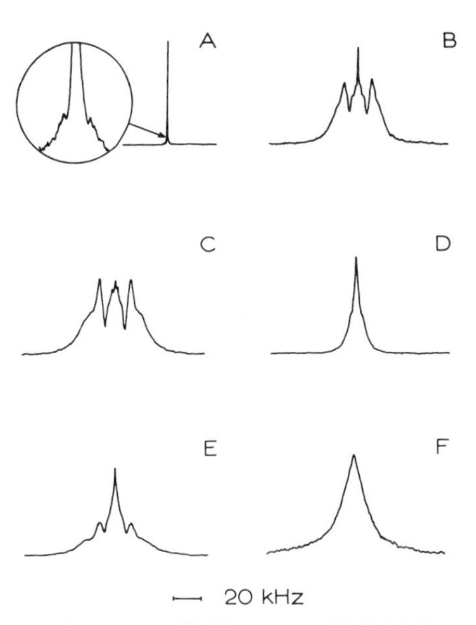

Fig. 1. Representative proton NMR spectra of DDA9-L at various temperatures, on cooling. (A) Isotropic phase with a small nematic fraction, at $T \gtrsim T_{IN}$; (B) nematic phase with a small isotropic fraction, at $T \lesssim T_{IN}$; (C) pure nematic phase homogeneously oriented; (D) unoriented nematic phase, or solid phase above the cold crystallization point; (E) solid phase with some nematic fraction; (F) solid phase slightly above T_g.

temperatures may differ, however, upon heating[3] or cooling the sample. This effect is more pronounced with T_{KN} and may be related to some deviation of the system from thermodynamic equilibrium.

It has also been noticed[6] that within some temperature interval around the transition points T_{KN} and T_{NI} the two neighbouring phases are not strictly homogeneous, but each phase contains embryos of the other phase. The relative amount of the phase existing in an embryonic form remains very small up to temperatures very close to the transition point, but it is detectable by NMR. Figure 1 shows several proton NMR spectra of DDA9-L representative of (A) the isotropic phase with a small nematic fraction, (B) nematic phase with a small isotropic fraction (1–2 %), (C) pure nematic phase homogeneously oriented, (D) unoriented nematic phase, or solid phase above the cold crystallization temperature, (E) solid phase with some nematic fraction, and (F) solid phase near room temperature. It is interesting to note that the sharp peak at the centre of spectrum B, which could not be obtained in the line shape simulation procedure reported in

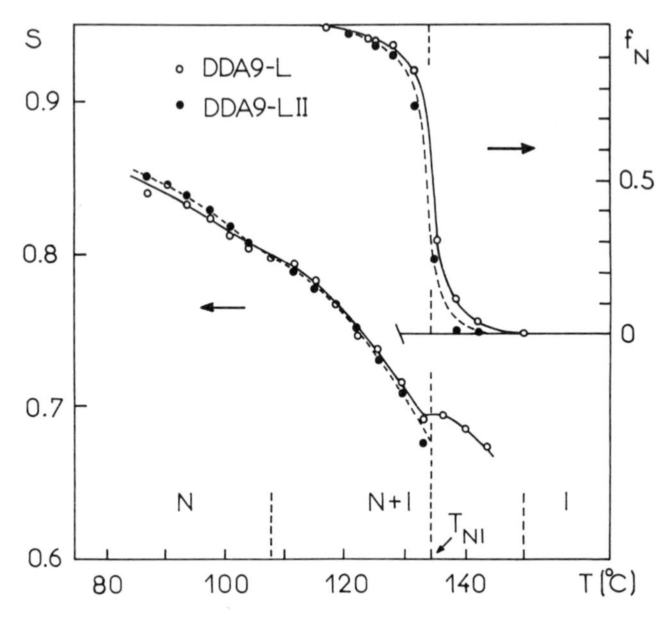

Fig. 2 Nematic order parameter S, and nematic fraction f_N in the N + I biphase, for polymer DDA9-L (open circles) and polymer fraction DDA9-LII (closed circles).

Ref. 3, is precisely the contribution from the isotropic embryos in the nematic phase, and this contribution was not considered in that previous work. Simulation of spectrum C by the same procedure requires that the central line of the spacer component of the spectrum be less important than shown in Fig. 8 of Ref. 3, i.e. the alignment of the spacer is further emphasized (note that spectrum C was recorded at a temperature lower than spectrum B). Spectrum B can be simulated from A and C; spectrum E can be simulated from C and D. The relative fractions of the two components of each biphase can be estimated in this way.

The temperature interval over which the heterogeneous phase behaviour is observed depends on the polydispersity of the sample. Note that T_{NI} for a given polymer is predicted to be an increasing function of the degree of polymerization before levelling off;[7] the same is true for the entropy of the transition, the degree of order $S(T_{NI})$, etc. This trend has been observed with polydisperse samples of DDA9 having different values of $\overline{DP}$.[6]

Figure 2 shows the range of the nematic–isotropic heterophasic behaviour and its consequences on the nematic order parameter $S(T)$ for two samples of DDA9. The polymer DDA9-LII, with $\overline{\mathrm{DP}} \simeq 10$, is a fraction of DDA9-L separated on a chromatographic column through removal of the longest and shortest species initially present in DDA9-L. It can be seen that the heterophasic range is narrower, and the values $S(T)$ are smaller for DDA9-LII than for DDA9-L, except in the pure nematic phase where $S(T)$ becomes higher for DDA9-LII, probably due to its higher average molecular mass. The curve $S(T)$ for DDA9-L (broader molecular weight distribution) shows an undulation at T_{IN} (defined as explained above) and another one at the lower limit of the biphasic range. The same undulations, but less pronounced, seem to exist for DDA9-LII. To explain this effect on $S(T)$ it has been proposed[8] that as the temperature is lowered from the isotropic phase into the N + I biphase, the nematic embryos first appearing in the isotropic phase are built up with the longest macromolecular species, and that incorporation of the species with lower and lower degree of polymerization is gradually obtained as the temperature decreases. This assumption is based on a similar fractionation scheme proposed by Flory[9] for polydisperse systems of rigid rods. The undulations would result from the interplay of the natural increase of S with decreasing temperature, and its opposite variation caused by a decrease in the average molecular mass of the nematic fraction upon incorporation of the shorter species. A quantitative interpretation of the data in Fig. 2 is not yet available. In this context one should also bear in mind that the temperature (108 °C) at which the isotropic fraction in the nematic phase vanishes corresponds to the critical temperature T^* of the magnetic birefringence observed in the isotropic phase.[10]

Finally, let us remark that the high values of $S(T)$ found for DDA9-L, as compared to PAA,[3] may not be a characteristic of all polymers of this type. Measurements on another polymer with a shorter —CO—$(CH_2)_7$—CO— spacer but otherwise identical chemical structure and comparable degree of polymerization, gave values of $S(T)$ about 20 % lower than those reported for DDA9-L.[11] Similar results were reported for a different polymer of the same (semi-rigid) type.[12] More generally, the degree of order $S(T)$ taken, for example, at T_{NI}, is expected to show an odd–even effect according to the parity of the number of chain bonds in the spacer.

3.2. Flexible Spacer Conformation in DDA9 Compared to 9DDA9

One of the most striking conclusions suggested by our earlier work[3] was that high nematic order of the mesogenic moiety and of the spacer were both necessary to simulate the proton NMR spectrum of DDA9-L. The implications of this result prompted us and others[13] to more detailed

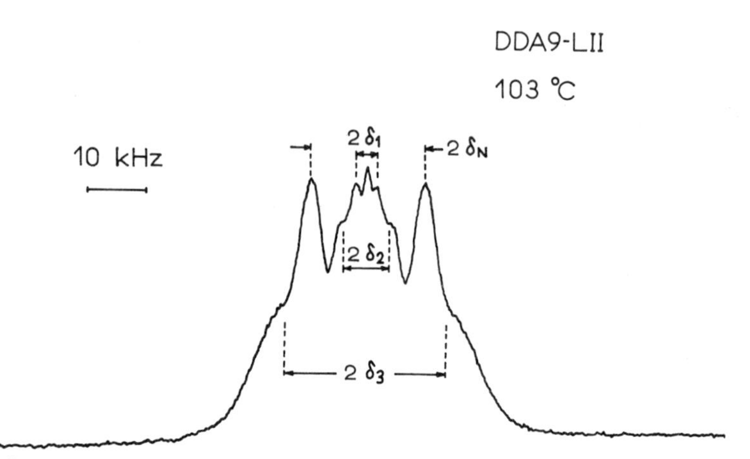

Fig. 3 Proton NMR spectrum of polymer DDA9-LII, in the well-oriented homogeneous nematic phase, at 103 °C, on cooling (see Section 3.2 for details and the meaning of the δ_i).

studies on the conformation of the spacer in semi-rigid polymers, as well as in model compounds such as 9DDA9.

It was remarked above that the experimental spectrum used earlier[3] as a reference for line shape simulation contained an isotropic component. Figure 3 shows a more representative spectrum of the nematic phase of homogeneously oriented DDA9-LII, recorded at 103 °C on cooling. Simulation of this spectrum confirms earlier assignments of the splittings $2\delta_N$, $2\delta_1$, $2\delta_2$, and $2\delta_3$ to dipolar interactions between (roughly speaking) mesogen *ortho*, terminal methyls, mesogen methyls, and alkyl spacer protons, respectively. The new simulation requires that the inner gaussian line of the triplet used in Ref. 3 to model the contribution of the spacer to the overall spectrum be of vanishing amplitude, which means that we should *not* assume, in this case, lower order at the middle of the spacer[3] but, instead, a similar degree of local order over the entire spacer (compare Fig. 3 here and Fig. 8 in Ref. 3). This suggests that the alkyl chain of the spacer is, on the average, confined to a cylinder whose diameter is comparable to the diameter of the mesogenic rod, presumably between one- and two-times the rod diameter.†

† With one more or less link in the alkyl chain of the spacer this strong cylindrical confinement is very improbable and we should expect a lower degree of order for both the mesogen and the spacer (odd–even effect). This model also predicts that the ratio of the quadrupole NMR splittings of the inner CD_2 to the α-CD_2 in a deuterated DDA9-L spacer should exceed $\simeq 0{\cdot}6$ at any temperature in the homogeneously oriented nematic phase. (Martins, A. F., current work.)

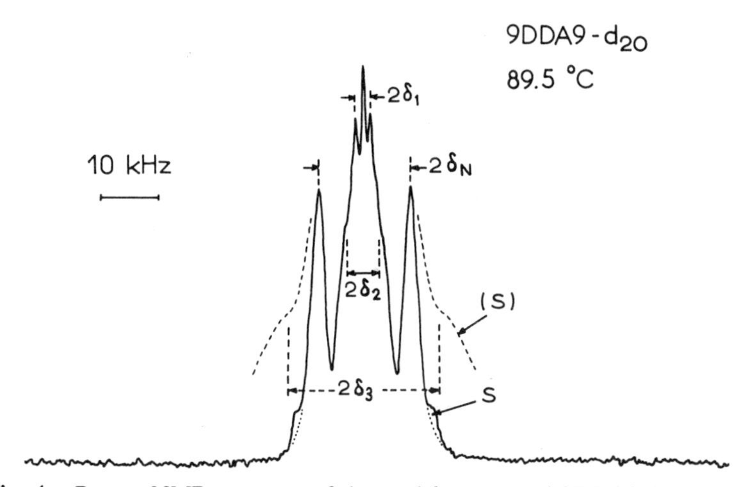

Fig. 4 Proton NMR spectrum of the model compound 9DDA9 deuterated on the alkyl chain (spacer), 9DDA9-d_{20}, in the nematic phase at 89·5 °C, on cooling. (S) represents the main contribution of the spacer to the proton spectrum, which has disappeared as a consequence of deuteration (a small contribution, S, remains due to incomplete deuteration). (For further details see Section 3.2.)

Results in Ref. 3 and here for the alkyl chain order in DDA9 were confirmed by Samulski *et al.*,[13] using deuterium NMR on a partially deuterated sample of this polymer.

Figure 4 shows a proton NMR spectrum of the model compound 9DDA9-d_{20} deuterated on the alkyl spacer, and Fig. 5 shows a deuterium NMR spectrum of the same compound. The former spectrum gives information on the two mesogenic moieties of the molecule and the latter gives information on the spacer. Both were recorded in the homogeneous and well-oriented nematic phase of this compound. Comparison of Figs 4 and 3 shows that the main (relevant) consequence of alkyl chain deuteration is the disappearance of the main shoulders on the external lines of the proton spectrum. The missing shoulders are represented by dashed lines marked (S) in Fig. 4 and do appear in the proton spectrum of non-deuterated 9DDA9. This spectrum (not represented here) differs from that in Fig. 3 essentially by the intensity of the central line, which is irrelevant in our comparison because this excess intensity only reflects the different proportions of the terminal CH_3 groups in DDA9 and 9DDA9 as compared to the rest of the proton–spins system.† Excluding the contribution

† Other small irrelevant differences arise from eventually different orientations of the —N=N(O)— central groups in the mesogenic units, etc.

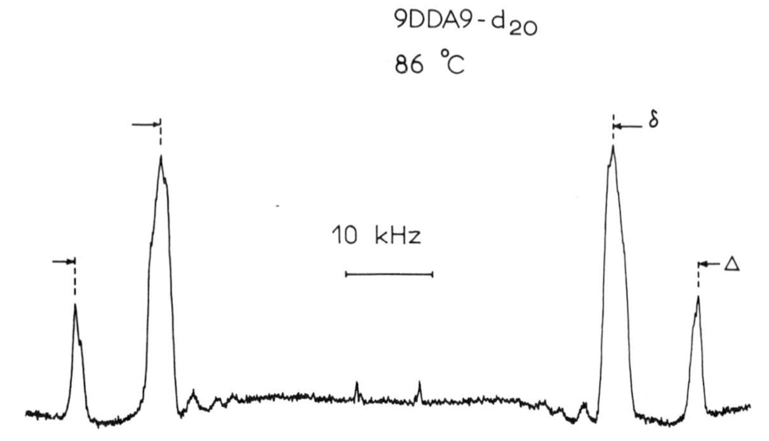

Fig. 5 Deuterium NMR spectrum of 9DDA9-d_{20} in the nematic phase, at 86 °C, on cooling. The external doublet is due to the deuterium linked to C_1 and C_{10} carbons, and the internal doublet to the deuterium linked to the remaining eight carbons in the alkyl chain.

of these terminal groups, the spectrum in Fig. 4 becomes similar to the form calculated in Ref. 3 for the mesogenic moiety of DDA9-L. These results not only confirm our previous analysis,[3] but also show that 9DDA9 is in fact an appropriate low molecular weight model of DDA9.

A more striking confirmation of this analysis is given by the deuterium NMR spectrum of the alkyl chain of 9DDA9 shown in Fig. 5. This spectrum is essentially equivalent to those given in Ref. 13 for the alkyl deuterated DDA9-d_{20}. It differs from the usual spectra of deuterated alkyl chains in conventional nematics by the appearance of only two doublets of resonance lines instead of (roughly speaking) one doublet per CD_2 group in the alkyl chain.[14,15] Considering that in both cases we have low molecular weight compounds and in both cases the alkyl chains are submitted to the same (on symmetry grounds) nematic potential, this essential difference is to be attributed to the fact that in 9DDA9 the alkyl chains are attached at both extremities to mesogenic units (dumb-bell type linkage) while in conventional nematics they are attached at only one extremity to the mesogen, having the other extremity free (pendent chain). As far as changes in alkyl chain conformation imply motions of the chain ends the freedom of a chain extremity is dependent on the inertia of the attached groups in addition to the constraints imposed by the symmetry of the nematic.[13] The areas under each doublet of lines in Fig. 5 are in the ratio 1:4. The external doublet is assigned to the two α-CD_2 methylenes. The inner doublet is

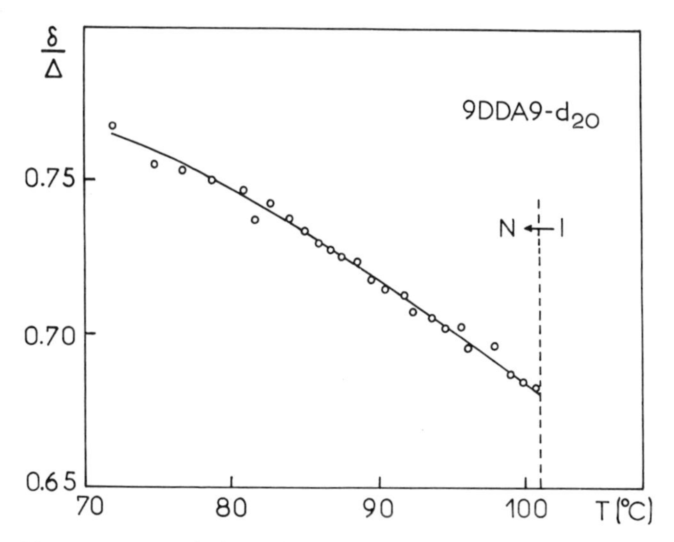

Fig. 6 Temperature variation of the ratio δ/Δ of the inner to the outer quadrupole NMR splittings shown in the spectrum of Fig. 5 (9DDA9-d_{20} in the nematic phase).

assigned to the remaining eight (internal) methylenes. This spectrum is thus direct evidence of the highly extended conformation of the spacer, with practically uniform local order of the various CD bonds. Only the α-CD_2 are slightly more ordered, their degree of order being near to that measured for the mesogenic moieties.

The nature of the spectrum in Fig. 5 does not change with temperature. The ratio δ/Δ of the inner to the outer splittings is slightly dependent on temperature, however, as shown in Fig. 6. This means that the mobility of the spacer increases with temperature, as expected, without losing the extended conformation (within the nematic phase). This same effect is revealed by the proton spectra, both in 9DDA9 and in DDA9-L, where δ_3/δ_N (Fig. 3) is also a decreasing function of the temperature. On the other hand the ratio δ_N/Δ or S/Δ is roughly independent of the temperature.

4. NMR OF A POLYMER–MONOMER NEMATIC MIXTURE

Mixtures of liquid crystals, either molecular or polymeric, are of great practical importance. Phase diagrams of mixtures of two nematogens A + B have been considered recently, on theoretical grounds, in three different situations, namely ($M_A + M_B$), ($M_A + P_B$) and ($P_A + P_B$), where M_α and P_α stand for molecular and polymeric nematic liquid crystals, respectively.[16] The order parameters S_A, S_B of each component in the mixtures are important ingredients of the theory, and measurements of these parameters in relevant situations should be helpful.

NMR is a particularly well-suited technique for the direct measurement of the order parameters of each component in a mixture, without disturbing the thermodynamic equilibrium of the mixture. We use this technique on a study of binary mixtures of PAA and DDA9-L currently under way.

The mixtures of PAA and DDA9-L are attractive because these two compounds are of similar chemical nature, have been extensively studied by NMR in their pure nematic phases,[3,17] are miscible in all proportions,[18] and have nearly the same clearing point T_{NI}, which should simplify the theoretical analysis of the data. Besides, the use of (available) deuterated samples, e.g. perdeuterated PAA (called PAA-d_{14}), as one of the components of the mixture preserves its physical properties and makes it easier to get the relevant information from the NMR spectra.

Figure 7 shows our (preliminary) results on a mixture of PAA-d_{14} (5%

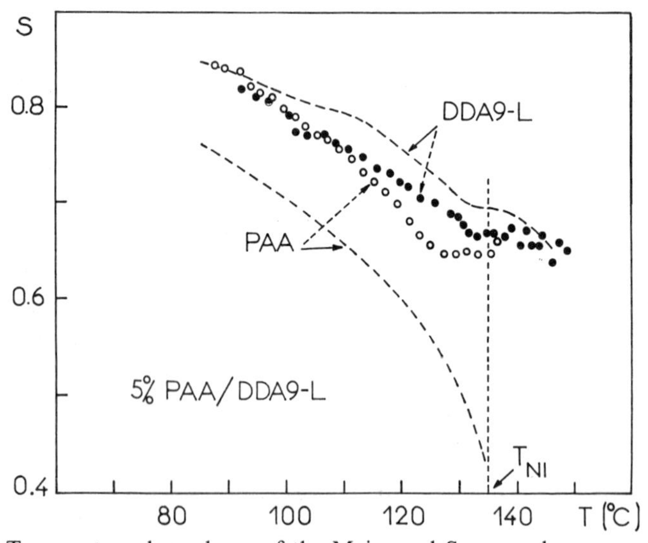

Fig. 7 Temperature dependence of the Maier and Saupe order parameters of polymer DDA9-L and 'monomer' PAA in pure melts (dashed lines), and in a 5% PAA + 95% DDA9-L, by weight, mixture (DDA9-L, closed circles; PAA, open circles).

by weight) and DDA9-L. Other mixtures with 10% and 50% PAA-d_{14} have also been studied.[19] The order parameter of PAA-d_{14} in the mixture, S_{mix}^{PAA}, was measured by deuterium NMR using the method described in Ref. 17, and $S_{mix}^{DDA9\text{-}L}$ was measured by proton NMR using the method described in Ref. 3. We have found, as expected, that any given temperature, in the nematic range, including the N + I biphase,

$$S^{PAA} < S_{mix}^{DDA9\text{-}L} < S^{DDA9\text{-}L}$$

where the symbols without subscript refer to unmixed compounds. Probably, the degree of order of the polymer in the mixture, $S_{mix}^{DDA9\text{-}L}$, varies continuously between the two extremes S^{PAA} and $S^{DDA9\text{-}L}$ as a function of concentration. Besides, we have $S_{mix}^{PAA} \simeq S_{mix}^{DDA9\text{-}L}$ in the pure nematic phase of the mixture (below 108 °C), and $S_{mix}^{PAA} < S_{mix}^{DDA9\text{-}L}$ in the N + I biphase, with S_{mix}^{PAA} being close to $S_{mix}^{DDA9\text{-}L}$ at the I to N + I transition. As shown in Fig. 7, we have observed that in the upper temperature limit of existence of the N + I biphase the values of $S_{mix}^{DDA9\text{-}L}$ tend to remain constant and near to the value $S^{DDA9\text{-}L}$ of the unmixed compound. The order parameter of PAA, S_{mix}^{PAA}, also remains nearly constant in this region. This might indicate that the longest molecules in DDA9-L, which, as remarked, are selectively transferred into the anisotropic phase at the I → I + N transition, favour the nematic ordering of the PAA molecules.

5. THE SOLID STATE OF DDA9 COMPARED TO THE UNORIENTED NEMATIC STATE

Besides the composite proton NMR line shape which is observed just below T_{KN}, on cooling (Fig. 1(E)), we can observe at least three different line shapes in the solid state of the polymer DDA9: namely, one below T_g, and two others, particularly meaningful, between T_g and the cold crystallization temperature[20] and above this temperature, respectively. The cold crystallization temperature, T_{cx}, of DDA9 occurs about 50 °C below T_{KN}.

The proton NMR spectra observed between T_g and T_{cx} have a 'super-Lorentzian' shape, i.e. one in which the wings spread out more than in the true Lorentzian, and consequently the ratio $\Delta_{1/2}/\Delta_{1/4}$ of the line widths taken at half- and quarter-height is lower than 0·577. In the present case we have found $\Delta_{1/2}/\Delta_{1/4} \simeq 0{\cdot}54$, nearly independent of the temperature within the interval $T_g < T < T_{cx}$ referred to above.

More interesting is the line shape observed in the *solid* phase above T_{cx}, which is qualitatively similar to that observed in the *unoriented nematic*

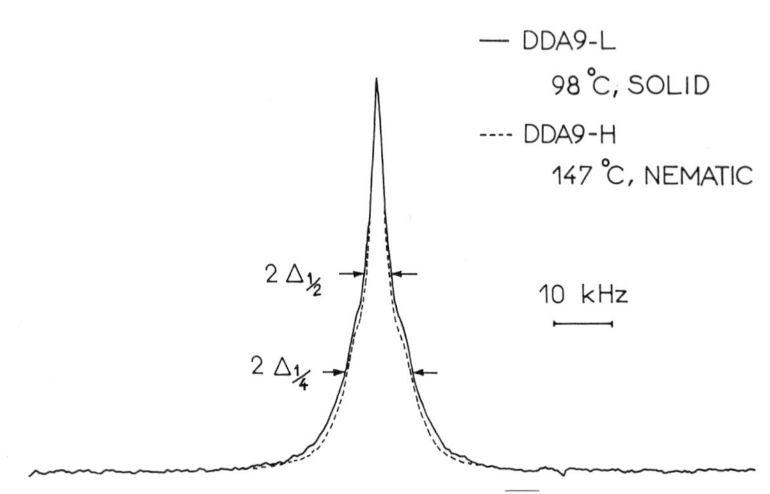

Fig. 8 Proton NMR spectra of polymer DDA9-L ($\overline{DP} \simeq 9$) in the *solid* phase, just below T_{KN}, on heating (full line), and of DDA9-H ($\overline{DP} \simeq 42$) in the unoriented *nematic* phase, at 147 °C (dashed line).

phase of DDA9-H. We remember that this high polymer sample ($\overline{DP} = 42$) does not orient in the ~10 kG magnetic field of the NMR spectrometer.[3] The similarity of both line shapes is illustrated in Fig. 8, and should imply some similarity in the microscopic structural organization of DDA9 in the solid just below T_{KN} and in the unoriented nematic phase. These line shapes can indeed be related to the line shape observed in the oriented nematic phase (Fig. 3) by the orientational averaging method described quantitatively in Ref. 3.

Taking, for example, the high polymer sample (DDA9-H) we can see (Fig. 9) that, although the line *shape* remains qualitatively invariant over a large temperature interval and through the solid–nematic transition, the line *width* is rather strongly dependent on temperature and shows a discontinuity at the transition point T_{KN}. We conclude that in this case the organization of the macromolecules is not changing very much at the transition, but their dynamics do change essentially. In the case of low polymer samples, such as DDA9-L, which are orientable in their nematic phases, the transition from this oriented nematic state to the solid state is furthermore accompanied by a strong distortion of the director field $\mathbf{n}(\mathbf{r})$, which acquires a (nearly) random configuration in space.

The temperature variation of the line widths, both at quarter- and half-height, ($2\Delta_{1/4}$ and $2\Delta_{1/2}$) follows the variation of δ_N or the order parameter $S(T)$ in the nematic phase and is of very different nature in the solid phase

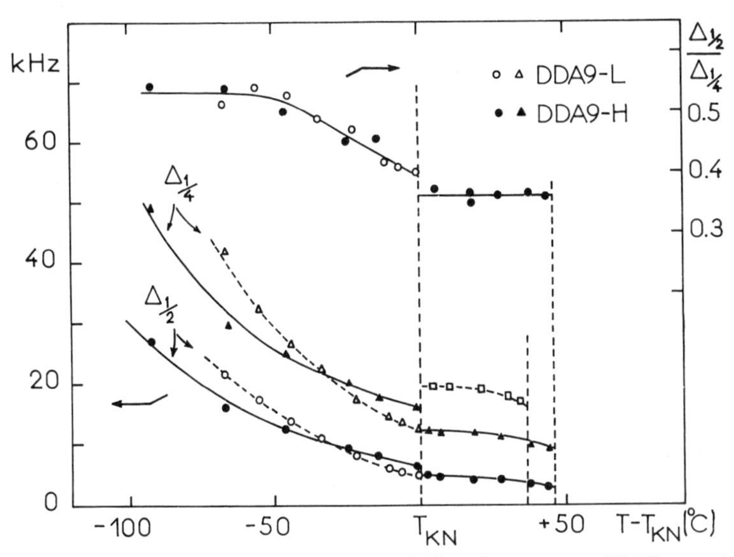

Fig. 9 Temperature variation of the line widths of the proton NMR spectra of DDA9-L and DDA9-H (see Fig. 8), at quarter-height ($2\Delta_{1/4}$) and half-height ($2\Delta_{1/2}$), and their ratio. Temperatures are relative to the solid–nematic transition T_{KN}. The splitting $2\delta_N$ (see Fig. 3) for DDA9-L in the nematic phase is shown for comparison. Open symbols and dashed lines refer to DDA9-L; closed symbols and full lines refer to DDA9-H.

(see Fig. 9). The ratio $\Delta_{1/2}/\Delta_{1/4}$ is constant in the nematic phase of DDA9-H (it is not defined in the nematic phase of DDA9-L, where the meaningful parameter is $2\delta_N$), and increases almost linearly with decreasing temperature below T_{KN}. The line width is thus apparently controlled by microscopic orientational fluctuations in the nematic phase, and by rotational and intramolecular motions (activated process) in the solid phase below T_{KN}. With further decrease in temperature we observe a levelling of the ratio $\Delta_{1/2}/\Delta_{1/4}$, which takes place around the cold crystallization temperature[20] and which we associate with it. The ratio $\Delta_{1/2}/\Delta_{1/4}$ probably varies again with T around the glass transition temperature T_g (see Table 1) but we did not investigate this temperature region.

6. CONCLUSION

The results reported here should allow some insight into the microscopic structure of liquid crystalline polymers and its relation to their properties, both in the melt and in the solid state. We have referred systematically to one particular polymer material, called DDA9, but we believe that these results are, at least qualitatively, applicable to many other liquid crystalline polymers of the same type, i.e. linear, semi-rigid, with mesogenic elements and flexible spacers in the main chain. Important to bear in mind, in the study of these materials, is the distinction between *orientation*, which may be characterized by the spatial configuration of the director field $\mathbf{n}(\mathbf{r})$, and *order*, which is well characterized by the Maier–Saupe order parameter S or, if needed, a second order tensor whose principal values are proportional to it.[21] For example, we can show[3] that the values of $S(T - T_{NI})$ are nearly the same in the well-oriented nematic phase of DDA9-L and in the unoriented nematic phase of DDA9-H (the small difference being related to the different degrees of polymerization).

The approach here has been descriptive, in part because the essentials of the quantitative approach to some of the problems treated here were given before.[3] The behaviour of the mesogenic rods, and of the flexible spacer, in the nematic as well as in the various regions of the solid phase, merits further attention. To this end a close comparison of the NMR data with other, e.g. thermodynamic data,[20] will be helpful. In relation to the fact that high polymers do not orient in moderately high magnetic fields, measurements of the rotational viscosity γ_1[21] in a series of sharp fractions of DDA9 with various (low) degrees of polymerization would be of great practical and theoretical importance. We believe that here again NMR will prove to be a very suitable technique.

ACKNOWLEDGEMENTS

Financial support of this work under NATO Research Grant No. 475.83 is gratefully acknowledged. AFM and RBB were further supported by JNICT (Portugal) under Research Contract No. 4248268, and by NSF (USA) under Research Grant DMR-8308939, respectively.

REFERENCES

1. Blumstein, A. and Hsu, E. C. (Eds), *Liquid Crystalline Order in Polymers*, Academic Press, New York, 1978; Ciferri, A., Krigbaum, W. R. and Meyer, R. B. (Eds), *Polymer Liquid Crystals*, Academic Press, New York, 1982.
2. Volino, F., Martins, A. F., Blumstein, R. B. and Blumstein, A., *J. Phys. (Lett.), Paris*, 1981, **42**, L-305.

3. Martins, A. F., Ferreira, J. B., Volino, F., Blumstein, A. and Blumstein, R. B., *Macromolecules*, 1983, **16**, 279.
4. Blumstein, R. B. and Stickles, E. M., *Mol. Cryst. Liq. Cryst. Lett.*, 1982, **82**, 151.
5. Blumstein, A. and Vilasagar, S., *Mol. Cryst. Liq. Cryst. Lett.*, 1981, **72**, 1.
6. Blumstein, R. B., Stickles, E. M., Gauthier, M. M., Blumstein, A. and Volino, F., *Macromolecules*, 1984, **17**, 177.
7. Ten Bosch, A., Maissa, P. and Sixou, P., *J. Phys.* (*Lett.*), *Paris*, 1983, **44**, L-105.
8. Volino, F., Allonneau, J. M., Giroud-Godquin, A. M., Blumstein, R. B., Stickles, E. M. and Blumstein, A., *Mol. Cryst. Liq. Cryst. Lett.*, 1984, **102**, 21.
9. Flory, P. J. and Frost, R. S., *Macromolecules*, 1978, **11**, 1126; Frost, R. S. and Flory, P. J., *Macromolecules*, 1978, **11**, 1134.
10. Maret, G., Volino, F., Blumstein, R. B., Martins, A. F. and Blumstein, A., *Proc. 27th Int. Symp. Macromolecules*, Strasbourg, 1981, **II**, 973.
11. Blumstein, A., Blumstein, R. B., Gauthier, M. M., Thomas, O. and Asrar, J., *Mol. Cryst. Liq. Cryst. Lett.*, 1983, **92**, 87.
12. Mueller, K., Hisgen, B., Ringsdorf, H., Lenz, R. W. and Kothe, G., *this book*, Chapter 13.
13. Samulski, E. T., Gauthier, M. M., Blumstein, R. B. and Blumstein, A., *Macromolecules*, 1984, **17**, 479.
14. Charvolin, J. and Deloche, B., in *The Molecular Physics of Liquid Crystals*, Luckhurst, G. R. and Gray, G. W. (Eds), Academic Press, London, 1979, Chapter 15.
15. Samulski, E. T. and Dong, R. Y., *J. Chem. Phys.*, 1982, **77**, 5090.
16. Brochard, F., Jouffroy, J. and Levinson, P., *J. Phys., Paris*, 1984, **45**, 1125.
17. Volino, F., Martins, A. F. and Dianoux, A. J., *Mol. Cryst. Liq. Cryst.*, 1981, **66**, 37; Dianoux, A. J., Ferreira, J. B., Martins, A. F., Giroud-Godquin, A. M. and Volino, F., *9th Int. Conf. Liquid Crystals*, Bangalore, India, 6–10 Dec. 1982 (Abstracts), to be published.
18. Billard, J., Blumstein, A. and Vilasagar, S., *Mol. Cryst. Liq. Cryst. Lett.*, 1982, **72**, 163.
19. Blumstein, R. B., Blumstein, A., Stickles, E. M., Poliks, M. D., Giroud, A. M. and Volino, F., *ACS Polym. Prepr.*, 1983, **24**, 275.
20. Grebowicz, J. and Wunderlich, B., *J. Polym. Sci., Polym. Phys. Ed.*, 1983, **21**, 141.
21. De Gennes, P. G., *The Physics of Liquid Crystals*, Oxford University Press, London, 1975.

PART IV

APPLICATIONS

20

ARAMIDS—BRIDGING THE GAP BETWEEN DUCTILE AND BRITTLE REINFORCING FIBRES*

M. G. NORTHOLT

Akzo Research Laboratories, Corporate Research Department, Arnhem, The Netherlands

This chapter presents in a concise form the relation between the microstructure and the mechanical properties of aramid fibres. At the same time an attempt is made to demonstrate that the rationale behind the design of composite materials can also be employed for the molecular design of the reinforcing fibre itself.

Composite materials have the advantage that they exhibit the best properties of their constituents and often have additional properties that none of the constituents can provide. Particularly nowadays fibre-reinforced composites having high strength-to-weight and stiffness-to-weight ratios are becoming increasingly important. Not only a high tenacity and stiffness are desired but also an adequate ductility or impact resistance, a low creep, a low specific weight, an appreciable compressive strength, a low thermal expansion coefficient and a proper thermo-mechanical stability.

Let us now draft some basic principles for the structure of a single component material which exhibits most of these composite material properties. A material of a high specific strength and specific modulus should consist mainly of elements of the first two rows of the periodic system linked together by covalent bonds. This postulate can easily be derived from a consideration of the single bond enthalpies between the various elements[1] and their specific densities.

However, a material merely composed of atoms covalently linked together in all directions will not only be very strong, like diamond, but also be extremely brittle since it does not have a mechanism that prevents crack propagation by eliminating local stress concentrations. These

* Lecture presented at the 22nd International Man-Made Fibre Conference in Dornbirn, Austria, 8–10 June 1983.

materials, being glass or crystalline substances like SiO_2, SiC, BN, B_4C, etc., mainly offer a high modulus of elasticity and a good thermal resistance. They largely absorb elastic energy before fracture, so that in order to decrease the brittleness, i.e. to increase the ductility, another energy absorbent mechanism should be introduced. On a macrostructural scale this is realized by designing a composite material consisting of structural components, with a high modulus of elasticity, which are embedded in a soft matrix with ductile properties. Anisotropic mechanical properties are introduced by giving the high-modulus component a needle-like or lamellar shape. These principles, being applied to, for example, glass fibre and carbon fibre reinforced composites, can also be used on a molecular level for the design of a fibre having the properties of a composite material.

The introduction of a plastic deformation mechanism in a single component material is achieved by incorporating bonds having a much lower dissociation energy than the covalent bonds. For materials consisting of first row elements this function is fulfilled by the van der Waals' intermolecular bond and/or the hydrogen bond. The presence of them also facilitates the manufacturing process since a liquid stage in the process is now possible. To ensure stiffness and strength, loadbearing molecular conformations in one or two dimensions provide these properties. For example, carbon fibre is made up of stacks of graphite planes which have an almost crystalline order and are oriented parallel to the fibre axis. Unlike the material graphite itself, the carbon fibre does not show a low yield stress for shear between the planes. This may be caused by some covalent bonding between the graphite planes.[2] So carbon fibres are brittle and have an almost linear stress–strain curve up to fracture with a fracture energy of about 35 $MNm\,m^{-3}$.

A polymer chain is a one-dimensional loadbearing conformation. By employing the most simple polymer chain with the smallest cross-sectional area per covalent bond, viz. polyethylene, it should be possible to obtain a large tensile modulus and strength. By gel spinning followed by hot drawing of polyethylene we achieved in our laboratory a tensile strength of 3 GPa with a modulus of 200 GPa.[3] However, the intermolecular adhesion in this fibre arises only from van der Waals' forces. Therefore extremely long chains are needed for an optimum load transfer between adjacent chains. In addition, the low melting point of 140 °C severely limits the application. These disadvantages are removed by the introduction of stiff polymer chains which are laterally linked by hydrogen bonds. A stiff polymer chain is obtained when free rotation around interatomic bonds is

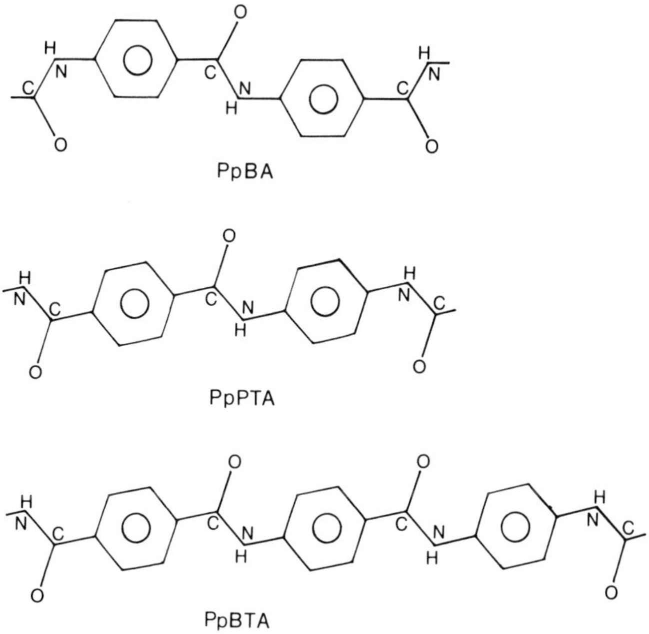

PpBA

PpPTA

PpBTA

Fig. 1. Molecular conformations of some aromatic polyamides.

absent and the chain adopts the linearly extended conformation. This goal is effectually realized in the aromatic polyamides or aramids. Sequential arrangements of phenylene and amide segments located at the *para* carbons yield rigid chains like poly-*p*-phenylene terephthalamide (P*p*PTA), poly-*p*-benzamide (P*p*BA) and poly-*p*-*p*′-benzanilidene terephthalamide (P*p*BTA) shown in Fig. 1.

The conformation of these polymer chains is primarily governed by competitive intramolecular interactions between the conjugated groups in the chain. They are the resonance effect trying to stabilize coplanarity of the amide groups and the phenylene groups, and the counteracting steric hindrance found between the oxygen and an *ortho*-hydrogen of the *p*-phenylene diamine segment and between the amide hydrogen and an *ortho*-hydrogen of the terephthalic segment. Due to these intramolecular interactions the interplanar angle between the phenylene and the amide segments ranges from 25 to 40°.[4] Figure 2 shows the chain packing in P*p*PTA in two projections.

Regularly positioned amide segments allow for medium strong intermolecular hydrogen bonds which ensure a proper load transfer between

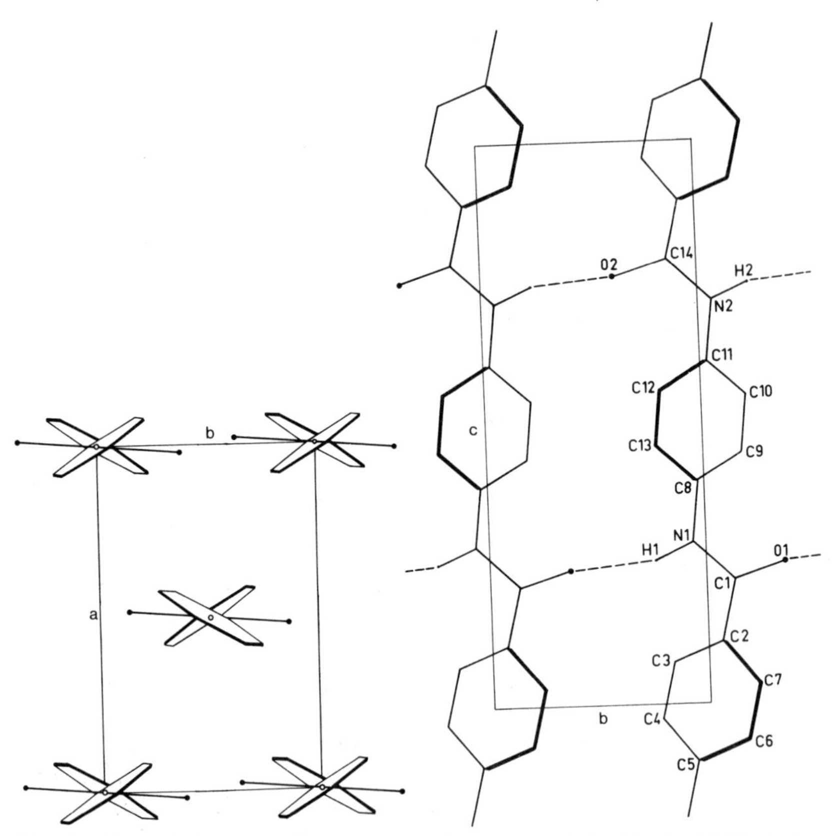

Fig. 2. Crystal structure of poly-*p*-phenylene terephthalamide (P*p*PTA). Left, the projection along the chain axis; right, the projection on the hydrogen-bonded plane. Hydrogen bonds are indicated by dashed lines. Reproduced from Ref. 4 by permission of Pergamon Press, Oxford.

chains as is found in a zipper. The hydrogen-bonded chains form sheets which are stacked parallel into crystallites. Between adjacent hydrogen-bonded planes the interaction largely takes place by van der Waals' forces with some π-bond overlap of the phenylene segments. This causes the hydrogen-bonded planes to act as slip-planes like the closest packed lattice planes in metals. To take full advantage of the chain properties, the chains are aligned in one direction, thereby limiting the macroscopic shape of the material to filament and films. The aromatic polyamides are therefore manufactured in a wet spinning process in which the chains are given a narrow orientation distribution around the filament axis. This leads to an

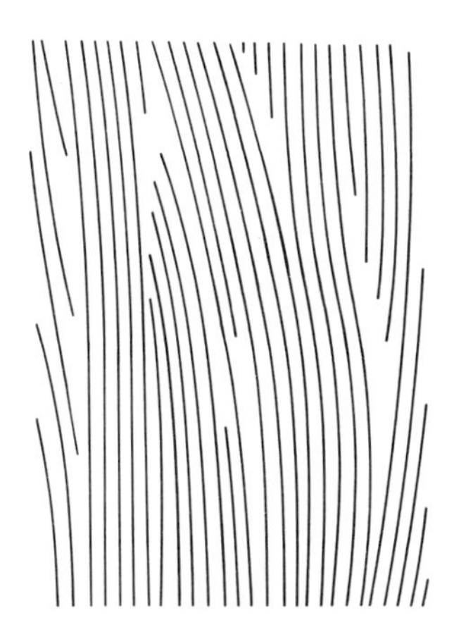

a b

Fig. 3. Schematic representation of the microstructure of (a) semi-crystalline polymers such as PET and nylon-6 and (b) poly-*p*-phenylene terephthalamide. Fibre axis vertical.

initial modulus of about 70 GPa which can be increased by subsequent hot drawing to 140 GPa. Figure 3(b) depicts a schematic representation of the one-phase paracrystalline structure. The filament strength for a gauge length of 2·5 cm is 4 GPa with an elongation at break of about 4% and a fracture energy of about 80 MNm m^{-3}.

The hydrogen-bonded planes may give rise to different modes of lateral texture in the fibre, viz. a radial, tangential or random orientation.[5,6] In addition the hydrogen-bonded planes may show a regular pleat along the fibre axis with a periodicity of about 500 nm.[7] The fibre can be regarded as consisting of parallel arranged fibrils. These fibrils are formed by chains of crystallites arranged end to end. The crystallite size in the fibre axis direction is 20–100 nm and normal to this direction, 4–10 nm. The perfection of the crystallites is rather high as shown by the lattice distortion parameter g_{II} which ranges from 1% for heat-treated fibres to 3% for as-spun fibres. Actually, it is the first organic polymer fibre which has shown equatorial as well as meridional lattice fringes.[8]

On the basis of the structure described, the series model can be used for the analysis of the mechanical properties.[9] Assuming a uniform stress distribution along the fibre, the elastic extension of the fibre is governed by the relation

$$S_{33} = \frac{1}{e_3} + A\langle \sin^2 \phi \rangle \tag{1}$$

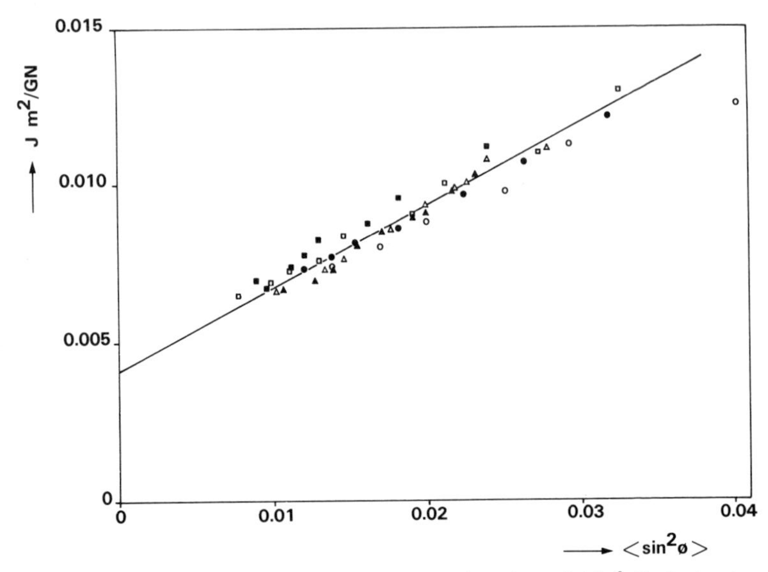

Fig. 4. The dynamic compliance, $S_{33} = J$ as a function of $\langle \sin^2 \phi \rangle$ during tensile deformation of the fibre. △, ▲, Fibre 1; □, ■, fibre 2; ○, ●, fibre 3. Open symbols represent the first extension of the fibre; closed symbols, repeated extension of the fibre. Reproduced from Ref. 9 by permission of Butterworth & Co. (Publishers) Ltd, London.

Here S_{33} is the fibre compliance, e_3 the modulus of elasticity of the chain and $\langle \sin^2 \phi \rangle$ the orientation distribution parameter of the crystallites with respect to the fibre axis, which is zero for perfect orientation and 2/3 for random orientation. Figure 4 shows the observations which confirm expression (1). The constant A represents a measure of the mechanical anisotropy of the crystallite:

$$A = \frac{1}{2g} - \frac{2(1 + \nu_{13})}{e_3} \tag{2}$$

where g is the shear modulus in the plane containing the chain axis and ν_{13} is the Poisson ratio for a stress along this axis. Dynamic moduli and crystallite orientation measurements during tensile deformation have confirmed the validity of expression (1) up to rupture for P*p*PTA fibres. For e_3 and g, values of 240 and 2 GPa have been found. Up to a stress of about 1·5 GPa the extension of the fibre is brought about mainly by elastic rotation, by some retarded as well as plastic rotation of the crystallites towards the fibre axis and by elastic extension of the polymer chain itself.

A relation between the stress and the orientation parameter has been derived and experimentally confirmed:

$$\langle \sin^2 \phi \rangle = \langle \sin^2 \phi_0 \rangle \exp(-C\sigma) \tag{3}$$

where $C = 2A + \lambda$. The parameter λ represents both the permanent and the retarded elastic rotation of the crystallites. For $\sigma > 1{\cdot}5$ GPa the fibre extension increases by axial flow, which is, presumably caused by slip between adjacent chains. Without this contribution the stress–strain relation is given as a first approximation by

$$\varepsilon = \frac{\sigma}{e_3} + \tfrac{1}{2} \langle \sin^2 \phi_0 \rangle [1 - \exp(-C\sigma)] \tag{4}$$

where $\langle \sin^2 \phi_0 \rangle$ is the initial value of the orientation parameter. So the ductility of the fibres is primarily determined by the initial orientation distribution of the crystallites and by the ability of the crystallites to rotate their symmetry axis towards the stress direction as a result of the relatively low value of the shear modulus g. During extension this distribution contracts and near rupture the chains are oriented almost parallel to the stress direction, as shown by the observed dynamic S_{33}^{-1} of 190 GPa.

Owing to the crystallite orienting mechanism governing the entire tensile deformation process, the presence of structural irregularities may hamper the alignment of the crystallites along the stress direction, which can give rise to premature rupture of the fibre. However, the effect of these inhomogeneities is probably partly mitigated by the fact that local slip of the hydrogen-bonded planes occurs. Especially as far as the ductility is concerned, the aramid fibres distinguish themselves from the carbon fibres and E-glass fibres.

In Figs 5 and 6 the tensile and creep behaviour of P*p*PTA fibre is compared with that of semi-crystalline fibres like nylon 66 and polyethylene terephthalate. The difference is striking and completely caused by the difference in microstructure as visualized in Fig. 3. Semi-crystalline fibres have a two-phase structure consisting of a series arrangement of amorphous and crystalline domains, whereas aramids have a one-phase paracrystalline structure. As a result of this highly oriented structure aramid fibre fractures according to an axial splitting process. Because of the high aspect ratio of the split fibres, they will still contribute effectively to the strength of an assembly, e.g. a multifilament yarn embedded in a matrix.

Special attention should be given to the important property of the strength in axial compression, commonly designated as the compressive

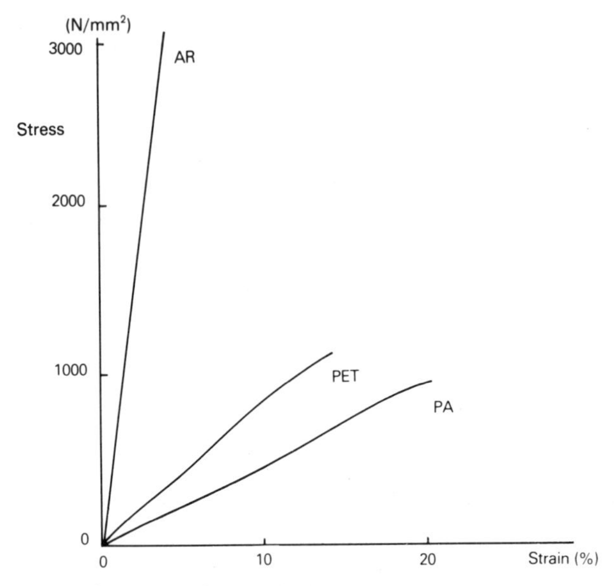

Fig. 5. Stress–strain curves of P*p*PTA (AR), polyester (PET) and nylon 66 (PA) yarns.

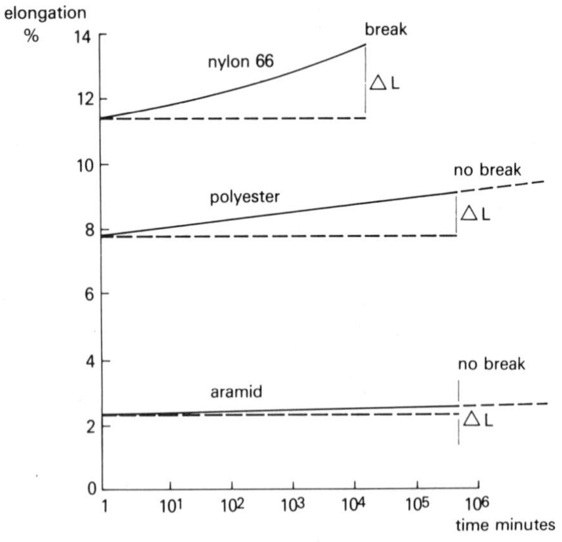

Fig. 6. Creep of nylon 66, polyester and aramid cords loaded with 60 % of their breaking load for well over a year.

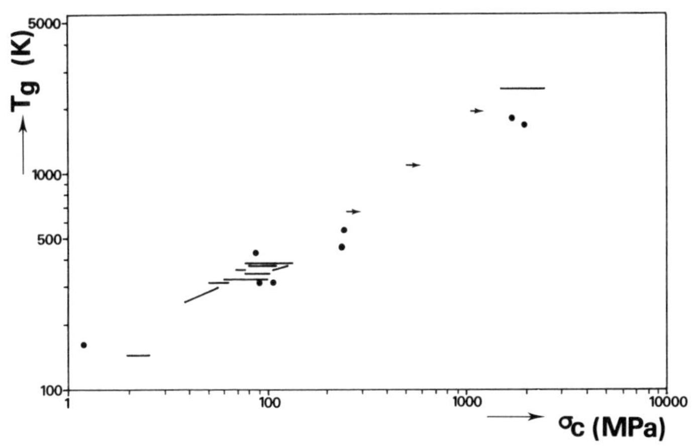

Fig. 7. Glass transition temperature (K) as a function of the strength in axial compression σ_c. Reproduced from Ref. 11 by permission of Chapman and Hall Ltd, London.

stress at which yielding occurs. In many applications the materials are subjected to compressive stress as in the case of a bending deformation. The phenomena which usually accompany compressive yielding of fibres are the well-known kink and slip bands.[10] In polymer fibres these distortions are macroscopic manifestations resulting from the buckling of chains. In places where the buckling occurs the chains adopt a very different conformation which may extend over a large part of the chain. The resistance to buckling of chains under compression is related to the glass transition temperature of the polymer T_g. Below T_g a polymer has glassy properties; above this temperature it has rubber-like properties. In the molecular interpretation of the glass–rubber transition of polymers this relaxation process is associated with the onset of large-scale segmental motions in the chain. Figure 7 depicts the relation between the glass transition temperature and the compressive strength σ_c for materials composed of mainly first-row atoms. From this figure it follows that

$$\sigma_c \propto T_g^2 \tag{5}$$

This relation can also be derived on the basis of the assumption of the equivalence of the work done in compression up to yielding and the activation energy of the glass transition.[11]

According to eqn (5) the ratio of the compressive strength values of aramid fibre to carbon fibre is 1:8. This seems to be an almost prohibiting constraint for the use of aramids in high-performance composites. However,

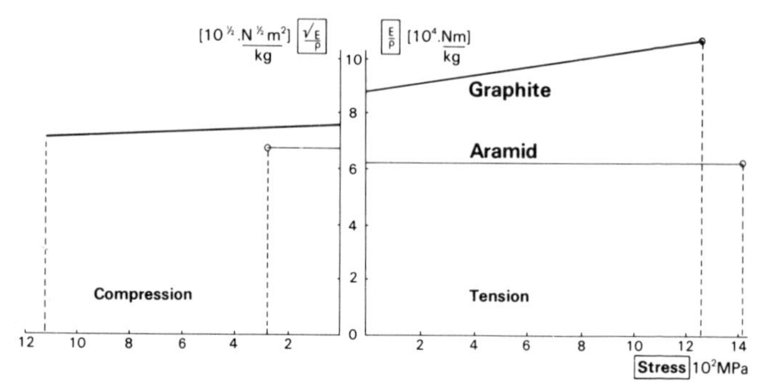

Fig. 8. Comparison of compressive and tensile properties of unidirectional carbon and aramid fibre reinforced composites.[12]

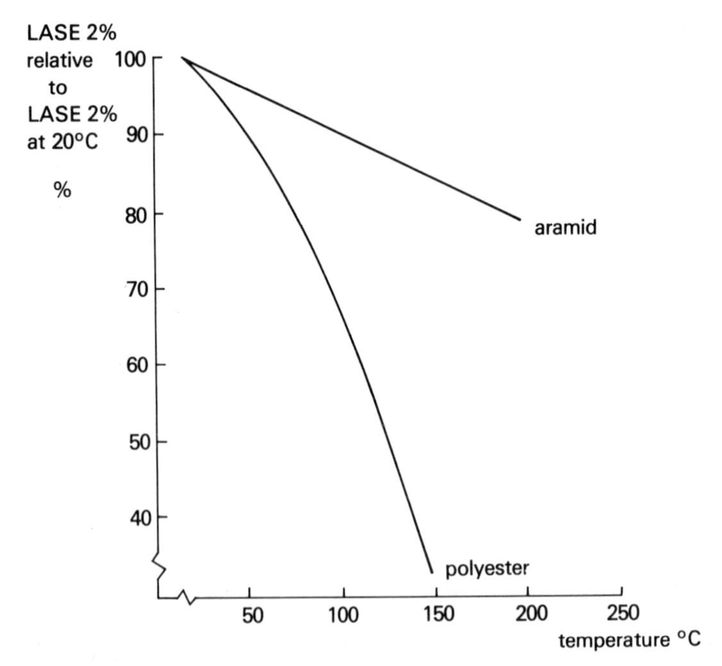

Fig. 9. The effect of temperature on the LASE (load at standard elongation) of 2% for aramid and polyester fibre.

as Van Dreumel[12] has pointed out, this constraint has only a modest effect. As a matter of course the application of aramid fibre in compression loaded structures is ultimately restricted by the compression strength limit, but in many cases instability of the structure will occur before this limit is reached. The efficiency of columns and panels under compression loads is given by the ratio of the load carried and the weight of the structure. It is proportional to the product of (1) the 'structure loading coefficient', which is $P^{1/2}/L^2$ for columns and $P^{2/3}/L^{5/3}$ for panels and (2) 'the material efficiency criterion', which is $E^{1/2}/\rho$ for columns and $E^{1/3}/\rho$ for panels.[13] Here P is the load on the column or panel, L its length, ρ the specific weight and E the compression modulus of elasticity. The material efficiency criterion depends thus only on material properties and Fig. 8 shows that this criterion for columns made from aramid almost equals that of carbon fibre reinforced composites.

Due to the stiffness of the chain the thermo-mechanical stability of aramid fibre is appreciably better than that of the conventional synthetic fibres, as shown in Fig. 9. The coefficient of linear thermal expansion of P*p*PTA fibre is negative, viz. $-4 \times 10^{-6}\,°C^{-1}$, which can be explained from the anisotropy of the thermal motion of the atoms in the rigid chain. The longitudinal component of the thermal vibration will increase the length of the chain, while the transverse components shorten the chain. Because of the chain stiffness the longitudinal vibration is small compared with the transverse vibrations resulting in a negative linear thermal expansion coefficient.

ACKNOWLEDGEMENT

The author is indebted to W. H. M. van Dreumel of the Department of Aerospace Engineering of the Delft University of Technology, for his permission to present here the results of his investigation of the compressive behaviour of aramid reinforced materials.

REFERENCES

1. Pauling, L., *Nature of the Chemical Bond*, 3rd edn, Cornell University Press, Ithaca, 1960.
2. Fitzer, E. and Hüttner, W., *J. Phys.* (*D*), 1981, **14**, 347.
3. Maatman, H., To be published.
4. Northolt, M. G., *Eur. Polym. J.*, 1974, **10**, 799.

5. Ballou, J. W., *ACS Polym. Prepr.*, 1976, **17**, 75.
6. Hagege, R., Jarrin, M. and Sotton, M., *J. Microscopy*, 1979, **115**, 65.
7. Dobb, M. G., Johnson, D. J. and Saville, B. P., *J. Polym. Sci., Polym. Phys. Ed.*, 1977, **15**, 2201.
8. Dobb, M. G., Hindeleh, A. M., Johnson, D. J. and Saville, B. P., *Nature*, 1975, **253** (5488), 189.
9. Northolt, M. G., *Polymer*, 1980, **21**, 1199.
10. Dobb, M. G., Johnson, D. J. and Saville, B. P., *Polymer*, 1981, **22**, 960.
11. Northolt, M. G., *J. Mat. Sci.*, 1981, **16**, 2025.
12. Van Dreumel, W. H. M., Report LR-341, Department of Aerospace Eng., Delft Univ. of Techn., The Netherlands, 1982.
13. Gordon, J. E., *Structures*, Pelican Books, Harmondsworth, England, 1978.

21

TOWARDS A PHOTOCONDUCTIVE LIQUID CRYSTAL: CARBAZOLE-CONTAINING SYSTEMS*

L. L. CHAPOY, D. K. MUNCK, K. H. RASMUSSEN,
E. JUUL DIEKMANN and R. K. SETHI

Instituttet for Kemiindustri, Technical University of Denmark, Lyngby, Denmark

and

DEREK BIDDLE

Institute of Physical Chemistry, University of Gothenburg, and Chalmers University of Technology, Gothenburg, Sweden

INTRODUCTION

The motivation for this work is to investigate the role of long-range order on the electrical properties of organic conductors through the use of a liquid crystalline vehicle.

Conducting polymers with conjugated backbones such as polyacetylene and poly-*p*-phenylene are generally quite intractable. Their morphology is difficult to study and is the subject of some debate at this time. It is thus not feasible to systematically investigate the electrical properties of this interesting class of materials with respect to long-range order. In the present context, it is significant that by stretching doped polyacetylene, appreciable conduction anisotropy has been achieved. An enhancement of conductivity in the stretching direction, presumably that which is parallel to the chain axis, was noted without any sizeable decrease in that of the perpendicular direction.[1] The simple averaging process, which would imply that the conductivity for the isotropic material should be $\frac{1}{3}\,\mathrm{Tr}\,\mathbf{\Omega}$, where $\mathbf{\Omega}$ is the conductivity tensor for the highly aligned material, does not appear

* Based on the paper presented at the Symposium on Order in Polymeric Materials, Waltham, Mass., USA, 25–26 August 1983, and published in *Molecular Crystals and Liquid Crystals*, **105**, 353–74 (1984).

to apply. This could be interpreted in terms of a sensitive co-operative dependence of the conductivity on the long-range order.

Order has also been postulated as a precondition for obtaining high levels of conductivity in one-dimensional organic conductors. For these materials, e.g. TCNQ–TTF complexes, the appropriate order is achieved via a specific crystal form in which the donor and acceptor component molecules form stacked structures of various types.[2] For these systems it is difficult to study conductivity in terms of modification of the chemical constituents because such changes are often accompanied by subtle changes in crystal structure so that more than one parameter is being changed in any given experiment. Effects due to varied spatial disposition and order for a given chemical composition are likewise not available since one generally cannot alter the predisposed crystal structure.

To surmount these difficulties we have elected to study the photoelectric properties of oriented liquid crystal systems containing carbazole. In this way it is possible to attain both well defined and variable arrays of carbazole, the active photoelectric chromophore.

The working hypothesis of this investigation is that long-range molecular order is crucial to the photoelectric properties, and that a suitable orientation will be conducive to both increased charge carrier formation through enhanced absorption and increased charge carrier mobility facilitated by the spatial disposition of orbitals on adjacent chromophores. Of course, variations in chemical structure will ultimately be responsible for the attainable level of conductivity. Thus, this approach could permit the unique possibility of investigating these two important parameters independently and unclouded by the effects of each other, escaping from a dilemma often encountered while studying properties of polymers in the solid state.

In order to evaluate this hypothesis, three approaches have been pursued:

1. The orientation of amorphous polymers such as poly(vinyl carbazole).
2. The use of low molar mass thermotropic nematic liquid crystals to orient photoconductive chromophores by a guest–host mechanism.
3. The use of a liquid crystalline polymer containing photoconductive chromophores which can be brought into a stacked array by alignment of the liquid crystal.

In the following section, previously obtained results will be briefly described and referenced, and new work will be presented in its entirety.

RESULTS AND DISCUSSION

1. The Orientation of Amorphous Polymers such as Poly(vinyl carbazole), PVK

This polymer in its commercial form exhibits sufficient photoconductivity to be used in imaging processes in spite of its stereochemical impurity and conformational disorder. Simple uniaxial creep experiments on thin films above the glass transition temperature were shown to produce only small orientations, i.e. order parameters $\langle P_2 \rangle$ of $<0{\cdot}2$ when defined in the usual way:

$$\langle P_2 \rangle = (3\langle \cos^2 \Theta \rangle - 1)/2 \tag{1}$$

where Θ is the angle between the stretching direction and the vector describing a chain backbone segment, or the long axis of a probe molecule and $\langle \ldots \rangle$ indicate an ensemble average. The order parameter was determined by the dichroic infrared absorption at 925 cm^{-1} corresponding to the symmetric out-of-plane bending mode of the carbazole ring.[3] It should be pointed out that according to the previously proposed hypothesis such a uniaxially oriented film might be expected to have diminished conductivity normal to the film plane since the carbazole rings will be tipped out of this plane when the chains are in the planar zig-zag conformation. The photoconductive measurements remain to be performed at this time.

The photoconductive phenomenon has long been associated with excimer emission. This is emission from a mono-excited dimer, resulting from two chromophores in intimate contact. This intimate contact is presumably closely associated with the mechanism of π-orbital coupling necessary for electronic conduction (hole transport in the case of carbazole). This has been elegantly demonstrated for a series of copolymers in which excimer emission and photoconductivity had essentially the same dependence on copolymer composition.[4] Features of excimer emission, thus, might be of interest in helping to elucidate the mechanism of photoconductivity in these systems. Figure 1 shows the dependence of the emission anisotropy for excimer emission on the order parameter for PVK films. The emission anisotropy, r', is given by:

$$r' = \frac{I_{\parallel} - I_{\perp}}{I_{\parallel} + I_{\perp}} \tag{2}$$

where I is the emission intensity and $\parallel$ and $\perp$ refer to parallel and perpendicular relative to the stretching direction. The samples were excited in the film plane at 344 nm and the emission was observed emanating from the thin edge, i.e. the standard 90° fluorescence geometry, at 404 nm.

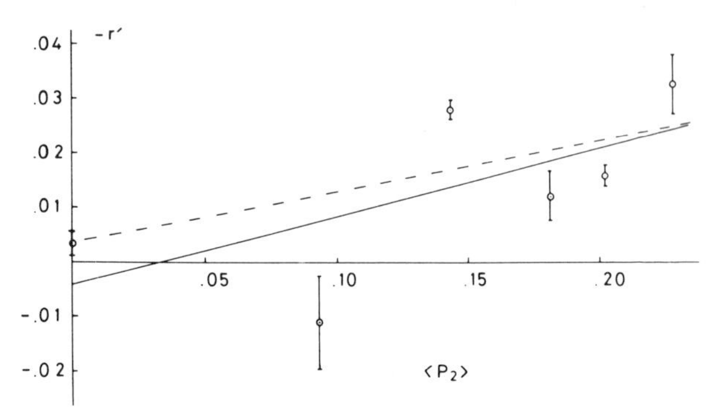

Fig. 1. The emission anisotropy as a function of order parameter for uniaxially oriented poly(vinyl carbazole). The solid line is a linear regression line for all points, while the dashed line omits the point falling below the horizontal axis. The lines have no physical significance in as much as r' should not be a simple function of $\langle P_2 \rangle$.

For isotropic bulk polymers r' is generally equal to zero due to the high concentration of chromophores and the extensive energy migration prior to the formation of the excimer state. For $\langle P_2 \rangle > 0$, r' is non-zero since the vector characterizing the excited state will not be randomized during excited state migration. Thus oriented samples offer the possibility of studying polarization effects associated with excimer emission that have not previously been observed and their possible consequences for photoconductivity. Note that theory does not predict r' to be a simple function of $\langle P_2 \rangle$ and the purpose of the figure is only to show that r becomes non-zero in the oriented state. Simple theory requires r to be a function of $[1 + B(\langle \cos^2 \Theta \rangle - \langle \cos^4 \Theta \rangle)]^{-1}$, where B is a parameter of the system. Data on $\langle \cos^4 \Theta \rangle$ are unfortunately not available.[5] The shape of the emission spectra was not affected by the orientation.

2. The Use of Low Molar Mass Thermotropic Liquid Crystals to Orient Photoconductive Chromophores by a Guest–Host Mechanism[6]

Carbazole was the chromophore of choice not only because of its good photoconductive properties in PVK, but also because the molecule is in general well characterized. This, however, presented certain technical difficulties which had to be overcome. Since carbazole absorbs strongly at 335 nm, glass was precluded as the material for cell construction. Indium–tin oxide coated 3 mm quartz plates (Spectrosil®) used as semi-transparent electrodes were found, however, to give good optical transmission at 335 nm. Subsequent treatment with a thin non-insulating SiO

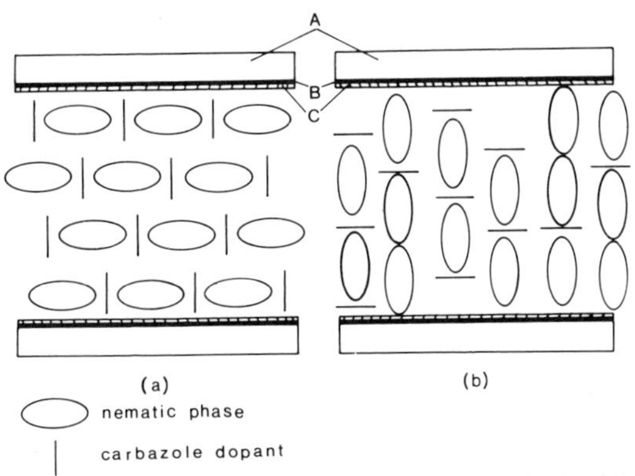

Fig. 2. A schematic cross-section of a thermotropic nematic liquid crystal doped with carbazole: (a) homogeneously aligned with an applied voltage less than the critical voltage; (b) homeotropically aligned with an applied voltage greater than the critical voltage. (A) Quartz plate; (B) indium–tin oxide layer; (C) aligning SiO layer. The light impinges from above.

layer with an ion gun at 30° to the plane ensures that homogeneous orientation of the nematic phase in the rest state will result, as shown schematically in Fig. 2(a). The requirement that the liquid crystal be a passive, transparent, orientating matrix is somewhat more difficult to achieve, since many thermotropic nematic phases are based on aromatic molecular components and because the molar concentration of the liquid crystal will be high. For the sake of experimental convenience only room temperature nematics were considered. A survey of commercially available potentially useful materials is given in Table 1.

Doping of the liquid crystals with carbazole was limited by the solubility of carbazole, i.e. about 10^{-1} M. Colourful microcrystals could easily be observed when the solubility was exceeded by viewing the oriented samples in a polarizing microscope with the liquid crystalline phase giving a dark background. The carbazole utilized here was purified to remove anthracene by a Diels–Alder reaction with maleic anhydride.[7]

Dichroic ultraviolet absorption measurements performed on the carbazole absorption band for homogeneously aligned samples showed, surprisingly, that the long axis of the carbazole ring plane is perpendicular to the nematic director, i.e. negative order parameters, as shown schematically in Fig. 2(a). The transition to the first excited state is known to be short-axis polarized for carbazole.[8] This fact is disturbing in as much as probe techniques have been used previously to determine both transition moment directions[9] as well as order parameters.[10] This perpendicular orientation has been

TABLE 1
Characteristics of Some Ultraviolet Transparent Liquid Crystals

Company	*Designation*	*Type*	$\Delta\varepsilon$	*Absorption edge (nm)*
Merck	NP-1083	Eutectic mixture of three phenyl cyclohexanes	+10·1	285
Merck	NP-1132	Mixture of three phenylcyclohexanes and one biphenylcyclohexane	+10·3	320
Hoffmann–La Roche	RO-TN-651	Hydrogenated cyanophenyl pyrimidine	+12·5	320
Chisso	Lixon K-0327	Cyclohexane carboxylic esters	−1·1	300

observed previously, however, with other carbazole systems.[3] Actual quantitative determination of the order parameters is complicated by the fact that the carbazole absorption is a shoulder on the red edge of the absorption from the conductive layer and/or liquid crystal.

For those liquid crystals having dielectric anisotropies $\Delta\varepsilon > 0$, an applied field of the order of a few volts d.c. over a film thickness of 23 μm led to homeotropic alignment as schematically indicated in Fig. 2(b), i.e. a Fréedericksz transition. In all cases the thickness of the cell was maintained with Mylar® spacers. The long carbazole axis was also found to lie perpendicular to the nematic director in the case of homeotropic alignment above the critical field, since the order parameter approached zero and the system appeared to be pseudo-isotropic. All absorption measurements are made normal to the plane of the quartz plates. For K-0327 for which $\Delta\varepsilon < 0$, and $\langle P_2 \rangle > 0$, i.e. parallel alignment, there is no change in order

TABLE 2
Order Parameters for Carbazole in Some Selected Nematic Phases as a Function of Applied Voltage

Designation	$\Delta\varepsilon$	$\langle P_2 \rangle$ *at V (volts)*		
		0	*5*	*10*
NP-1083	>0	−0·3	—	—
NP-1132	>0	−0·25	−0·11	−0·08
RO-TN-651	>0	−0·50	−0·20	−0·12
Lixon K-0327	<0	0·30	0·30	0·30

parameter with applied voltage, as would be expected. These data are summarized in Table 2.

Given these well-defined morphologies, experiments were performed to determine the low voltage d.c. photoconductivity of the above mentioned samples. Photocurrent measurements were carried out with a stabilized low voltage d.c. power supply, a Keithley 616 electrometer as the ammeter, a medium pressure mercury lamp (Philips MPK 125 W) and a 12 nm band-pass filter at 340 nm resulting in a light intensity of about $2\,W\,m^{-2}$. Cells were about $1\,cm^2$ in area. The introduction of a guard-ring in one of the electrodes to minimize interference from surface currents was not found to have any significant effect on the measurements and was not pursued further.

Figures 3 and 4 show the photocurrents for NP-1132 and RO-TN-651, respectively. The lower curve is for the undoped liquid crystal and the upper for that containing $8{\cdot}6 \times 10^{-2}$ M and $9{\cdot}0 \times 10^{-2}$ M carbazole, respectively. The numbers to the right on the figures refer to three consecutive runs and demonstrate the reproducibility. The increase in conductivity occurred immediately on opening the mechanical shutter. The measurements here are limited by the response time of and the ability to read data from the

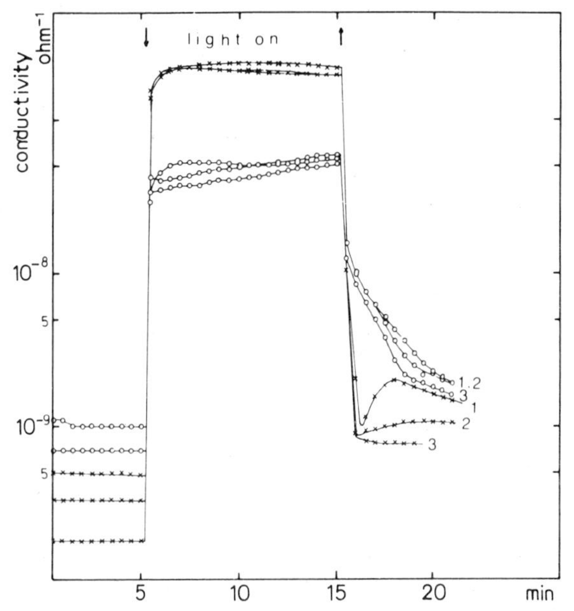

Fig. 3. Photoresponse of the conductivity for NP-1132.

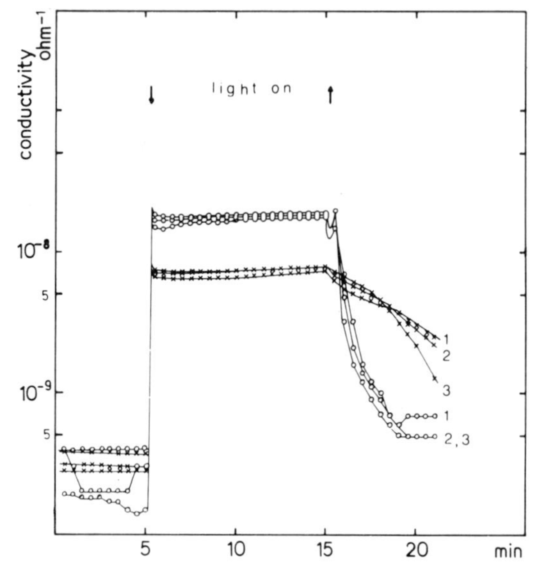

Fig. 4. As in Fig. 3, for RO-TN-651.

Keithley 616. Good values of the dark current were difficult to obtain and were no doubt complicated by the slow relaxation after the cessation of illumination.

Many questions remain to be answered: What is the conduction mechanism? Why does the liquid crystal give such a large photoresponse in the undoped state? This is perhaps not so surprising in light of experiments done years ago on samples of commonly available polymers, showing them all to be more or less photoconductive due to the presence of a variety of naturally occurring impurities.[11] Is the indium–tin oxide electrode being photostimulated to promote electrochemical reactions? Similar carbazole doping of an isotropic solvent such as paraffin resulted in no measurable photocurrent.

Figures 5 and 6 show the photoconductivity, $\Omega_L - \Omega_D \cong \Omega_L$ where Ω is the conductivity and the subscripts L and D refer to the illuminated and dark states respectively, as a function of voltage for NP-1132 and RO-TN-651, respectively. The vertical line indicates the occurrence of the Fréedericksz transition as observed in the polarizing microscope. The transition for NP-1132 is much sharper than for RO-TN-651 giving rise to a discontinuity in the photoconductivity when viewed as a function of voltage. It has been pointed out[12] that in the latter case, a direct

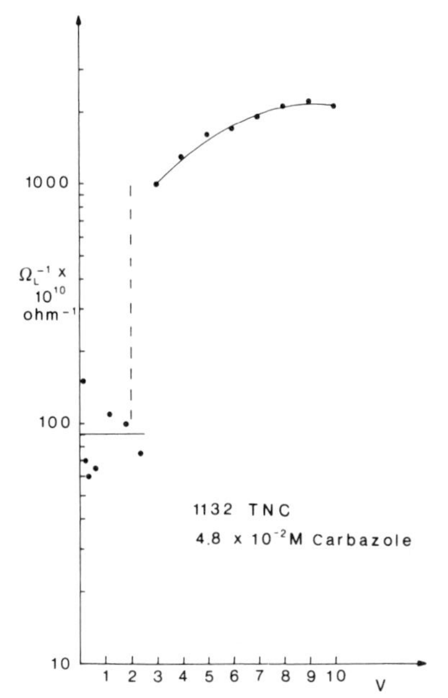

Fig. 5. The photoconductivity, $\sim\Omega_L$, as a function of applied voltage for NP-1132. The vertical line is the critical voltage for the Fréedericksz transition.

Fig. 6. As in Fig. 5, for RO-TN-651.

determination of the conductivity anisotropy $\Omega_\| - \Omega_\perp$ at the critical voltage is possible.

Figures 7 and 8 show the photoconductive gain Ω_L/Ω_D as a function of voltage for NP-1132 and RO-TN-651, respectively. The increase in conductivity with voltage for NP-1132 might be explained in terms of more carriers being detrapped under the influence of the higher field. We are at a loss to explain the decrease for RO-TN-651. The scatter in the curves results largely from the difficulty in determining the dark current. The continuous nature of both curves indicates a similar factor affecting both Ω_L and Ω_D at the critical voltage for the Fréedericksz transition.

Figure 9 shows the concentration dependence of the photocurrent, $\Omega_L - \Omega_D \cong \Omega_L$, for RO-TN-651 as a function of the carbazole concentration. Note that there appears to be a critical threshold concentration, $1{\cdot}0 \times 10^{-3}\,\text{M} < c^* < 6{\cdot}0 \times 10^{-3}\,\text{M}$, below which the presence of carbazole does not enhance the photoconductivity. A crude estimate of the carbazole intermolecular distance at this threshold from the $\frac{1}{3}$ power of the reciprocal

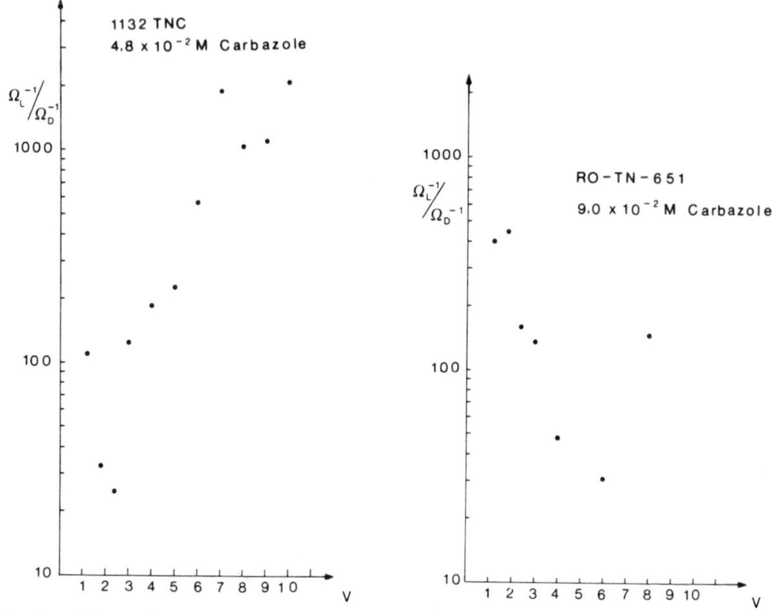

Fig. 7. The photoconductive gain, Ω_L/Ω_D, as a function of applied voltage for NP-1132.

Fig. 8. As in Fig. 7, for RO-TN-651.

concentration gives a value of 50–100 Å. Above the critical concentration there is an approximately linear relationship between dopant concentration and photoconductivity.

To our knowledge this is the first report of photoconductivity in liquid crystalline materials. Other photoelectric effects, however, involving liquid crystalline materials have been previously observed.[13,14]

3. The Use of a Liquid Crystalline Polymer Containing Photoconductive Chromophores Which Can be Brought into a Stacked Array by Alignment of the Liquid Crystal

In the first instance, poly-α-amino acids with carbazole-substituted side chains were chosen for this purpose.[15] Poly-α-amino acids are well known to be capable of forming lyotropic cholesteric phases due to their α-helical conformation. The formation of a homeotropic nematic phase, as shown schematically in Fig. 10, would give the desired morphology with the stacking of the carbazole rings in the plane of the cell. This is in essence the same morphology obtained with the carbazole-doped thermotropic

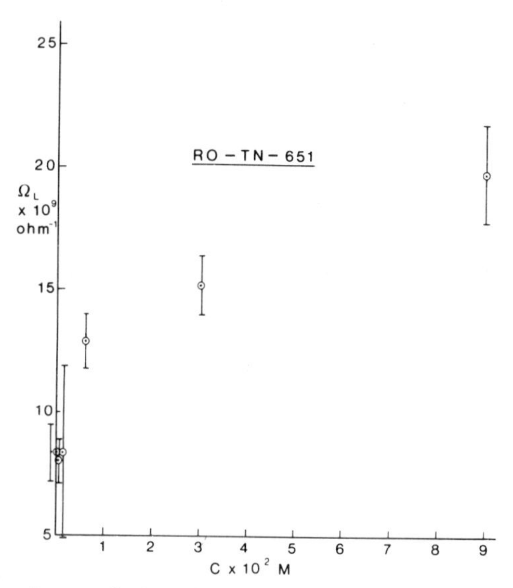

Fig. 9. The dependence of the photoconductivity, $\sim\Omega_L$, as a function of carbazole concentration for RO-TN-651.

nematic phases as shown previously in Fig. 2(b). The possibility of obtaining such a morphology is currently being explored. Possible approaches are the exploitation of the magnetic anisotropy to produce macroscopic alignment after which the solvent can be evaporated to give an oriented film or the use of the so-called thin cell technique.[16]

It should be pointed out that the polymer in question[15] has been shown to give highly structured excimer emissions in dilute solution.[17] This work is being continued to include lifetime studies, studies in the liquid crystalline state and studies of model compounds, in order to correlate these spectral properties with forthcoming measurements of the photoconductivity.

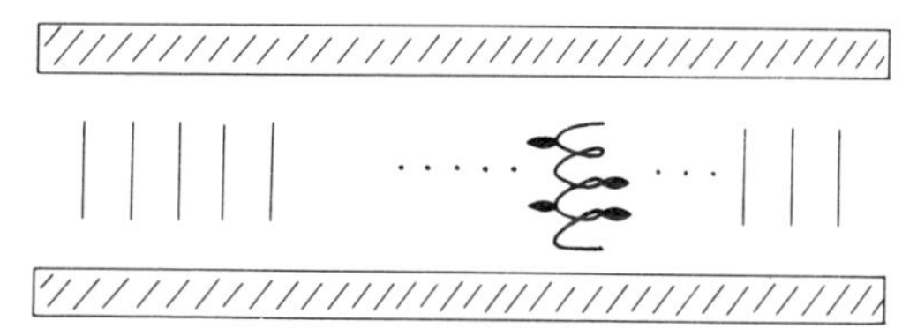

Fig. 10. A schematic cross-section, as in Fig. 2, for a poly amino acid with carbazole side-chain substituents homeotropically aligned so as to stack the carbazole rings in the plane of the cell and into the path of the impinging light.

ACKNOWLEDGEMENTS

The Danish group gratefully acknowledges financial support from Statens Teknisk Videnskabelige Forskningsråd, Danida and The BP Energifond.

Derek Biddle gratefully acknowledges the financial support of the Swedish Natural Science Research Council.

The authors would also like to thank Dr Åke Hörnell, Gagnef, Sweden, for his help in preparing the quartz plates for the liquid crystal experiments.

REFERENCES

1. Park, Y. W., Druy, M. A., Chaing, C. K., MacDiarmid, A. G., Heeger, A. J., Shirakawa, H. and Ikeda, S., *J. Polym. Sci., Polym. Lett. Ed.*, 1979, **17**, 195.
2. Shacklette, L. W., Eckhardt, H., Chance, R. R., Miller, G. G., Ivory, D. M. and Baughman, R. H., *Conductive Polymers*, Seymour, R. B. (Ed.), Plenum Press, New York, 1981, p. 115.
3. Chapoy, L. L., Sethi, R. K., Ravn Sørensen, P. and Rasmussen, K. H., *Polym. Photochem.*, 1981, **1**, 131.
4. Slobodyanik, V. V., Naidyonov, V. P., Pochinok, V. Ya. and Yashchuk, V. N., *Chem. Phys. Lett.*, 1981, **81**, 582.
5. Chapoy, L. L., Spaseska, D., Rasmussen, K. and DuPré, D. B., *Macromolecules*, 1979, **12**, 680.
6. Chapoy, L. L. and Munck, D. K., *J. de Physique, Colloque C3*, 1983, **44**, 697.
7. Kihara, K., Ishii, Y., Suzuki, Y. and Takeuchi, T., *Kogyo Kayaku Zasshi*, 1970, **73**, 2630.
8. Chakravorty, S. C. and Ganguly, S. C., *J. Chem. Phys.*, 1983, **52**, 2760.
9. Thulstrup, E. W. and Eggers, J. H., *Chem. Phys. Lett.*, 1968, **1**, 690.
10. Chapoy, L. L. and DuPré, D. B., *J. Chem. Phys.*, 1979, **70**, 2550.
11. McGibbon, G., Rostron, A. J. and Sharples, A., *J. Polym. Sci. A-2*, 1971, **9**, 569.
12. Bauernmeister, F., private communication.
13. Tien, H. Ti, *Nature* (*London*), 1970, **227**, 1232.
14. Aizawa, M., Hirano, M. and Suzuki, S., *Electrochim. Acta*, 1978, **23**, 1185.
15. Chapoy, L. L., Biddle, D., Halstrøm, J., Kovács, K., Brunfeldt, K., Qasim, M. A. and Christensen, T., *Macromolecules*, 1983, **16**, 181.
16. Uematsu, Y., Tomizawa, J., Kodohora, F. and Sasahi, T., *Acad. Rep. Tokyo Inst. Polytechn.*, 1980, **2**, 53.
17. Chapoy, L. L. and Biddle, D., *J. Polym. Sci., Polym. Lett. Ed.*, 1983, **21**, 621.

22

ELECTRO-OPTIC EFFECTS IN A SMECTOGENIC POLYSILOXANE SIDE-CHAIN LIQUID CRYSTAL POLYMER

H. J. Coles and R. Simon

Schuster Laboratory, Department of Physics, University of Manchester, UK

INTRODUCTION

The recent work of Krigbaum on main-chain[1] and Finkelmann,[2] Ringsdorf,[3] etc., on side-chain polymer liquid crystals has generated much interest in the potential of these systems for use in electro-optic devices. The combination of polymeric specific and monomeric liquid crystal specific properties leads to an interesting range of potential materials for new display devices. The majority of the research over the last five years has concentrated on synthesis and the establishment of the basic property–structure relationships. However in the last year or so papers have started appearing where the electro-optic properties of some of these materials have been examined.[4]

The main emphasis of the previous work both from a synthesis and a device point of view has been to concentrate on nematic or cholesteric polymer materials in the hope of producing displays analogous to the monomeric systems. As observed by Ringsdorf and Zentel[4] and Finkelmann *et al.*[5] the operating parameters of such polymer liquid crystals tend to be worse than those observed for equivalent monomeric systems. The threshold voltages are higher, the response times are slower (2–10 times), and the operating temperatures are not always convenient. This would seem to be a penalty of the increased viscosity imposed by the existence of the polymer main chain and the close proximity of the glass transition. The inescapable conclusion is that the device performance of such polymer liquid crystals will always be worse than the equivalent monomeric material.

However, as Finkelmann has pointed out[6] for side-chain systems, the operating voltages, etc., are not impossible and one can use the existence of T_g to effect optical storage in the nematic or cholesteric systems. If T_g is to be just above ambient temperature then the resultant high viscosity near to T_g implies very close response times unless high operating temperatures ($\sim 200\,°C$) are used. Our approach reported recently[7,8] has been somewhat different. We also believe that the main potential for side-chain polymer liquid crystals is for electro-optic storage devices. However recognizing that the problem in response times comes from the proximity of T_g to the operating or writing temperatures we have studied materials synthesized for us with lower T_g values.[9]

The question is then how to maintain optical storage? We have taken the novel approach in polymers of looking for storage effects in smectogenic materials. Optical storage effects are known in monomeric smectic liquid crystals and the object of our recent researches has been to look for similar effects in polymer liquid crystals. It is interesting to note that for monomeric storage systems the compounds used were based essentially on the cyanobiphenyls (originally synthesized at Hull University[10]) that had such a marked effect on the monomeric display industry. In an interactive programme with Hull we therefore studied cyanobiphenyl-based side-chain polymers. Preliminary results are being published elsewhere that concentrate on various electro-optic effects in a range of side-chain solution liquid crystals and in this report we will concentrate on electro-optic effects in a smectic side-chain polysiloxane liquid crystal polymer.

EXPERIMENTAL

(a) Apparatus

The electro-optic measurements were carried out using an Olympus BH-2 transmission polarizing microscope adapted to give direct sample observations, photodiode detection and photographic facilities (Fig. 1(a)). The sample cell (Fig. 1(b)) was heated in a thermostatically controlled hot stage system LIN-KAM 600. The temperature stability was better than $0{\cdot}1\,°C$ and temperatures could be adjusted in the range -20 to $+600\,°C$. Voltages were applied to the cell using a function generator output amplified by an Electro-Optic Developments LA10A linear amplifier capable of producing voltages up to 400 V (rms). This output was pulsed across the cell as required using a reed relay (not drawn) between the Amplifier (AMP) and the Cell. The photodiode output was recorded using a suitable chart recorder.

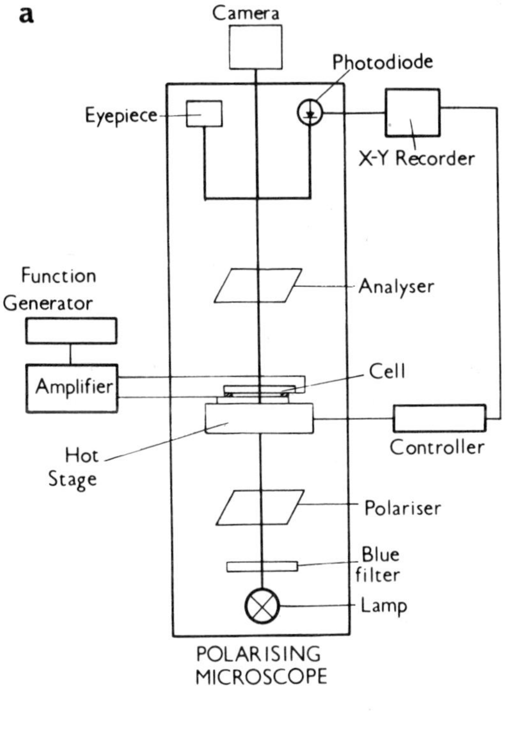

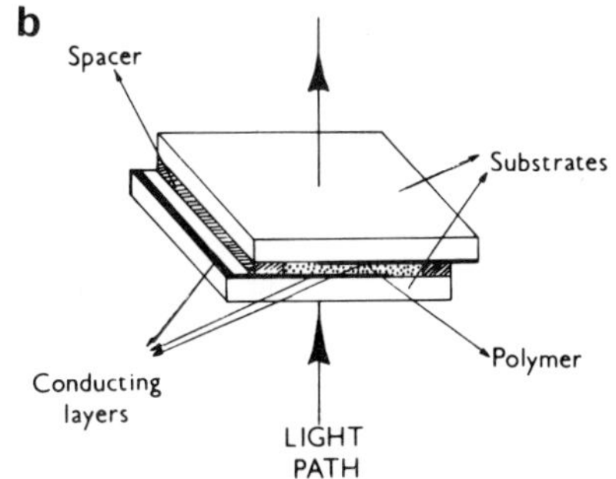

Fig. 1. (a) Schematic diagram of experimental apparatus and (b) assembly of polymer liquid crystal cell.

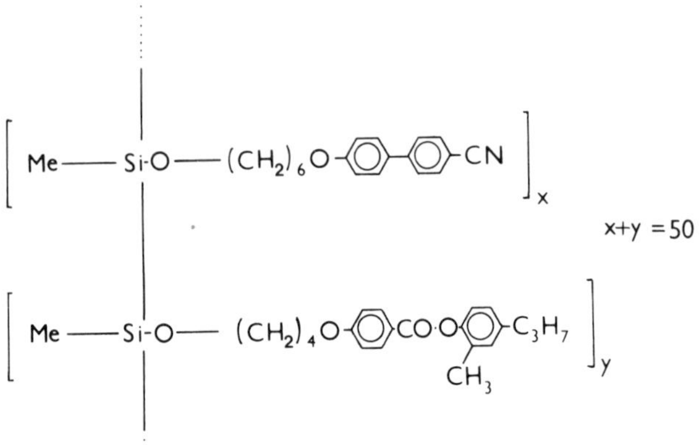

$$\left[\text{Me}-\text{Si-O}-(CH_2)_6O-\langle\bigcirc\rangle-\langle\bigcirc\rangle-CN\right]_x$$

$x+y=50$

$$\left[\text{Me}-\text{Si-O}-(CH_2)_4O-\langle\bigcirc\rangle CO{\cdot}O\langle\bigcirc\rangle-C_3H_7\right]_y \quad (\text{with } CH_3 \text{ substituent})$$

PG 296 g 4·0 s 85·9 i

Fig. 2. Phase transitions and structure of polymer PG 296. g = glass transition, s = smectic phase and i = isotropic phase (from DSC Ref. 9).

The sample cells (Fig. 1(b)) were constructed using In/SnO_2 coated glass substrates etched to give a 2 mm square electrode configuration. The two electrodes were spaced using a non-conducting epoxy (glue) laid down in narrow tracks and typical cell spacings were $\sim 40\,\mu$m. The cells were then heated using a hotplate set to $\sim 10\,°$C above the polymer clearing point and filled by capillary action. This resulted in a uniform and air bubble free texture. No surface alignment procedures were used in the current work but cells were thoroughly degreased and washed in isopropyl alcohol before assembly.

(b) Materials

The copolymer under investigation in the current work was a cyano-biphenyl and methylated benzoic ester side-chain substructural polysiloxane of high positive dielectric anisotropy ($\Delta\varepsilon$). The structure and transition details are given in Fig. 2 and the polymer is called PG 296 to relate to other work being published.[7,8]

RESULTS

With nematic or cholesteric monomer liquid crystals surface alignment effects are a precondition of their use for electro-optic displays. With smectic monomeric materials addressed using low or high frequency fields

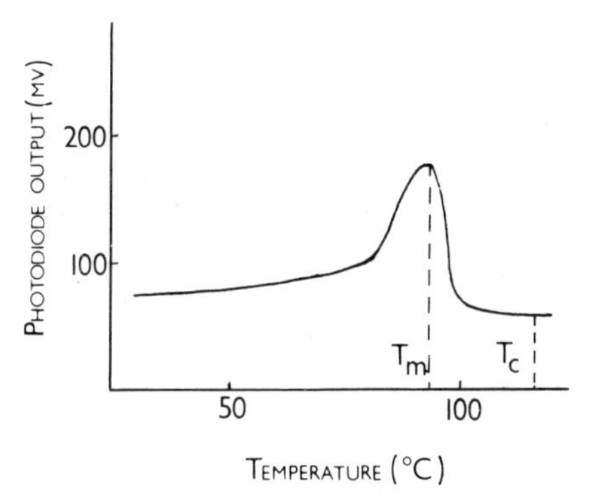

Fig. 3. Thermo-optic analysis (TOA) curve for polymer PG 296.

surface alignment techniques may be used to obtain the desired optical texture. Similar preconditions exist for nematic and cholesteric polymers. However in the case of smectic polymers this condition is not evident. The existence of the polymer main chain appears to be sufficient to give a highly scattering non-aligned texture. Surface alignment is not therefore a prerequisite before electro-optic effects might be observed, and in the work reported herein no surface alignment techniques were used.

(a) Thermo-optic Analysis

The thermo-optic curve (Fig. 3) is established without using aligning fields across the sample. The sample is observed between crossed polars and the transmission through the system measured as a function of temperature. At low temperatures (30–80 °C) the scattering texture is immobile and the transmission is low due to the turbidity of the sample and only varies slowly with temperature. At around 80 °C the texture becomes optically mobile and the transmission increases by a factor of ~3. At T_m the sample starts melting as evidenced by small dark regions appearing in the texture (Fig. 4). These regions are optically isotropic and therefore appear black through crossed polars. Further increases in temperature lead to an increasing dark area until the final traces of the focal–conic structure disappear at T_c, the so-called clearing temperature. We believe that this is due to the inherent polydispersity of the main-chain siloxane units used in the synthesis. The polydispersity, Z-number, of these samples is ~2. Following Finkelmann[6] we would expect low molecular weight (M_w) material to melt up to 20–30 °C before the high M_w compounds. This would explain the width of the biphasic region in our material. This polydispersity which does not appear to have raised serious comment in previous work will, we believe, be an

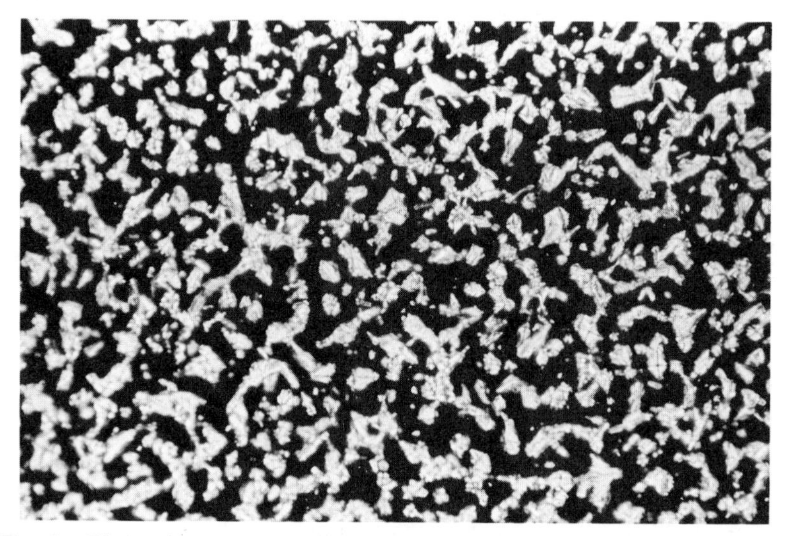

Fig. 4. Photomicrograph showing optical texture of PG 296 above T_m. Crossed polars, 99 °C, × 100.

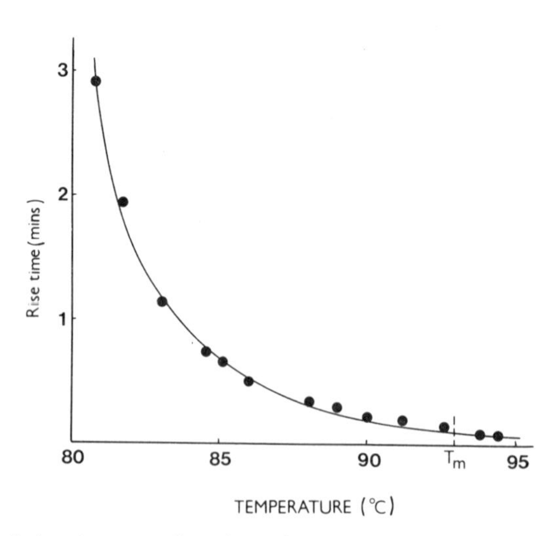

Fig. 5. Optical rise time, as a function of temperature, to an applied sinusoidal alternating electric field of 300 V_{rms}, 3 kHz.

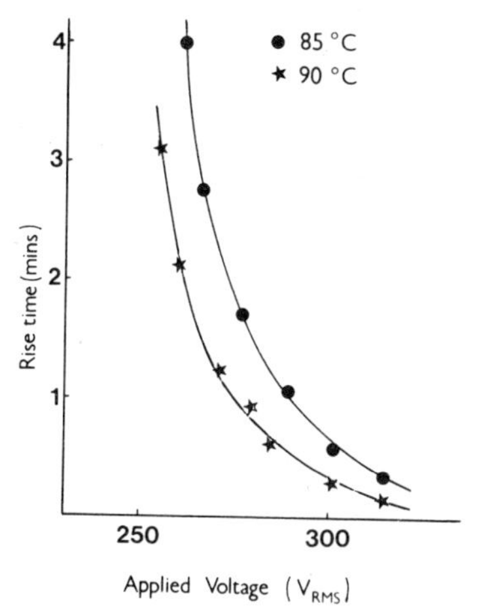

Fig. 6. Rise time as a function of applied voltage (2 kHz, sinewave) at temperatures just below T_m.

important factor in controlling both the magnitude and time response of the electro-optic effects.

(b) Response Time

In the data given in Figs 5, 6 and 7 the response or rise time is defined as the time for the light transmission to drop to 50% of its initial value on application of an electric field across the system. Below the biphasic region the response times are of the order of minutes at 80 °C and increase to several hours at room temperature. The data presented in Fig. 5 correspond to measurements in and just below the biphasic region for a fixed voltage (300 V_{rms}) and frequency (3 kHz sinewave). It is significant that, despite the fact that the samples are smectogenic in nature, the response times are on the timescale of seconds in this biphasic region. These times are easily comparable with those observed in nematics at less accessible temperatures. As can be seen from Fig. 6 increasing the rms voltage applied to the sample reduces these response times further and for ease of comparison we have shown two voltage–response time curves for temperatures 8 °C and 3 °C below the start of the biphasic region. As can be seen from Fig. 7 for a fixed V_{rms} and temperature the response time is very critically dependent on (a) the frequency of the applied field and (b) its waveform. Although DC fields may be used to induce changes in the scattering texture the response times are very slow, i.e. from minutes to hours. As the frequency increases the apparent response time decreases and reaches a

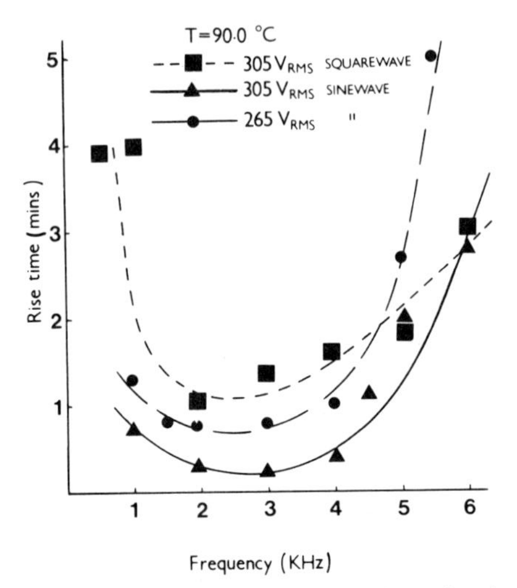

Fig. 7. Frequency, and waveform, dependence of the rise time, at a temperature of 90 °C ($T_m - 3$ °C).

minimum at about 3 kHz. Below this minimum frequency the response time corresponds to a turbulent dynamic scattering observed as a swirling motion in the optical texture. This turbulence disappears in the region of the critical frequency and above it is replaced by effects due to director reorientation.

The use of a square waveform shows how the response times vary for

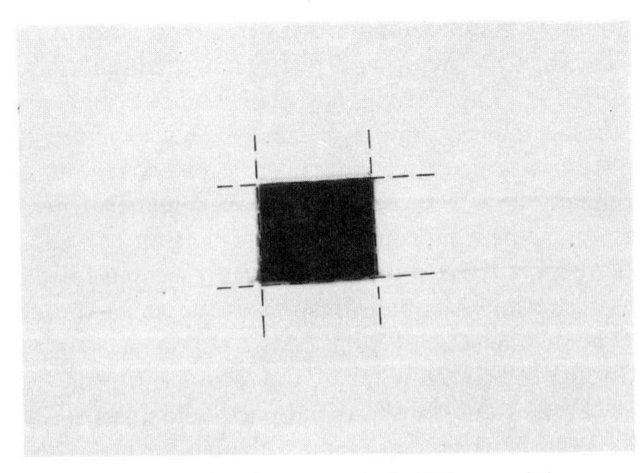

Fig. 8. Photomicrograph showing electric field induced homeotropic texture within the cell area (dark texture). Crossed polars, × 100. $T = 90$ °C.

different waveforms. It seems that the response time is then governed by the power spectrum of the applied waveform. We are currently carrying out measurement to quantify this effect further. It is evident from Figs 5, 6 and 7 that a combination of temperatures, voltage, frequency and waveform critically control the response times in these polysiloxane side-chain liquid crystals and that a judicious combination of these parameters leads to response times of a few seconds or less.

It is important to note that for frequencies below the minimum the optical response corresponds to a change from one scattering texture to another. Therefore the sample always appears opaque but with varying degrees of optical density. For frequencies above this minimum the response corresponds to a change from a scattering texture to a homeotropically aligned system. This is shown in Fig. 8 where the dark area (as observed through crossed polars) implies an optically isotropic texture in the region of the electrodes marked by a dashed line. This suggests a well-ordered homeotropic texture. If the sample is cooled the texture remains clear in the region of the electrodes. In Fig. 9 we have shown the optical texture observed at room temperatures. The blue inner region is the isotropically

Fig. 9. Photomicrograph showing the lighter field induced homeotropic texture stored at room temperature (blue region) as compared to the opaque scattering texture (black region). Crossed polars, × 100.

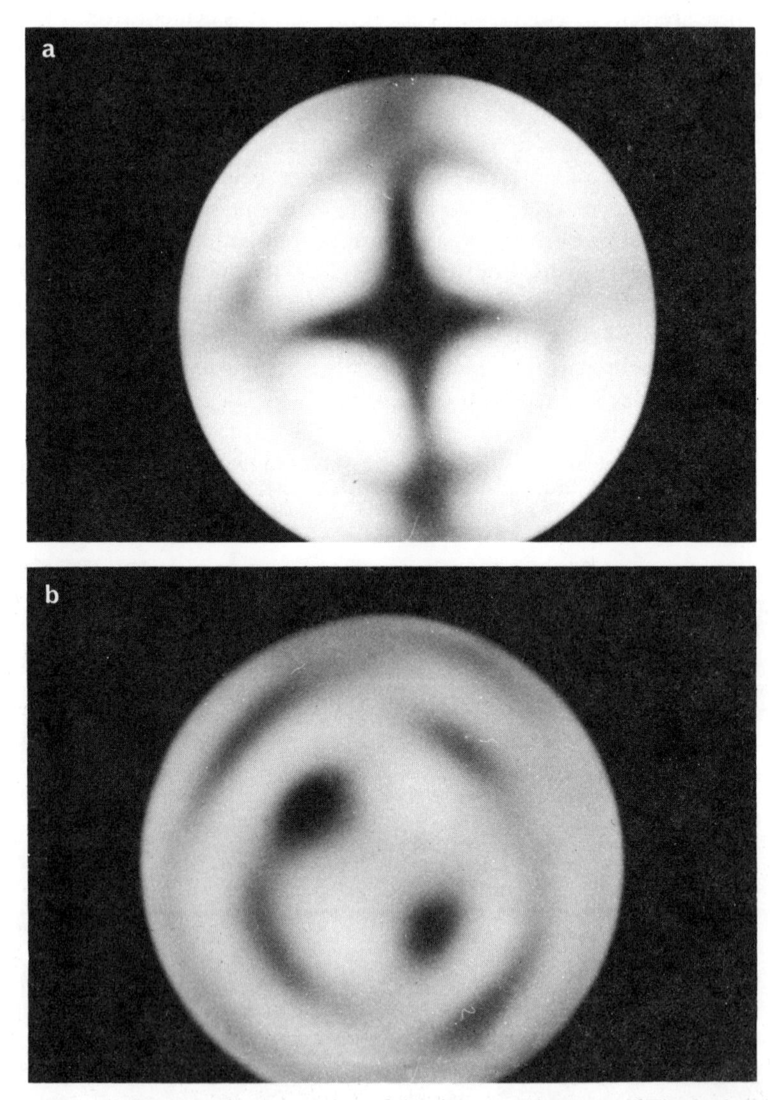

Fig. 10. (a) Conoscopic observation of the clear stored texture of Fig. 9, showing homeotropic alignment of mesogenic groups, and (b) positive uniaxial behaviour as evidenced by the insertion of a $\lambda/4$ plate.

clear texture and the outer region the strongly scattering texture. We have used optical conoscopy (Fig. 10(a)) to establish that the texture is optically isotropic and insertion of a $\frac{1}{4}$ waveplate (Fig. 10(b)) confirms that the system is uniaxial and positive. This means that the highly optically polarizable cyanobiphenyl side groups are oriented perpendicularly to the glass surface in the region of the applied field. It is important to remember that these photographs are taken at room temperature when the polymer is in the smectic phase some 20 °C above the glass transition temperature. Textures of this type have remained stored with no observed deterioration in their optical properties now for at least 18 months.

CONCLUSIONS

It is evident from the above results that electro-optic effects are manifest in smectic side-chain siloxane polymer liquid crystals. Although the response times are slower than for equivalent monomeric smectic materials where response times between 10 ms and 100 ms have been observed, these times may be approx. seconds or less under suitable conditions of field, voltage, waveform and frequency. These response times may be further reduced by making thinner cells. Although we do not envisage these materials, in their present form, competing with conventional monomeric nematic or cholesteric materials, these new smectic polymers do have several advantages over monomeric smectics. First, no aligning agents are required in order to obtain a suitable scattering texture. Secondly, the polymers have not been doped with ions to produce the dynamic scattering effect. This could be done and might improve the device response times; however we feel that this could also have the disadvantage of reducing the cell lifetime. Thirdly, by writing in the biphasic region and thus cooling back into the smectic phase optical textures may be readily stored. These textures have a very high contrast between the on and off states. Finally and importantly the texture is stored at room temperature above T_g, and indeed suppression of T_g to as low a value as possible would appear desirable as this also improves the response times further. We believe that these novel liquid crystal polymer systems will lead to a new generation of optical storage media.

ACKNOWLEDGEMENTS

The authors are grateful to Professor G. W. Gray, Dr D. Lacey and Dr P. A. Gemmell for the provision of the polymer sample and DSC data, and to the SERC for the award of a research grant (HJC) under the electro-active polymer scheme. RS also thanks the SERC for the research assistantship.

REFERENCES

1. Krigbaum, W. R. and Lader, H. J., *Mol. Cryst. Liq. Cryst.*, 1980, **62**, 87.
2. Finkelmann, H. and Ringsdorf, H., *Makromol. Chem.*, 1978, **179**, 273.
3. Ringsdorf, H. and Schneller, A., *Br. Polym. J.*, 1981, **13**, 43.
4. Ringsdorf, H. and Zentel, R., *Makromol. Chem.*, 1979, **180**, 803.
5. Finkelmann, H., Kiechle, U. and Rehage, G., *Mol. Cryst. Liq. Cryst.*, 1983, **94**, 343.
6. Finkelmann, H., *Phil. Trans. Roy. Soc.*, 1983, **309**, 105.
7. Simon, R. and Coles, H. J., *Mol. Cryst. Liq. Cryst. Lett.*, 1984, **102**, 43.
8. Coles, H. J. and Simon, R., *Mol. Cryst. Liq. Cryst. Lett.*, 1984, **102**, 75.
9. Gray, G. W., Lacey, D. and Gemmell, P. A., private communication.
10. Gray, G. W. *et al.*, *Electron. Lett.*, 1973, **9**(6), 130.

INDEX